JOHN LANKFORD is adjunct professor of history and assistant to the provost at Kansas State University. He is editor of the *Encyclopedia of the History of Astronomy* and coeditor of *Essays on American Social History*.

AMERICAN ASTRONOMY

✦

AMERICAN ASTRONOMY

COMMUNITY, CAREERS, AND POWER, 1859–1940

JOHN LANKFORD

with the assistance of Ricky L. Slavings

THE UNIVERSITY OF CHICAGO PRESS
CHICAGO & LONDON

JOHN LANKFORD is assistant to the provost and adjunct professor of history at Kansas State University. He is editor of *The History of Astronomy: An Encyclopedia* (1996) and former chair of the Historical Astronomy Division of the American Astronomical Society. He has also served as cochair of the History of Astronomy Interest Group in the History of Science Society. RICKY L. SLAVINGS is director of institutional research and associate professor of sociology at Radford University, Radford, Virginia.

The University of Chicago Press, Chicago 60637
The University of Chicago Press, Ltd., London

Printed in the United States of America
06 05 04 03 02 01 00 99 98 97 1 2 3 4 5

ISBN: 0-226-46886-0 (cloth)

Library of Congress Cataloging-in-Publication Data

Lankford, John, 1934–
American astronomy : community, careers, and power, 1859–1940 / John Lankford.
p. cm.
Includes bibliographical references and index.
ISBN 0-226-46886-0 (cloth)
1. Astronomy—United States—History—19th century. 2. Astronomy—United States—History—20th century.
QB33.U6L36 1997
306.4′5—dc20 96-31461
CIP

The paper used in this publication meets the minimum requirements of the American National Standard for Information Sciences—Permanence of Paper for Printed Library Materials, ANSI Z39.48-1984.

FOR JIM McCARTNEY

✦

Friend and colleague

for a quarter of a century

without whom this book never

would have been written

✦

CONTENTS

✦

TABLES AND FIGURES

TABLES

FIGURES

✦

PREFACE

This book deals with unfinished business: the examination of certain possibilities and options first explored by historians of science in the early 1970s, but, for whatever reasons, only briefly considered. These possibilities and options may be expressed as a series of questions. Using the tools of collective biography, is it possible to quantitatively reconstruct a scientific community across the lifetimes of several cohorts of scientists? Would cost/benefit analysis justify the investment of time and money needed to carry out such a reconstruction and the analysis of the data? What would the quantitative reconstruction of a scientific community tell us about its demographic composition and institutional development, as well as the changing nature of scientific careers and the forms and uses of power? What light would such investigations throw on the status of women and the nature of the reward system in a scientific community? On the basis of the material presented in this book, I believe these questions can be answered in ways that advance our understanding of the history of science.

Both sociology and history provide rich precedents for quantitative investigations of modern communities. From the pioneering sociological studies of Middletown by Robert and Helen Lynd (1929, 1937), to the landmark investigations of Robert Park (1925, 1952) and W. I. Thomas (1966), through the classic monograph by historian Merle Curti (1959), the quantitative reconstruction and sociologically informed analysis of communities has been a compelling line of research for several generations of scholars.

However, in the 1970s, just as sophisticated methodological (e.g. Shorter 1971; *Social Science History* 1976–) and historiographical (Shapin and Thackray; 1974; Pyenson 1977) resources became available to historians of science, the winds of scholarly fashion abruptly shifted. In part, shrill and bitter controversy surrounding the quantitative analysis of American slavery (Fogel 1974) may have dampened the enthusiasm for quantitative history. From another perspective, many who called themselves members of the social science history movement became mired in quantitative studies of voting behavior that generated a great deal of data which often had little historiographical significance.

For whatever reasons, after a few pioneering studies, historians of science did not continue to examine scientific communities using quantitative techniques. Seminal work such as Morrell and Thackray (1981) on the early years of the British Association for the Advancement of Science or the study by Thackray et al. (1985) and his University of Pennsylvania team on the American chemical community apparently marked an end rather than a beginning.[1] Most historians of science continued to deal with community in epistemic terms, rather than as a social construct worthy of quantitative analysis.

Despite changes in historiographical fashion, my fascination with community as a fundamental unit of social/historical analysis has remained undiminished. This interest can be traced to my teachers, J. Milton Yinger (Simpson and Yinger 1953), Maurice Stein (1960), and Robert Samuel Fletcher (1971), whose Oberlin lectures often focused on community as a key aspect of American social and historical development. My teachers envisioned communities as dynamic assemblages. Populations changed and their members created new institutions. In turn, these institutions helped define the nature and scope of political, social, economic, and cultural life for individuals and groups that made up communities.

If quantitative studies of communities play a significant role in several fields of sociology and history, why not utilize the approach in the history of science? So I reasoned in the late 1970s and began exploring the feasibility of such an investigation. Across the years, through uncounted hours in the archives and computer center, this fascination has remained undiminished.

An understanding of communities as populations, institutions, and complex social processes can help scholars explain science as a many-sided social activity. While the laboratory and observatory are appropriate sites for certain kinds of studies, they are not the only venues for investigation. Scientists do more than design experiments, build instruments, collect data, and develop explanatory hypotheses. Their professional careers are located within scientific communities.

As social beings, scientists depend on and interact with others of their kind. Careers and power within a scientific community are not free-standing entities, but intricate social processes that can only be understood in context. The same is true for the reward system and other

1. Other classic examples of the quantitative reconstruction of scientific communities include Forman et al. 1975; Kohlstedt 1976; Kevles 1979; Weart 1979 and Rossiter 1982.

institutions that scientists construct. In turn, institutions help define community; they provide resources that scientists use to shape the future of communities.

At one time I believed the work of Robert K. Merton (Storer 1973) would be of great value in analyzing a quantitatively reconstructed scientific community. However, when confronted by rich data and a growing realization of the complexity of a single scientific community, I quickly discovered the limits to Merton's structural functional approach. The structural functional school of social theory (e.g., Parsons 1949), in which Mertonian sociology of science is rooted, provides an elegant but ahistorical edifice that obscures the role of contingency and volition in human affairs. Nor am I interested in perpetuating the Mertonian model of science as a self-regulating meritocracy that transcends those political and social dialectics to which other institutions in a democratic society are subject.

As for Merton's students, their attempts to defend his theoretical constructs using sophisticated statistical methods that obscured complex patterns of changing human behavior (e.g., Cole and Cole 1973; Cole 1979) most certainly contributed to the disillusion with quantitative studies and provided impetus for the rise of the Edinburgh school of social studies of science and its concern for carefully researched historical case studies (e.g., Edge and Mulkay 1976).

The study of communities provides conceptual and analytical perspectives that may lead to a richer and more nuanced reconstruction and explanation of the past. In this book, questions concerning the demographic and institutional aspects of community take center stage. Between 1859 and World War II, who made up the American astronomical community? What were their geographical and social origins? Where were they educated and what were their highest earned degrees? What was the gender balance? And how did gender affect careers? What forms did the scientific career take in the American astronomical community? How did the institutional landscape grow and change over the professional lifetimes of three cohorts of scientists? And what of power? How was it defined? Who held power and how did they use it? And finally, how did the reward system work? Only in the context of community can these questions be answered.

The conceptual vocabulary employed in this study has been generated (perhaps hammered out would be the better phrase) at the fluid and often chaotic intersection of data (both quantitative and literary), curiosity, historiographic tradition, and imagination. My fascination with communities, with the quantitative reconstruction of groups that aggregate to form communities and the dynamics within and between the components of community, represents the deep foundations on which this study rests.

The pages that follow are not cast in traditional narrative form. I find the literary conventions of narrative useful only when dealing with the reconstruction and analysis of events that are of limited chronological scope. This book is analogous to a large-scale canvas on which is depicted the activities of three cohorts of scientists at work between 1859 and 1940. In this reconstruction, quantitative data are at least as important as literary materials collected in the archives. The numbers help us understand patterns of collective behavior while individual biography provides exemplars, giving faces to the numbers. This technique might be called sociological narrative: the story alternates between quantitative and qualitative representations of human behavior in the past.

Communities are far too complex and dynamic for single-factor analysis. This book contains no grand theory or powerful interpretative thesis. Readers will not come away with an intellectual sound bite similar to the famous frontier thesis of Frederick Jackson Turner or John Maynard Keynes' theory of the relationship between government expenditures and economic growth. A grand theory would destroy myriad subtle connections between various components of the American astronomical community and wash out highly nuanced but extremely significant interactions between the community and American culture and society.

The reconstruction and analysis of the American astronomical community is presented here as a series of interlocking chapters that examine the traditional fields of celestial mechanics and astrometry and the new field of astrophysics in a variety of contexts, including education, the structure and nature of careers, power, the reward system and gender. The final product is analogous to a large-scale action painting representing the major elements of the American astronomical community and their underlying dynamics.

I have been told that projects like this are too time consuming, labor intensive, and expensive, that the technical demands of such a project would frighten away researchers. As a mainframe user who relied on a team of assistants, I admit that in the past there was some validity to these criticisms. But the revolution in personal computing provides scholars with new ways to study communities. Desk top machines can carry out operations once reserved for the university mainframe. Optical data entry systems scan documents directly into the memory of a desktop machine in the historian's study. New software helps organize and analyze data stored in the hard drive of a personal computer. These innovative technologies make the quantitative study of scientific communities much more researcher-friendly than in the age of the mainframe. Those who begin such investigations today will not have to punch cards or deal with reels of magnetic tape. They will have much more control over data entry and analysis, both of which can be carried out at the historian's desk rather than in the computer center. When increased computing capacity is required, data can be uploaded from a personal computer to the mainframe through the use of a modem. New computing technologies make the quantitative study of scientific communities an attractive research field for Ph.D. students and professors alike.

Beyond these obvious technological advantages, the historiography of science is moving in new directions.[2] The social constructivist school seems to have reached its upper limits and some of its followers are moving toward the philosophy of science (Shapin 1994). Historians of science are showing interest in the emergence of interdisciplinary (Tatarewicz 1990; Doel 1996) or local scientific communities (Kohler 1994), focusing on such problems as patronage or the development of scientific practice. While these investigations do not attempt a quantitative reconstruction similar to this discussion of American astronomy, there is, nonetheless, a sense of convergence. New conceptions of scientific communities play an important role in a number of recent investigations. Innovative research on scientific communities may come to play a significant role in expanding the frontiers of our discipline.

In addition to historians and sociologists of science, one other group will find this book of interest. Astronomers will learn something about the recent past of their science, about what has changed and what has remained the same since 1859. The Historical Astronomy Division of the American Astronomical Society attests to the fact that astronomers are interested in history. Its annual meetings promote dialogue between scientists and historians on topics ranging from archaeoastronomy to space exploration. But I have observed preferences; astronomers are attracted by the technical challenge of mapping an ancient astronomical site, biographies of great leaders, or the construction of famous telescopes. In the pages that follow, I offer astronomers something different: the social history of a modern science. Astronomers know a great deal about their intellectual roots. I want to provide an understanding of the social roots of modern astronomy; hence the emphasis on community, careers, and power.

The history of American science in the nineteenth and twentieth centuries is a vital and growing field. I hope this study will contribute to a deeper understanding of the development of American science in the modern period and provide historians, sociologists, and astronomers with fresh perspectives on one of America's most interesting scientific communities.

2. Readers alert to trends in the social and behavioral sciences as well as tides in American political culture also realize that there is a resurgence of interest in community by scholars and politicians alike (Sassen 1990; Tickamyer and Duncan 1990; Marks 1991; Baldassare 1992; Walton 1993).

✦

ACKNOWLEDGMENTS

This book, a dozen years in the making, would not have been possible without the patient help of archivists and librarians across the United States. I can not hope to acknowledge each and every one, but to all go my thanks and appreciation for courteous, professional service. Brenda Corbin, librarian at the United States Naval Observatory in Washington, D.C., has, for more years than either of us would like to admit, been a continuing source of assistance and advice. She has tracked down the answers to obscure bibliographical questions in at least three languages. Judy Lola Bausch, librarian at the Yerkes Observatory, guided me through the rich archives of that institution and did not loose her temper when I jammed the Xerox machine on a Friday afternoon. Jean St. Clair, retired archivist at the National Academy of Sciences, placed her unparalleled knowledge of the Academy's archives at my disposal and made my visits to 2100 Constitution Avenue seminars in the history of the Academy. Dorothy Schamberg, who presides over the Shane Archives of the Lick Observatory, located at the University of California at Santa Cruz, taught me how to use one of the great resources for the history of American astronomy and has answered many questions by mail since that wonderful summer I spent at Santa Cruz. Carolyn V. Miller, retired science librarian at Santa Cruz helped me get to know both the collections and my way around the town and campus. To Anne Graver Edwards, reference librarian at the Ellis Library of the University of Missouri at Columbia, I owe an immense debt accumulated over a dozen years and more.

In more than one instance, these librarians and archivists arranged for me to meet informally with senior scientists at their institutions. These discussions, while hardly qualifying as formal oral history interviews, provided important perspectives and enhanced my ability to make sense of the archives.

Support from the program in the history and philosophy of science and the program in sociology at the National Science Foundation (two awards) and from the Trustees of the Dudley Observatory (two grants) released me from teaching and provided funds for travel to archives and to pay research assistants. I am also indebted to the Graduate Research Council of the University of Missouri at Columbia for funding at several critical junctures, especially to Dr. John McCormick, Associate Dean for Research and former Graduate Dean Don Blount. A project like this is labor-intensive and without support from the NSF, the Trustees of the Dudley Observatory, and the Graduate Research Council at the University of Missouri at Columbia, the work could never have been done. I am also deeply grateful to successive chairs of the Department of Physics and Astronomy at Missouri, Sam Warner and Henry White, for providing space for my research team and my ever-increasing hoard of notes and printouts. For many years, the penthouse on the Physics Building was my workshop and intellectual laboratory. Missouri astrophysicists Charles J. Peterson and Terry W. Edwards were always available to answer technical questions.

At West Virginia University, I thank the staff of the Wise and Evansdale Libraries for assistance, especially with interlibrary loan requests. My colleague, Dr. David A. Wisner, was especially helpful with advice on the history of nineteenth-century French education. Kathy Fletcher of Academic Computing provided invaluable assistance as we tried to link the WVU and Radford University computers.

Colleagues in the history of astronomy and American science have been free with advice and criticism. Marc Rothenberg, editor of the Joseph Henry Papers at the Smithsonian Institution has, since 1981, read drafts of my work, made suggestions, and critically watched the development of this project. Also at the Smithsonian David H. DeVorkin provided advice and important suggestions concerning manuscript collections. Robert Smith, who divides his time between the Smithsonian and The Johns Hopkins University, has been especially helpful in keeping me focused on the meaning of the historical experience rather than on its details. Practitioner historians LeRoy Doggett, Owen Gingerich, and Donald E. Osterbrock provided technical help and warm encouragement over the years. I deeply regret that LeRoy did not live to see the final product. Since our first meeting in 1980, David Edge has been a constant source of inspiration and wise counsel. To historians Ron Brashear, Steve Dick, Karl Hufbauer, Peggy Kidwell, Robert Kohler, Howard Plotkin, D. J. Warner, George Webb, and Tom Williams I owe immense debts for shared knowledge and critical perspectives. For almost fifteen years, Kim Lacey Rogers has been a source of encouragement and advice, sharing her perspectives on American social history, as well as discussions on methodology and interpretation. James P. Wittenberg of the College of William and Mary guided me through the earliest stages of this project with technical advice and patient encouragement. I treasure his memory.

This book would never have been written without the assistance of Ricky L. Slavings of Radford University. Soon after Rick joined the project as my graduate research assistant in the fall of 1984, it became clear that he was far more than a gopher solving computer problems and writing programs to analyze the data. Over the years our collaboration has deepened. I thank Rick for his wisdom, patience, and good humor even in the face of ill-tempered mainframes and glitch-ridden programs.

This book is dedicated to James L. McCartney. Since the mid-1960s our scholarly lives and personal histories have run parallel. Across the years, Jim has freely shared his knowledge of the sociology of science, criticized drafts of papers and chapters of this book, and looked over my shoulder as I wrote grant proposals. His fingerprints are on every page.

My thanks to those archivist and librarians who granted permission to quote from materials in their collections. Permission to cite from Maria Mitchell letters in the Rare Books and Manuscript Department of the Vassar College Library has been granted by the Nantucket Maria Mitchell Association.

Finally, warmest thanks to Susan Abrams of the University of Chicago Press, who presided over the production of this book.

✦

ABBREVIATIONS

The following is a list of the archives, libraries, and manuscript depositories in which materials cited in this study are located and the abbreviations used in citations. In the interests of space, I have not listed each individual collection in an archive or depository. To do so would make this list excessively long and duplicate information contained in the citations.

AIP — American Institute of Physics, Niels Bohr Library. College Park, Maryland.

BLUCB — University Archives, Bancroft Library, University of California at Berkeley.

DOA — Dudley Observatory Archives, Dudley Observatory. Schenectady, New York.

FML — Henry Ford Museum and Greenfield Village. Dearborn, Michigan.

HCA — Hamilton College Archives. Hamilton, New York.

HPMFE — George Ellery Hale Papers, Microfilm Edition. California Institute of Technology.

HUA — Harvard University Archives, Pusey Library. Cambridge, Massachusetts.

LC — Library of Congress, Manuscripts Division. Washington, D.C.

LOA — Lowell Observatory Archives. Flagstaff, Arizona.

MTHCA — Mount Holyoke College Archives. South Hadley, Massachusetts.

MTWA — Mount Wilson Observatory Archives, The Huntington Library. San Marino, California.

NAS National Academy of Sciences Archives. Washington, D.C.

NA National Archives. Washington, D.C.

NWUA Northwestern University Archives, University Library. Evanston, Illinois.

SALO Mary Lea Shane Archives of the Lick Observatory, University Library. University of California at Santa Cruz.

SCA Smith College Archives. Northampton, Massachusetts.

UCA University of Chicago Archives, Regenstein Library. Chicago, Illinois.

UCYOA University of Chicago, Yerkes Observatory Archives. Williams Bay, Wisconsin.

UPAIS University of Pittsburgh, Archives of Industrial Society, Hillman Library. Pittsburgh, Pennsylvania.

UVA University of Virginia Archives, Alderman Library. Charlottesville, Virginia.

UWMA University of Wisconsin at Madison Archives, University Library. Madison, Wisconsin.

VCA Special Collections, Vassar College Libraries. Poughkeepsie, New York.

WCA Wellesley College Archives. Wellesley, Massachusetts.

WHMC Western Historical Manuscripts Collection. Ellis Library. University of Missouri at Columbia.

YUA Yale University Manuscripts and Archives. New Haven, Connecticut.

1 ✦

ON WRITING THE HISTORY OF A SCIENTIFIC COMMUNITY

In the mid-1850s, New York University professor Elias Loomis suggested that the number and quality of telescopes in the United States should permit American astronomers to compete with European observatories (Loomis 1856: 50–51). Loomis, however, implied that there were more telescopes than qualified observers. The message was clear. Philanthropists should not endow new observatories; they should provide resources to guarantee the effective use of those already in existence.

Scarcely a decade later, Naval Observatory superintendent James Melville Gilliss lamented that American astronomers had not lived up to their potential; American astronomy lagged behind Europe. For Gilliss, the problem lay deeper than the disruption caused by the Civil War. Telescopes and auxiliary research instrumentation gathered dust because there were not enough astronomers to use them. The solution was obvious: create a viable scientific community around existing telescopes.

This book traces the development of an astronomical community in the United States. Demographic patterns, the education of astronomers, the structure of scientific careers, the reward system, power and institution-building, as well as conflict are examined across the professional lifetimes of three cohorts of astronomers. By 1940, American astronomy had attained world-class status. This achievement rested on more than the construction of giant telescopes. It involved the creation of a scientific community capable of effectively utilizing those instruments.

WHAT IS A SCIENTIFIC COMMUNITY?

Defining a scientific community is neither simple nor straightforward. It involves both theoretical and empirical considerations. This book uses a model of a scientific community based on its social system. Frequently, scientific communities have been treated as presumptive entities (DeVorkin 1981) whose primary functions are epistemic (Kuhn 1970: 176–78; 1977: 296). But social historians require different conceptual perspectives. One fruitful approach is to suggest that scientific communities operate on two distinct levels: the generic community and the specialist research community. Just as modern urban communities are best understood as collections of neighborhoods, so the modern scientific community can be viewed as a collection of specialist research communities. The American astronomical community operated at the generic level (all astronomers) as well as at the level of specialist research communities. External to the generic scientific community and related specialist research communities are larger social entities including the national and international scientific communities as well as the all-encompassing context of American culture.[1]

More formally, scientific communities are bounded social systems serving identifiable populations. The social system of a scientific community operates through a set of institutions. I have used data from the collective biography to reconstruct the population of the American astronomical community and its institutions. Both the generic and research communities reflect levels of consensus and shared meanings. In times of rapid change, however, consensus and shared meanings are disputed. Many of the conflicts discussed in the chapters that follow hinged on differing views of the aims and goals of astronomy as a discipline and the kind of institutions that should reflect and, in turn, define and enforce consensus and shared meanings.[2]

Two caveats must be entered. There is a great deal of overlap between the two levels of community. And, in studying the social history of science, we must not overlook the fact that one component of the role of scientist involves adding to the store of natural knowledge. While this study is not directly concerned with epistemic activities, it does not ignore the intersection of intellectual and social history. Relations between the social and epistemic aspects of astronomy are discussed in chapters 3 and 7. These caveats remind us that our understanding of science can not be neatly packaged. Scientists work in overlapping social contexts. In short, we are dealing with human behavior in all its rich complexity.

The generic scientific community is the arena in which the primary social institutions of modern science are located. Here we find education and the reward system, and it is here that recruitment takes place. It is at the level of the generic community that scientists struggle for scarce re-

1. E. C. Lindeman's classic analysis of community (1931: 102–5) remains a challenging point at which to begin any discussion of the topic. Barnard (1968) deals with more recent perspectives.

2. This paragraph is based on Barnard (1968). See also Traweek (1988: 6).

sources and the generic community legitimates the victors in major epistemic and political conflicts. The professional organizations of American astronomy are located here, and it is the generic community that provides access to the national and international scientific communities as well as to American culture and society.

The specialist research community is, by definition, the location in which astronomers add to the store of natural knowledge. Here the work place (observatories) is located. It is also the research community that provides professional identity and related values. Identity and values developed at the level of the specialist research community are the functional equivalents of political ideology or worldviews; they drive many behaviors when scientists move into the generic community.

It is in the context of the generic community that specialist research communities meld. While each specialist research community had its own publications outlet, astronomers also presented their findings at the twice-yearly meetings of the Astronomical and Astrophysical Society of America (later the American Astronomical Society). These meetings symbolize the way in which the generic community integrated American astronomy. Through the 1940s, meetings of the American Astronomical Society were not organized as concurrent sessions. Rather the society sat as one body, in a single auditorium or conference room, and listened to every paper, no matter what the topic. Speakers were open to questions from members of each research community.

The individual career also serves as a bridge between the two levels of community. For example, success at the level of the specialist community was not enough to guarantee access to the reward system. Individuals had to secure patronage in the generic community and the national or international scientific communities as well.

Beyond the generic and specialist research communities lies American culture. In this context are located major patrons including the federal government, which supported activities in the specialist research community concerned with celestial mechanics and astrometry, as well as private donors and philanthropic foundations that funded research in astrophysics and were responsible for the creation of major astrophysical research institutions. Powerful astronomers reached out to the national culture and appropriated models, roles, and values and then gave them institutional form in the specialist research community. This process involved conflict that occurred in the generic community. Male astronomers also took their perceptions of women from the larger culture.

Sometimes, it is difficult to separate the functions performed by specialist research communities and the generic community. The reward system provides the clearest example of relationships between the two levels of community. The specialist research community is where contributions to knowledge are made and initially evaluated. Here, too, are mentors and sponsors. Individuals must achieve a significant level of success and develop important contacts in the specialist research community if they hope to compete for national and international honors. At the level of the generic community we find patrons who dispense a wide range of favors, as well as the astronomy section of the National Academy of Sciences and astronomers who vote on the distribution of stars in the *American Men of Science* series (*AMS*). The National Academy itself is located beyond the generic scientific community, at the level of the national scientific community. Further out lies the international scientific community; for American astronomers its most important institutions included the Royal Society of London, the Royal Astronomical Society, and the French Academy of Sciences.[3] Early in the twentieth century, international astronomical organizations emerged.

In order to understand the dynamics of a scientific community, social historians need to think in ecological terms. Both biological and social communities are composed of complex interrelated structures that form dynamic systems. To understand the dynamics of a scientific community, the following aspects must be examined.

Boundaries are important in defining a community. Who is part of a community and who is not? To be sure, this problem can be dealt with as a semantic formalism, but that would miss the nature of community as social process. A community is dynamic; it changes over time. If the concept of community is to have maximum explanatory power, it should be used in a comprehensive way. Internalist historiography, whether produced by scientists or professional historians, is generally written from the perspective of "best science elitism" (Moyer 1985: 174). It focuses on great men and their intellectual activities in almost complete isolation from the rest of the scientific community. This is far too narrow a conception of the scientific enterprise.

I prefer to define community inclusively. In this study astronomers are grouped into three classes. The elite include those who were elected to the National Academy of Sciences (NAS) and/or who earned stars in editions of *AMS*. This group was a power elite, controlling resources and defining the cognitive activities of the discipline. Its members also acted as mediators between the community and American culture as well as between astronomy and other sciences. The rank and file, those males who never

3. Compared to the process by which honors and awards are allocated, the education of scientists seems to operate in reverse. Undergraduate education takes place at the level of the national scientific community (chap. 4), with students moving into the generic community for graduate training and then into the specialist community for doctoral research and the acquisition of a mentor.

achieved elite status, and women complete the social map. While the elite stand at the center, the social location of the rank and file and women place them closer to the periphery. During this period, American astronomers often worked closely with mathematicians, physicists, and engineers. However, I have, limited membership in the astronomical community to those whose career patterns clearly identified them as astronomers.

How was the American astronomical community integrated? What gave the community cohesion? Integration involved at least four mechanisms. First, there were shared scientific concerns, either with the old astronomy (celestial mechanics and astrometry) or with the new astronomy (astrophysics). Second, there were organizations, including the Mathematics and Astronomy Section (1882), of the American Association for the Advancement of Science (AAAS), the Astronomical and Astrophysical Society of America (1899) and the Mathematics and Astronomy Section (1899) section of the NAS. The American Philosophical Society and the American Academy of Arts and Sciences also provided astronomers with opportunities for contacts. Third, there were journals that represented the major cognitive divisions of American astronomy: the *Astronomical Journal* (2nd series, 1886) published papers dealing with celestial mechanics and astrometry, and the *Astrophysical Journal* (1895) served the interests of those working in the new astronomy. The *Publications of the Astronomical Society of the Pacific* (1889) started as a regional journal and later attained national stature. And last, but hardly least, there was higher education: colleges and universities with astronomy departments offering undergraduate and graduate degrees.

Membership in the American astronomical community offered many advantages. A sense of professional identity was one benefit; access to resources, ranging from patronage to membership in the NAS, was another. To be sure, there were levels of membership. Amateurs played a significant role in American astronomy (Lankford 1981b, 1988; Williams 1988), providing observational data and developing new instrumentation useful to professionals. But these activities did not lead to full membership in the community. Amateurs rarely had access to resources, power, or the reward system.

The astronomical community did not exist in a vacuum; it was connected to American culture and society. These linkages ranged from patronage and the American system of higher education through models for large-scale research institutions borrowed from the industrial economy. Members of the American astronomical community were also in contact with scientists in other communities. Meetings of the AAAS, NAS, American Philosophical Society, and American Academy of Arts and Sciences provided opportunities for the exchange of information between scientists from different disciplines. Popular journals such as the *Sidereal Messenger*

(1882) and *Popular Astronomy* (1893) linked astronomy and the larger culture.

The historical development of the American astronomical community can be understood through an examination of its demography, institutional growth, and cognitive development. A community existed before 1859, but it would be hard to trace its roots to a time earlier than the 1840s (Warner 1979; Bruce 1987). Most of the major institutions and cognitive divisions of the community developed after 1859.

Did the participants in the historical drama speak of an astronomical community? Occasionally, to be sure, but certainly not early or often. Does this mean there was no community? Some scholars argue that unless the historical actors use a given term or concept themselves, historians have no right to employ it. Such historiographic asceticism would place unacceptable limits on the possibility of explaining the past (Danto 1968; Jardine 1994). Historians should make every effort to employ the most powerful conceptual and analytical tools available to explain the past. Community provides an effective conceptual device that helps us describe and explain the social history of modern science.

ORGANIZING RESEARCH ON THE HISTORY OF A SCIENTIFIC COMMUNITY

The move from theories and concepts to data collection and analysis involves defining the parameters of a research project and selecting appropriate research tools. The following discussion examines the parameters, methods, and data on which this study rests.

The year 1859 was chosen as the beginning date because in that year Bunsen and Kirchhoff announced empirical laws for the physical interpretation of spectra. Since the time of Newton, the meaning of spectra had been an unsettled topic. After 1859 the field of experimental spectroscopy developed rapidly. A few astronomers were attracted to spectroscopy and began research that would transform their science. First the sun and then the stars were subjected to spectroscopic investigation. Astronomers working in the traditional fields of celestial mechanics and astrometry were often critical of these new areas of investigation. Soon a split developed between workers in the traditional fields and those using spectroscopy to examine the physical and chemical composition of the cosmos. The new science of astrophysics emerged and came into competition with celestial mechanics and astrometry.

I selected 1940 as the ending date for this study because the coming of World War II marks a watershed. After the war, new patrons, problems, and instrumentation redefined the social structure and cognitive dimensions of American astronomy. Wartime experiences provided younger astronomers with exposure to a wide range of new electronic techniques, from radar to computers. After the war they took these methods and applied them to astronomical research. New patrons appeared: the military and the National Science Foundation provided lavish support for astronomy. In the 1960s the National Aeronautics and Space Administration joined the list of patrons. After 1945, a growing number of astronomers worked to integrate the findings of quantum mechanics and atomic physics into astrophysics. Soon, many recruits to astronomy were drawn from the ranks of physics Ph.D.'s. New research agendas and technologies eroded distinctions between the old and new astronomy. To carry the story beyond 1940 would be to enter a new social, political, and cognitive landscape.

American astronomy was selected as the research site for several reasons. Archival and manuscript sources are readily accessible. Most of these collections are well organized and some have published finding aids. There are few restrictions on the use of most collections. The history of American science is experiencing a renaissance. The history of astronomy in the United States no longer stands in historiographic isolation. Further, the history of astronomy itself has enjoyed rapid growth. Most recently, we have seen a move away from the study of the ancient, medieval, and early modern periods in favor of the nineteenth and twentieth centuries. Finally, my graduate training was in American social and intellectual history and it seemed prudent to capitalize on these strengths.

In order to study a community, its population and institutions have to be reconstructed. It simply will not do to talk about representative individuals or typical institutions. There is no way to recognize the typical (i.e., average) except on the basis of an examination of the *whole*. This is a quantitative problem and can only be answered numerically. Thus collective biography was selected as the primary research methodology. However, with a database of this size, it would be patently impossible to transcribe all the information on traditional note cards. The result would be a sea of paper! Data were coded and analyzed on the mainframe computer at the University of Missouri at Columbia and later West Virginia University. The machine-readable collective biography provides detailed information on the backgrounds, education, careers, and publications of 1205 astronomers active between 1859 and 1940. Data that constitute the collective biography were drawn from a variety of sources including biographies, bibliographies, obituaries, biographical encyclopedias, oral history interviews, and manuscripts.

In identifying members of the American astronomical community I have cast a wide net. Since the 1960s, social historians and feminist scholars have urged the importance of moving beyond the study of elites (Stearns 1985). We cannot achieve a deep understanding of a scientific community by concentrating only on its leading members. The pages that follow present a scientific community in the round, including its less powerful members and those who stood far from the center, as well as elite astronomers.

No simple formula can specify who should or should not be included as members of a scientific community. And, to complicate matters, selection criteria must be applied across the better part of a century. The study opens as the age of preprofessional science was coming to an end. Many individuals still made significant contributions to more than one field of science. Professional identity was fluid. Clearly defined disciplines and research fields were just coming into being. Eighty years later, the situation was very different.

No definition of membership in the American astronomical community can be applied in a rigid manner. Criteria must be flexible enough to reflect the changing historical context. The criteria include one or more of the following (Rothenberg 1974: 3–7): employment in an observatory or a college or university astronomy department, publications in the field, or recognition by other scientists as an astronomer.[4] In the American astronomical community, educational credentialling, especially graduate education, is an unsatisfactory method for identifying scientists. It is hardly applicable for the preprofessional history of any science and of limited value in astronomy well into the twentieth century. Further, lines of demarcation between amateurs and professional are extremely difficult to draw. Even in the first quarter of the twentieth century a few amateurs earned stars in *AMS*.[5]

The identification of astronomers is complicated by the range of activities practiced in observatories. In Europe it was often the case that astronomical and geophysical research frequently were conducted under the same roof. Indeed, in late-seventeenth and eighteenth-century France, astronomers at the Paris Observatory may have spent more time working in geodesy than on what we would define as astronomical pursuits.[6] Euro-

4. These criteria exclude most technical staff (machinists, opticians, electricians, and the like) unless they moved from the shop to the telescope. Clearly, some did so, e.g., George Willis Ritchey or Milton Humason.

5. Amateurs in the world of professionalizing science have been given serious attention by several scholars. For an introduction to the literature and a discussion of amateurs in nineteenth-century astronomy, see Lankford (1981a and 1981b). Yeo (1989) provides an interesting comparison through a discussion of authority in nineteenth-century British science.

6. See for example Pellitier (1990) for a discussion of eighteenth-century geodesy or Wolf (1902) for a review of activities at the Paris Observatory. An examination of the Uccle reports on world astronomy by Lancaster (1886) and Stroobant et al. (1907, 1931) indicate that geophysical observatories remain important in European astronomy down to World

pean astronomers were often called on to take part in expeditions to determine arcs of the meridian or to assist in trigonometric surveys. In a sense, this was the price they paid for state support. In the United States, mapping was the province of the Coast and Geodetic Survey and the Army Corps of Topographic Engineers.

While American observatories participated in studies of geomagnetism, seismography, meteorology, and provided local time to neighboring communities, these activities were secondary to their primary mission and seldom demanded the full-time commitment of staff members. Geophysics and meteorology generally involved reading instruments at stated intervals and keeping accurate records that would later be forwarded to others for reduction and analysis. As geophysics matured, it left the observatory for an independent institutional and professional identity.[7] The same pattern can be seen in meteorology.

Activities related to the determination and transmission of exact time fit comfortably with the work of most American observatories. Observatory clocks were continually monitored by the use of transit observations. It was not difficult to carry this practice a step further and provide at least one accurate civil time-signal each day. These activities were not burdensome, nor did they distract astronomers from their primary goals.

While the Coast and Geodetic Survey employed astronomers as consultants in the 1840s and 1850s, the practice declined after the Civil War. As American astronomy matured, its practitioners found employment and patronage elsewhere and did not need the Coast Survey. Put another way, from about 1860 the process of professionalization led astronomers to concentrate on astronomical research and drop other scientific interests.

In addition to organizing the population according to status, type of astronomy to which individuals devoted their careers, and gender, I have grouped astronomers into cohorts. Unlike birth cohorts, used by population demographers, the population of the American astronomical community has been organized according to shared experiences.[8] The first cohort includes those who entered astronomy in 1859 or before. At this time astronomy was dominated by celestial mechanics and astrometry. The or-

War II and that in many observatories there were staff members whose designated activities involved geophysical rather than astronomical research.

7. Reingold's view (1964: 59–62) that geophysics was a primary activity in American observatories is difficult to substantiate after mid-century.

8. The problem of generations is much more complex for historians than demographers. Anyone who deals with generations in numerical terms alone misses the point of shared experiences. See Spitzer (1973) for an important discussion. I am indebted to Professor John Modell of Carnegie-Mellon University for his comments on this problem at a session on "Collective Biography and Life Course Studies" held at the 1986 Social Science History Association meeting in St. Louis.

bits of solar system objects and the publication of astrometric catalogues were the primary concerns of astronomers. This cohort often did research in astronomy and at least one other science, a characteristic of preprofessional science.

The second cohort entered astronomy between 1860 and 1899. This cohort witnessed the development of astrophysics and a debate over its legitimacy. The second cohort also experienced a revolution in instrumentation, including the construction of large refracting telescopes and the application of photography to astronomy as well as the incorporation of solar and stellar spectroscopy into the research agenda.[9]

The third cohort, those who entered between 1900 and 1940, experienced the resolution of the debate over the legitimacy of astrophysics and the acceptance of the reflecting telescope as the primary research tool for most astronomers, as well as demographic changes that included a long period (ca. 1900 through the 1920s) in which the demand for trained astronomers exceeded the supply. By the 1930s, many astronomers realized that the infrastructure of their science needed to be modernized. From telescopes to learned journals, the astronomical community was in need of renewal.[10]

Rich data, organized in machine-readable form for rapid and often complex manipulation, can be analyzed in a number of ways. These range from simple frequency tables to multivariate analysis. The variables discussed above (status, cohort, gender, basic research interests) make comparative analysis possible, e.g., between those who practiced the old and new astronomy or the elite and rank and file, and women.

From the perspective of traditional internalist historiography, it may be a matter of concern to find that scientists presented in these pages spent much of their time on career management, the quest for honors and awards, searching for resources, or in power struggles with peers, rather than in the production of scientific knowledge.[11] A scientist's life, however, involves a variety of activities other than research. Indeed, this book fo-

9. Nineteenth-century developments are covered by Clerke (1908). This discussion should be supplemented by Lankford (1984a) and Osterbrock (1985) on new forms of instrumentation. Osterbrock's life of James E. Keeler (1984a) provides a good overview of this cohort through the eyes of one of the pioneers in astrophysics. Meadows (1970) deals with the rise of solar physics, and Hearnshaw (1986) traces the development of stellar spectroscopy.

10. The period after 1900 can be followed in Gingerich (1984) and its broad intellectual contours traced in Struve and Zebergs (1962).

11. Forman (1991) deals with these complicated historiographical problems in a major paper. In brief, he argues that historians of science should not devote themselves to the glorification of scientists, but report their findings independently of scientist practitioners whose historiography is dedicated to providing a privileged status for science and scientists. In a sense, Foreman is asking professional historians of science to disassociate themselves from the Sarton model of the history of science (Reingold 1991).

cuses on those aspects of the scientific career that are not directly related to the production of knowledge.

Andrew Pickering (1992: vii) has argued that the historiography of science can be divided into two broad domains: studies of scientific knowledge and studies of scientific practice. To this typology I would add a third category: the study of the large-scale social organization of science, especially the population and institutions of the scientific community, scientific careers, and the exercise of power in the community.

Unlike the problematics of science-as-knowledge or science-as-practice, the problematics addressed in this book are not epistemic. My interest is in social institutions and social processes; the focus involves social rather than intellectual history.

THE ORGANIZATION OF THIS BOOK

The following chapters explore the development of the American astronomical community between 1859 and the coming of World War II. Each chapter rests on quantitative data and manuscript sources. I have tried to make clear the human meaning of the numbers through examples from the lives of individual astronomers. Where possible, biographical materials are drawn from manuscripts rather than better-known printed sources.

In order to provide a benchmark against which to measure change, chapter 2 examines the population and structure of the American astronomical community in 1859. I also discuss the place of astronomy in the culture of modern science at mid-century. The careers of several astronomers illustrate aspects of the community in 1859.

Conflict between the old astronomy (celestial mechanics and astrometry) and the new astronomy (astrophysics) is the focus of chapter 3. Of special importance is understanding the mind-set of workers in the old and the new astronomy. What assumptions did they make about the nature and scope of the discipline? How did they value the process of discovery in astronomy? How did they define the role of astronomer? The chapter also examines common ground between the two groups and traces the emergence of a research agenda for astrophysics.

The education of American astronomers is the subject of chapter 4. Here quantitative data drawn from the collective biography are especially useful in illuminating patterns of education within the community, while manuscript sources help us explain findings suggested by the numbers. The content and structure of undergraduate and graduate education are examined and the collective experiences represented by quantitative data brought to life through biography. The chapter ends at the point astronomers moved to a first job.

Chapter 5 takes up the topic of the changing scientific career. My goal here (as elsewhere in the book) is to move beyond impressions based on the lives of a few elite astronomers. Thus the task was to model the scientific career as a social process responding to changing historical circumstances. This approach places the career in a market context and considers both the supply and demand sides of the equation. Collective biography permits the quantitative discussion of careers and illuminates the process of career development in ways that anecdotal evidence never could. Questions such as career duration, the pools from which Ph.D. and non-Ph.D. astronomers were drawn, employment patterns, and the complex nature of career paths, as well as the careers of those who left astronomy are considered.

Career management is the subject of chapter 6. The focus shifts from quantitative models developed in the previous chapter to qualitative considerations based on extensive use of manuscripts. Individual careers are examined in order to demonstrate the role played by mentors, patrons, and sponsors. Failed careers are also discussed. Both quantitative and manuscript sources are used to examine the role of chance and choice in fundamental career decisions such as the choice between the old or the new astronomy.

Chapters 4, 5, and 6 form a unit, focusing on preparation for a career in astronomy and the ways in which careers unfold. Chapter 7 returns to problems related to the rise of astrophysics. Conflict in the American astronomical community was, in large part, centered around two closely related areas: tensions between the old and new astronomy and the status and power of observatory directors, who formed a powerful group within the community. Chapter 3 explores the epistemic and social differences between workers in the old and the new astronomy. Chapter 7 discusses structural and ideological factors that define the nature and uses of power. This chapter also explores linkages between the astronomical community and American culture and society, including the ways in which astronomers borrowed models and roles from the industrial economy.

Discussion of the reward system in the American astronomical community in chapter 8 is an extension of problems considered in the previous chapter. Data from the collective biography provide a quantitative portrait of the power elite and allow us to trace the routes by which they attained power, while archival materials illuminate the process of nomination and election to the National Academy of Sciences, the American Philosophical Society, and the American Academy of Arts and Sciences. The chapter closes with a statistical analysis of the reward system. This provides insights into a long-debated problem in the history and sociology of science: How fair is the reward system? Does the reward system reflect patronage or merit? The answer to this question mirrors the complexity of the scientific career discussed in earlier chapters.

Chapter 9 examines the role and status women in the American astronomical community. In a sense, this chapter is a recapitulation of topics examined earlier. The education and careers of women are discussed and illustrated through biographical examples. The relation of women to the reward system is also explored. Materials presented in this chapter stand in sharp contract to the experiences of male astronomers. Economic differences are explained in part by the existence of a dual labor market in American astronomy.

The American astronomical community in 1940 is the subject of chapter 10. Here empirical data drawn from the collective biography invite comparison with the portrait of the community in 1859 developed in chapter 2. The chapter also discusses the political economy of American astronomy from 1859 to World War II.

The final chapter is comparative. Astronomy is discussed in the context of American science, especially physics, chemistry, and biology. Further, comparisons are developed between the growth of astrophysics and the rise of genetics. This strategy suggests important structural and cognitive constraints on the development of new fields of science in America. In addition, the chapter deals with cross-cultural comparisons. What was the structure and organization of astronomy in the United States, the United Kingdom, France, and Germany. How were these national scientific communities similar and how were they different? Were there national styles in astronomy? Finally, what was characteristically American about astronomy in the United States?

2 ✦

THE AMERICAN ASTRONOMICAL COMMUNITY IN 1859: A BENCHMARK

The year 1859 stands as an *annus mirabilis* in the history of modern science, witnessing the publication of Bunsen and Kirchhoff's first paper on spectrum analysis as well as the appearance of Darwin's *On the Origin of Species*. The one led to important developments in the physical sciences and the rise of astrophysics. The other transformed the biological sciences and, in time, much of Western thought.

In order to understand the American astronomical community in 1859, the decade of the 1850s merits synoptic review. The 1850s mark a watershed in the history of American science. Through the 1840s there was little differentiation between fields and the social role of scientist had not been created; scientific societies were local rather than national. By the 1860s the scientist had emerged as a legitimate social role and national rather than local scientific societies catered to the needs of the scientific community. In this new context, specialization became increasingly important.

AMERICAN ASTRONOMY AND THE CULTURE OF MODERN SCIENCE

Sociologist Joseph Ben-David has provided one of the most widely accepted models of the culture of modern science. Ben-David (1971: 159) sees the rise of the nineteenth-century university and the institutionalization of the role of the scientist within universities as key to understanding the culture of modern science. Historian Nathan Reingold (1976) has produced an influential analysis of the culture of science in America. Reingold examines changing relationships between three social groups, cultivators (a lay audience), practitioners (technicians and teachers), and researchers, within the nineteenth century demographic, economic and intellectual milieux. In principle, these models are compatible and reinforce one another at key points. From the perspective of discipline-specific history of science, however, they are necessarily limited. Ben-David is concerned with a comparative discussion involving England, France, Germany, and the United States, while Reingold seeks to integrate the history of the sciences in America. It will be instructive to review the sociological and historical models from a fresh perspective. The emergence of the culture of modern science appears somewhat different when examined from the vantage point of the history of American astronomy.

As Ben-David (1971: chap. 5) suggests, before science can flourish there must be a belief in its value to society. This belief can take many forms. In Anglo-American culture, science was justified both in terms of its moral and economic benefits. Astronomy, for example, was viewed as indispensable to the growth of ocean-borne trade and commerce, and astronomers were charged with developing various aids to navigation. As Dick (1983) points out, this was the reason behind the founding of the United States Naval Observatory. By the mid-nineteenth century, however, astronomy had achieved a unique position. While its practical aspects were held in great esteem, astronomical research of a decidedly non-utilitarian nature was being supported by private philanthropy as well as tax dollars.

After 1859 astronomical and astrophysical research in the United States became increasingly divorced from practical applications. Astrometry and celestial mechanics achieved levels of precision far in excess of the demands of practical navigation. Astrophysical research in solar and stellar spectroscopy or photometry offered virtually no practical returns. Yet private patronage grew rapidly without concern for the material value of astrophysical research.[1] While chemists, geologists and physicists were constantly reminded of the need to produce practical, economically beneficial results (Brush 1979: 51–55), American astronomers occupied a privileged position. Philanthropists and the scientifically literate public supported astronomical research without demanding an economic payoff.

Public support for astronomy was not entirely accidental. Starting in the 1840s, a number of textbooks and popular expositions of astronomical topics issued from the presses. English books on astronomical topics also found a ready market in the United States (Hetherington 1983). Gifted popularizers, such as Cincinnati astronomer Ormsby MacKnight Mitchel (1809–62), were in great demand on the antebellum lecture

1. Miller (1970) is the best introduction to the history of philanthropic support for astronomy. Astronomy's one income-generating activity, providing accurate time to local communities, generally did not have a direct impact on shaping the research programs of observatories. It was a relatively simple mechanical problem that did not involve innovative research.

circuit. Mitchel also published a popular astronomical monthly, the *Sidereal Messenger* (1846–48) and his lectures were collected and circulated widely in printed form (McCormmach 1966).

An important consequence of these activities was the development of a market for astronomical knowledge. The emerging urban middle class eagerly consumed a growing stream of astronomical books and lectures and there developed what Reingold (1976: 34–46) calls the cultivator class. The reasons behind the growth of interest in astronomy are complex. Many persons must have been stimulated by romantic religiosity, which found the wonders of the universe a source of deep spiritual inspiration. Others wished to see astronomical objects for themselves and supported the building of observatories (Rothenberg 1990). The interest of some may have been stimulated by a desire to develop astronomy as part of local or regional culture. For whatever reason, the increasing demand for astronomical knowledge expanded employment opportunities for scientists. Because astronomy developed a supportive public earlier than most other sciences, its subsequent growth would differ from theirs.

Once an astronomically literate public reached a certain critical mass, scientists were able to tap this resource. Ormsby McK. Mitchel gained financial support from citizens of Cincinnati (Goldfarb 1969) and built an observatory in 1843. Four years later, the Harvard College Observatory (Jones and Boyd 1971: chap. 2) installed a fifteen-inch refracting telescope paid for by public subscriptions. In 1856 the Dudley Observatory (James 1987) was dedicated at Albany, New York. This institution owed its existence to private philanthropy.

New institutions providing superior instrumentation allowed astronomers to pursue more challenging research programs. But, at the same time, the possibility of conflict between patrons and scientists increased. At Cincinnati, Mitchel was continually faced with demands from his supporters for virtually unlimited access to the telescope while at Harvard conflict was avoided by a system of public viewing nights. The Dudley Observatory became embroiled in a bitter struggle involving both personalities and conflicting views over the management of a scientific institution financed by public subscriptions.

Ben-David (1971: 40–41) reminds us that in the early nineteenth century, the role of research scientist did not exist. This role came into being in Germany. As it spread through the Western world, the role of research scientist was adapted to particular national circumstances. In the United States, research soon became institutionalized as part of the scientific community, where it served to insulate scientists from pragmatic and materialistic dimensions of American life. Further, within a given scientific community, researchers were accorded higher status and greater rewards than scientists involved only in teaching. The development of the research ethic underpins Reingold's distinction between practitioners, whose primary concern was teaching, and research scientists. Clearly the active researcher was separated from the cultivator (whether patron, interested layperson, or active amateur scientist) by a wide gulf. Those who aspired to the role of researcher sought institutional contexts that would provide them with the freedom and resources needed to pursue a research career.

This process of differentiation did not occur overnight. A viable institutional base for research careers was especially difficult to develop in a science like astronomy, with its demand for expensive instrumentation. In 1859 the single largest employer of astronomers was the federal government. Despite a tendency toward mission-oriented research, federal institutions seemed to offer the best available opportunities. After the Civil War there would be a shift away from federal service toward institutional locations that offered greater latitude for the individual researcher. Prior to 1860 (Dupree 1957b: chaps. 5–6), however, institutions like the U.S. Naval Observatory (1830), the Naval Academy (1845), the Nautical Almanac Office (1849), the reorganized Coast Survey (1843) and the various boundary and survey commissions provided numerous opportunities for careers in astronomy.

As research activities increased, American astronomers strengthened contacts with the international astronomical community. This was accomplished through personal correspondence, visits to European observatories and publication in international journals.[2] To be sure, in 1859 this process was just beginning. Of the one hundred and sixty papers that appeared in the *Astronomische Nachrichten* during 1859, only seven (4 percent) were by American astronomers. However modestly, Americans were beginning to make their presence felt on the international scene.

Driven by cultural nationalism, as well as frustration at the slow growth of astronomical research, leading scientists like Naval Observatory Superintendent James Melville Gilliss (1811–65) lamented, "Our country has done so little to redeem the noble promises implicitly made to astronomers of the old world." He enumerated the major research institutions of the day: the Naval Observatory (USNO), the Harvard College Observatory (HCO), the Cincinnati Observatory, and the Dudley Observatory. Gilliss noted that each of these institutions had "one or more magnificent instruments from which valuable results have justly been expected." Yet, because of political or administrative problems, lack of funds and the economic and social crisis brought on by the Civil War,

2. A discussion of American astronomers abroad would make an interesting and valuable contribution to our knowledge of nineteenth-century science. Reading through manuscript collections, I get the impression that a grand tour of European observatories was required for those who aspired to positions of eminence.

these observatories had accomplished little. "It is our duty," Gilliss concluded, "to strive for a new order of things and to apply these splendid [telescopes] . . . to the purposes for which they were constructed."[3]

With increased funding and improved facilities, American astronomers sought to establish domestic outlets for the publication of research and to create new institutions that would permit increased contacts between scientists. Benjamin A. Gould (1824–96) founded the first professional journal (Herrmann 1971) for American astronomers in 1849. Trained in Germany, Gould modeled his *Astronomical Journal* on the *Astronomische Nachrichten*. The *AJ* suspended publication during the Civil War and did not resume until its founder retired from the directorship of the Cordoba Observatory in the mid-1880s. From 1849 to 1861 and again after 1885 the *AJ* was the premier journal for American astronomers, although many papers continued to appear in the *American Journal of Science*. In the 1890s, its position was challenged by the *Astrophysical Journal*.

Astronomers like Alexis Caswell (1799–1877) joined Gould in working to create and nurture the American Association for the Advancement of Science (1848). In its original form the AAAS did not differentiate between cultivators, practitioners, and researchers (Kohlstedt 1976). It soon was captured by those committed to the research ethic and played a major part in defining the culture of modern science in America.

Gould was also active in organizing the National Academy of Sciences (Cochrane 1978). The Academy was created (Dupree 1957a) at a time of national crisis (1863), under the guise of providing scientific advice to the federal government. In fact, its founders wished to establish an elite organization, membership in which would confer honor similar to election to the Royal Society of London or the French Academy of Sciences. The founders of the Academy also took advantage of the opportunity to define the content of the sciences in America. Gould used his position as a member of the powerful inner circle that created the Academy, to exclude George Bond (1825–65), director of the HCO and John Draper (1811–82), whose research included chemistry, physics and astronomy. Gould believed (Jones and Boyd 1971:126–30) that the only worthwhile research in astronomy focused on celestial mechanics and astrometry using techniques and instrumentation developed in European observatories.

While a status hierarchy in American astronomy had emerged by 1859, it would not become as rigid as that in the chemical or physics communities. The process of professionalization in astronomy never completely excluded the amateur. For this reason it is necessary to examine carefully the class of cultivators described by Reingold. After 1859 this

3. James Melville Gilliss to George W. Hough, 26 December 1864. NWUA, Hough Papers.

Table 2.1 Employment of American Astronomers in 1859

	Status			
Institution	Elite	Rank and File	Women	**Total**
Government	14 (44%)	12 (36%)	1	**27 (41%)**
University	9 (28%)	8 (24%)		
Public	1	2		**17 (26%)**
Private	8	6		
College	6 (18%)	6 (18%)		**12 (18%)**
Public	2	0		
Private	4	6		
Private Research Institutions	3 (9%)	6 (18%)		**9 (14%)**
Polytech	0	1 (3%)		**1 (1.5%)**
Total	**32**	**33**	**1**	**6**

class would become stratified. It contained the great patrons of astronomy: women and men who gave money for instruments and the buildings to house them or to endow large-scale research projects. Cultivators also included an astronomically literate public who followed the science as reported in newspapers, popular books, and lectures. At yet another level came amateurs who were themselves divided into several classes. Amateurs included a large number who enjoyed learning the constellations and the names of the brighter stars or who took pleasure in scanning the sky with opera glasses. A much smaller number owned telescopes and a few employed them in serious investigations that might even be published in the *Astronomical Journal*. Frequently, recruits to the science of astronomy were drawn from those who made up this portion of the cultivator class. The class of cultivators (Reingold 1976: 47) remained an important factor that must be taken into account in order to understand the history of American astronomy.

In 1859 the American astronomical community already contained the seeds of remarkable diversity. As table 2.1 indicates, universities stood second to the federal government as an employer of astronomers, with private institutions outdistancing public by a considerable margin. It should be remembered, however, that virtually none of these institutions had or would have Ph.D. programs in astronomy for many years. The Harvard College Observatory remained aloof from graduate education until the 1920s. Active research institutions like HCO were virtually autonomous from the universities with which they were associated. Colleges rank third. Here, too, private institutions were preferred to public. For example, Christian H. F. Peters (1813–90), who spent most of his American career at Hamilton College, managed to secure excellent instrumentation and carry out important researches in solar system astronomy and star mapping. Private observatories stand last on the list of institutions in which astronomers found employment in 1859. These ranged from the Dudley with its world-class transit circle through the observatory maintained by New York lawyer turned astronomer, Lewis Morris Rutherfurd (1816–92), who pioneered new forms of instrumentation and the application of photography to spectroscopy and astrometry. One astronomer was employed at Philadelphia's Central High School, America's only equivalent of a European polytechnical institution.

This diverse pattern of employment meant that the American astronomical community did not become concentrated in one institutional setting. Several institutional locations were available. And there is an important corollary; American astronomers were probably not as Ph.D.-conscious as scientists in other fields. Entry into the community was never restricted to holders of the doctorate, and those with only baccalaureate degrees (or even less) frequently achieved a full measure of status and power.

In summary, the early development of the American astronomical community followed many of the patterns suggested by Ben-David and Reingold in their discussions of the culture of modern science, but there are also important differences. To use an analogy from economic history, local markets (Greene 1976; Gerstner 1976; and Shapiro 1976) developed as an astronomically literate public both consumed and supported the production and dissemination of astronomical knowledge. The success of these local markets led to further developments; astronomers became involved in creating national institutions that would help define and legitimate the role of scientist. Differentiation took place along the lines proposed by Reingold, but, at least for astronomy, the structure of the cultivator class became increasingly complex. Further, the role of researcher was realized in various institutional settings. In astronomy, the lines of demarcation that, according to Reingold and Ben-David, separated various statuses in a scientific community were neither clear-cut nor final. Mobility between Reingold's class of cultivators and the ranks of practitioners and researchers was common.

THE STRUCTURE OF THE AMERICAN ASTRONOMICAL COMMUNITY

The discussion of social structure begins with demographic considerations. Population data includes information on gender, geographic origins and levels of educational attainment. The distribution of the population by location of employment throws light on career patterns and suggests the range of professional opportunities. Finally, specific institutions including the publications network, professional organizations and obser-

Table 2.2 Professional Activities of American Astronomers in 1859

Activity	Status			Total
	Elite	Rank and File	Women	
Research	10 (31%)	17 (52%)	1	28 (42%)
Teaching & Research	10 (31%)	12 (36%)		22 (33%)
Research & Administration	5 (16%)	4 (12%)		9 (14%)
Teaching, Research & Administration	5 (16%)			5 (8%)
Teaching	2 (6%)			2 (3%)
Total	32	33	1	66

vatories are examined. The discussion of observatories also involves examination of research instrumentation and research programs.

In 1859 the American astronomical community numbered 66 individuals (65 men and one woman). Seventy-nine percent (49) of these astronomers were native-born, with New England contributing the largest share (47 percent), followed by the mid-Atlantic states (21 percent). Twenty-one percent (13) were foreign-born with the largest group (6) coming from the United Kingdom. Germany and Canada each contributed two and the rest were scattered between the Austrian Empire, Scandinavia, and Italy. Thirty-seven percent had fathers who were employed in white-collar occupations, while the fathers of 39 percent were professionals.[4] Only two members of the American astronomical community came from astronomical families (George Bond and Charles A. Young), while the father of one was a medical doctor. For 74 percent of the group the highest degree was an A.B. and for 15 percent it was the M.A. Two held the Ph.D. and three were M.D.'s. The federal government was the single most important employer (41 percent) for these scientists.

Table 2.2 reports data on the professional activities of astronomers in 1859. Forty-two percent of the American astronomical community were involved primarily in research. Fifty-four percent mixed research with teaching and/or administration. Only 3 percent devoted themselves exclusively to teaching.

4. These figures are generally compatible with the findings of Elliott (1982). However, the socioeconomic status of fathers of these astronomers differs from findings reported by Elliott. More fathers were employed in white-collar occupations than in the population of scientists studied by Elliott. Elliott (1982: 84) is correct in observing that "One of the most difficult kinds of historical evidence to acquire is the economic class of scientists." My classification is based on the reported occupations of fathers taken from various biographical sources.

The oldest member of the community was Rutgers mathematical astronomer Theodore Strong (1790–1869) who began his scientific career in 1812, while among the youngest was George W. Hough (1836–1909) who in 1859 moved to the Cincinnati Observatory after a year as an assistant at the Harvard College Observatory.

A hypothetical individual representing the demographic characteristics of the American astronomical community would have the following characteristics. He would be a white male, born in a small town in New England. His father would be a clergyman, doctor, lawyer, or teacher. The individual would hold the bachelor's degree. This composite figure would work at a government research facility. He would probably have entered into a scientific career in the late 1840s.

In 1859 the professional institutions of the American astronomical community were underdeveloped. It may be fruitful (Hetherington 1983) to consider the early history of American astronomy (Warner 1979) in the context of a developing nation. Many American astronomers relied on European journals to communicate research findings and looked to English and continental academies and learned societies for recognition. European honors and awards were virtually the only significant form of recognition available to American astronomers in 1859.

Central among institutions necessary for the growth of modern science are those that facilitate communication. The *Astronomical Journal* was the only specialized domestic journal for astronomers and it would suspend publication in 1861. The *American Journal of Science* (1818) included papers on astronomical subjects, but the *AJS* served all of American science. Some American observatories published volumes of annals, but these were slow to appear. The *Memoirs of the American Academy of Arts and Sciences* provided an outlet for the research of New England astronomers. Bureaus of the federal government also published astronomical research, but like the annals of observatories, these volumes appeared after long delays and in limited quantities.

The only national professional organization available to astronomers was the AAAS. Not until the very end of the nineteenth century would a professional astronomical society be formed. Astronomers were active in the affairs of the AAAS and attended its meetings. After 1849, the association included a section devoted to mathematics and astronomy, in which astronomers communicated through oral presentations as well as informal discussions. The published proceedings of the annual meetings of the AAAS included papers presented at each of the sections. Personal correspondence, however, remained the major vehicle of communication between astronomers.

Neither local nor regional scientific societies can be viewed as a significant source of honors or recognition for American astronomers. Beyond election to office, these organizations offered little that was unique or exclusive. In keeping with the democratic spirit of the day, most were inclusive: cultivators, practitioners and research scientists were thrown together. Despite the cultural nationalism of the antebellum period, American science still looked across the Atlantic both for ideas and approbation. Awards from the Royal Society of London or the French Academy of Sciences meant a great deal more than recognition by local or regional American societies. Election as a foreign associate of a European academy was viewed by many as the capstone of a scientific career.

Astronomy is a data-driven science. In order to collect data these scientists needed increasingly complex and expensive instrumentation. Each generation sought to improve on the work of its predecessors. This involved more refined instrumentation in order to make astrometric measurements with ever greater precision. Astronomers were driven to secure sophisticated forms of research instrumentation for social and political reasons as well. The culture of modern science places a premium on discovery. Scientists exchange knowledge (Hagstrom 1965) for recognition from the scientific community. In order to make significant observations and collect data that peers would think worthy of reward, astronomers needed access to the best available instrumentation.

The institutional context in which instruments are located is also of great importance. The most advanced instrumentation is of little value in a hostile environment. The early history of the Cincinnati and Dudley observatories suggests that if financial resources are lacking or the institution is divided by bitter quarrels, there is little chance that telescopes will be put to productive use. Further, observatories with small staffs deny scientists an opportunity to discuss research problems with knowledgeable colleagues. Observatories lacking assistants to help reduce and analyze data may do little more than accumulate observations. Lack of funds to support the publication of research will also limit an observatory's effectiveness. When heavy teaching demands were placed on astronomers, they could not, no matter how advanced their telescopes, produce much in the way of results. Federal facilities were committed to applied research programs and frequently did not permit investigators to select their own research problems. It is important, therefore, to consider the institutional context in which instrumentation is located.

In his remarks as retiring president of the AAAS in 1859, Alexis Caswell, Brown University astronomy and meteorology professor, reviewed the history of American astronomy. Caswell argued that until about 1840, Americans "had done almost nothing in the way of astronomical experiments." While Caswell (1860: 19–20) listed a number of telescopes at federal, collegiate, and private institutions, he concluded that none were suitable "for the exact determination of the elements of position," the essence of astronomical research.

After 1840 the situation began to change. In that year the observatory at Philadelphia's Central High School mounted a six-inch refractor from the Munich firm of Merz and Mahler, the premier makers of telescopes in Europe. New York University astronomer Elias Loomis (1811–89) suggested (1856: 29) that the opening of the observatory at Central High School marked "an epoch in the history of American astronomy" because of the "introduction of a class of instruments superior to any which had been hitherto imported."[5]

Caswell listed several refracting telescopes from which important research results could be expected. The largest of these, a fifteen-inch telescope, was located at the Harvard College Observatory. Thirteen-inch refractors were at the University of Michigan, Hamilton College and in the private research observatory of Lewis M. Rutherfurd in New York City. The Cincinnati Observatory was equipped with a twelve-inch refractor, but the director, Ormsby McK. Mitchel, spent much of his time as a consulting engineer to several major railroads "and the observatory has consequently been neglected." Elias Loomis (1856: 38) must have spoken for many in the American astronomical community when he chided, "We trust that he [Mitchel] will soon free himself from such groveling occupations, and again direct the gaze of his powerful telescope to study the movements of distant worlds." Mitchel, who died in the Civil War, found little time for astronomy during the last years of his life. The U.S. Naval Observatory in Washington had the smallest of these refractors, a nine-and-a-half-inch Munich instrument.

Large refracting telescopes were used to measure double stars, study the surface features of planets and the moon, as well as search for new satellites and asteroids, discover nebulae and star clusters and determine the position of comets, minor planets and planetary satellites so that their orbits could be calculated. In order to carry out these tasks, refractors had to be supplied with expensive auxiliary equipment.

By 1859 many observatories were equipped with other instrumentation as well. The transit circle is a highly specialized instrument that permits astronomers to measure the position of an object with more precision

5. There is no standard authoritative list of telescopes and observatories in antebellum American and little agreement among scholars on what such a list should include. For an introduction to the literature, see Warner (1979: 70, n. 24).

When astronomers compile what they call historical lists of telescopes they generally apply ahistorical (i.e., twentieth- century) standards to define large instruments. For an example see the list of "The World's Largest Telescopes, 1850–1950," which forms an appendix to Gingerich (1984: Ai-Avi). To be of any historical value, this list should have included refractors with objectives of 6 inches or larger. Because the compiler selected 15 inches as the lower limit, there are no references to refracting telescopes built by the American telescope-maker Henry Fitz (Lankford 1984b). These were major research instruments in mid-nineteenth-century American astronomy, if not by the standards of astronomy in the 1990s.

than is possible with a refracting telescope on an equatorial mount. Data collected with these instruments were used to compile star catalogues for the use of navigators, surveyors, and astronomers. These data also helped improve the value of several fundamental constants in astronomy.

Caswell (1860: 21) reported that in 1859 there were six transit circles "of the most refined construction" located at the Harvard Observatory, the Naval Observatory, the Observatory of Georgetown College in Washington, D.C., at West Point, the University of Michigan, and the Dudley Observatory in Albany. Caswell did not feel it necessary to elaborate on what his audience knew only too well: most of these instruments were either not being used or, because of small staffs and limited budgets, were underutilized. In the case of the Naval Observatory, publication of results was so far in arrears that its research was inaccessible to most astronomers.

Beneath these premier institutions there was a second tier. Caswell (1860:21–2) suggested in his AAAS address that "there are in the possession of collegiate institutions and in the hands of amateur astronomers, I think, not less than twenty instruments, which . . . from time to time will be able to furnish most valuable contributions to astronomy." Several of these instruments (Loomis 1856: 46–8) are worthy of mention. They provide another illustration of the pluralistic organization of astronomical research in antebellum America. An amateur, a Mr. Van Arsdale of Newark, New Jersey, used a four-inch comet seeker carried on a six-inch refractor (both instruments by the American telescope-maker Henry Fitz) to discover three comets. At his home on Sixteenth Street near Fifth Avenue in New York City, a Mr. Campbell remodelled the top story so that he could mount an eight-inch Fitz refractor with which he took some of the earliest solar eclipse photographs in 1854 (Loomis, 1856:46–48). While hardly qualifying as amateurs, Benjamin A. Gould and former Shelby College professor Joseph Winlock (1826–75), a staff member at the Nautical Almanac Office, borrowed the Shelby telescope and a transit instrument from the Coast Survey and set up the Cloverden Observatory (Loomis 1856: 49), where they measured the positions of comets and asteroids.

In a widely read article in *Harper's*, Elias Loomis (1856) painted a bright picture of American astronomy. Published a few years before Caswell's AAAS address, the Loomis paper is of great value for understanding conditions in the early 1850s. The New York University scientist suggested (Loomis 1856: 50–51) that "We now have instruments which permit us to engage in astronomical researches upon a footing of equality with the oldest establishments of Europe" and argued that the number of qualified observers "has kept pace with the increase of our instruments." However, from the tone and direction of Loomis's remarks, it was clear that he believed an adequate financial base was lacking. He devoted much of the *Harper's* article to justifying the role of astronomy in contemporary life and, by extension, why it was worthy of philanthropic support.

The arguments presented by Loomis can be organized under two broad headings: economic and cultural. He stressed the importance of exact time to navigation, the railroads, and businessmen. Observatories were the providers of time in nineteenth-century America and thus performed a significant economic and social service. In addition, the *Nautical Almanac* provided a major aid to navigators. It was important both for commercial shipping and the Navy.

Loomis (1856: 52) then turned to the cultural influences of astronomy. He argued that as the premier research science in America, "an astronomical observatory . . . is a center of genial influence, which directly or indirectly imparts life and efficiency to all subordinate institutions of education." At a more personal level, astronomy offered a form of recreation that could "inspire the mind with noble sentiments" and "direct the thoughts to the wonders of the material universe." With the help of an astronomer and telescope "men of business may acquire new ideas of the wonders of the material universe" and at the end of a day "spent toiling for the acquisition of wealth, may learn that there are mines of intellectual riches more inexhaustible than the mines of California."

In the year 1859, American astronomers possessed some world-class research instrumentation, but often lacked institutional settings conducive to their effective use. In the second half of the nineteenth century this problem would be solved as new research institutions emerged. These developments had a great impact on careers in the American astronomical community.

SCIENTIFIC CONCERNS OF AMERICAN ASTRONOMERS

In 1859 the scientific interests of American astronomers fell into three broad categories (Clerke 1908: 1–2; Young 1902: 2–3). Observational or practical astronomy included the study of instruments (especially instrumental errors), methods of observation, and mathematical reduction and analysis of observational data. Most workers in practical astronomy were engaged in making transit-circle observations and the production of astrometric catalogues giving the position of stars with great accuracy. Gravitational or theoretical astronomy was the most active and prestigious field in astronomy. It included the study of celestial mechanics—the calculation of orbits and ephemerides for the planets, their satellites, comets, and asteroids. The study of double stars was also part of gravitational astronomy. Astronomers working on problems in gravitational or theoretical astronomy employed refined tools of mathematical analysis to study the motions of solar system objects and state of the art instrumentation developed by workers in practical astronomy to measure the positions of solar system objects in order to compare prediction with observation. The final area, physical astronomy, would grow rapidly after 1859 as the field of astrophysics. As Agnes Mary Clerke reported (1908: 2), there was a clearly defined hierarchy in mid-nineteenth-century astronomy. "Orthodox astronomers of the old school [celestial mechanicians and those who worked in astrometry] looked with a certain contempt upon observers who spent their nights in scrutinising the faces of the moon and planets rather than in timing their transits, or devoted daylight energies, not to reductions and computations, but to counting and measuring spots on the sun. . . . [Astronomers not engaged in astrometry or celestial mechanics] "were regarded as irregular practitioners, to be tolerated perhaps, but certainly not encouraged."

Both astrometry and celestial mechanics developed rapidly in the century before 1859. While Isaac Newton (1642–1727) was the founder of celestial mechanics, the subject was brought to perfection by a distinguished group of French mathematical astronomers in the eighteenth century. The *Mécanique céleste* (1799–1825) of Pierre Simon de Laplace (1749–1827), stands as the monument of this school. In the early nineteenth century, the American mathematical astronomer Nathaniel Bowditch (1773–1838) produced a translation with commentaries of *Mécanique céleste*. Using Newton's theory of gravitation and the calculus, theoretical astronomers developed mathematical models for each planet. Because of mutual gravitational attractions of solar system objects, these models were very complex. Theory and observation were continually compared and improvements made in the models of mathematical astronomers. In the early nineteenth century, the center of research in theoretical and gravitational astronomy moved to Germany. Johann K. F. Gauss (1777–1855), provided important new quantitative tools for the study of gravitational astronomy. Other German astronomers soon followed Gauss into the field. By 1859, American astronomers looked to Germany for the tools and techniques needed to make progress in their research.[6] A knowledge of the works of European mathematical astronomers was fundamental for anyone who aspired to work in the field.

Pre–Civil War American workers in astrometry used transit circles to collect data on the position of stars and planets. Like the celestial mechanicians, these astronomers looked to Europe for methods, techniques, and inspiration. Their instruments came from the workshops of German and English instrument-makers, and the observational programs on which they embarked were often shaped by suggestions from European astronomers. Both the methods of making observations and their reduction and

6. For an introduction to the history of celestial mechanics see Norberg (1974: chap. 1).

analysis were carried out according to guidelines established by the German astronomer Friedrich W. Bessel (1784–1846), the father of modern astrometry.[7]

By the 1850s, Reingold's class of cultivators was dramatically separated from researchers in celestial mechanics and astrometry. Cultivators could hardly hope to understand much about these complex mathematical areas. According to Clerke (1908: 1), these fields could be "unraveled only by the subtle agency of an elaborate calculus." This meant that intending scientists needed to study mathematics. For students "interested in astronomy, the significant feature of a college was not the quality of the descriptive astronomy course, but the quality of the mathematics course" (Rothenberg 1974: 90). In order to get full value from such courses, students needed a reading knowledge of French and German. Many important publications in celestial mechanics and astrometry were available only in the form of original monographs or books written by European scholars. As Benjamin A. Gould was advised (Comstock 1922: 156) while an undergraduate at Harvard, "Study foreign languages, for thus alone can you keep pace with the progress of modern science."

In addition to a thorough grounding in advanced mathematics and the language skills needed to read the works of French and German masters of celestial mechanics and astrometry, there were other lines of demarcation between cultivators and researchers. Astronomers working in this field argued that their investigations were paradigmatic for astronomical science. As Gould put the matter (Comstock 1922: 164), observations of the surface features of the sun, moon, and planets or the structure of comets, and observations of nebulae and star clusters, may have some value, but "the study of the *motions* of the heavenly bodies is, nevertheless, the sole problem of astronomy."

Agnes Clerke observed (1908: 2), that emerging professionals sought to draw lines of demarcation between themselves and those who worked in other areas of astronomy. When one of America's leading observational astronomers, William C. Bond (1789–1859), first director of the Harvard College Observatory, died, his Harvard colleague, Benjamin Peirce (1809–80), the foremost mathematical astronomer in the United States, recalled that Bond (Rothenberg 1974:107) "had not received a special training in astronomy and mathematics and that his knowledge of theoretical astronomy was not extensive." Because of these limitations Bond did not engage in research that called for "long, intricate, and profound mathematical computations, but preferred those [lines of scientific investigation] which

7. For an introduction to the history of astrometry, see Clerke (1908, chap. 2). On Bessel, see Fricke (1970). Rothenberg (1974: 81–87) discusses the impact of European methods on American astronomy.

were merely dependent upon the thorough discipline of the senses." Mathematical astronomers sought to capture the Harvard Observatory directorship and were displeased when it went to George P. Bond, who was interested in the study of comets and nebulae and a pioneer in the application of photography to astronomy.

Another way of understanding the intellectual concerns of American astronomers is through an examination of Gould's *Astronomical Journal*. During the calendar year 1859, the *AJ* published ten numbers containing thirty papers. A few papers dealt with observational astronomy including a series of visual estimates of the maxima of bright variable stars made by an amateur, Stillman Masterman (1831–63) of Franklin County, Maine. Others explored theoretical aspects of celestial mechanics including a discussion by Simon Newcomb of the common origin of asteroids. Most, however, were data papers, containing the results of astrometric observations of stars, comets and asteroids.

Of the five major astronomical research institutions in the United States in 1859, two (the Dudley and Cincinnati observatories) were virtually closed. Political and economic difficulties reduced their operations to a bare minimum. Only in the 1880s would the Dudley Observatory realize its promise as a major center of astrometric research. The Cincinnati Observatory was equipped for observational astronomy and its director, Ormsby McK. Mitchel, devoted what time he could spare from fundraising, public speaking, and consulting, to the study of double stars. However, as contemporaries noted, the Cincinnati instrument was not being used to capacity.

The Naval Observatory was the main representative of astrometric research. Here a large staff of scientists was actively pursuing several lines of investigation designed to improve knowledge of the positions and motions of the stars and planets. At Harvard, equipped with one of the largest telescopes in the world, practical astronomy was favored. Major planetary discoveries had been made with the great Cambridge refractor, as well as pioneering studies of comets and nebulae. These investigations brought international recognition to the observatory.

A final institution deserves consideration. The Nautical Almanac Office (NAO), created in 1849, became the center for gravitational astronomy in the United States.[8] Charged with the production of an astronomi-

8. The most reliable history of the Nautical Almanac Office is found in Norberg (1974: 94–102). A history of the NAO would be a welcome addition to our knowledge of American science. Such a study would illuminate many aspects of both the history of astronomy and the development of science in the federal government. Rich archival and manuscript sources are available in the National Archives, the Newcomb Papers at the Library of Congress, and elsewhere. Doggett (1996) discusses the comparative history of nautical almanac offices in Europe and America.

cal ephemeris for the use of navigators, surveyors, and explorers in the American West, the annual volume was also of great value to professional scientists. *The American Ephemeris and Nautical Almanac,* first published in 1852, quickly took its place along side the German, English and French astronomical almanacs.

The NAO provided employment for mathematical astronomers and gave them a congenial environment in which to work. America's first professional woman astronomer, Maria Mitchell (1818–89), was a computer on the NAO staff. From the beginning, its director, Admiral Charles Henry Davis (1807–77), encouraged the staff to carefully evaluate European research and, where necessary, develop new theories of planetary motion. On the basis of these models, computers constructed tables predicting the positions of the sun, moon, and planets. Observers supplied data with which to test positions computed by the NAO. Using these observations, mathematical astronomers worked to improve their models of the motions of solar system objects.

In 1859, no American observatory was devoted to the study of astrophysics. In a few years, this would change. Soon amateurs and then professionals (Lankford, 1981a) would begin to devote time and energy to the new field. Activities at private university and college observatories were generally divided between research in astrometry and celestial mechanics.

REPRESENTATIVE INDIVIDUALS

Three individuals serve to illustrate aspects of American astronomy in 1859. Denison Olmsted (1791–1859) represents much that was characteristic of the culture of preprofessional science. The lives of Charles Augustus Young (1834–1908) and Simon Newcomb (1835–1909) exemplify aspects of career development open to younger members of the cohort that entered astronomy in the 1850s.

Denison Olmsted was born into a Connecticut farm family (Dexter 1912, 6:592) of "moderate circumstances." After graduation from Yale, Olmsted taught in a private school. He accepted a tutorship at Yale in 1815 and began studying for the ministry. At this time Olmsted also became involved in the movement to reform secondary education. While he abandoned plans to enter the ministry, educational reform remained a lifelong interest.

The scientific career of Denison Olmsted apparently began by accident. He was elected to the chair of chemistry at the University of North Carolina in 1817 and accepted with the provision that he spend a year at Yale studying the subject. Olmsted remained at North Carolina until 1826. From 1822 he acted as state geologist as well. Returning to Yale in 1826 as professor of mathematics and natural philosophy, Olmsted spent the rest of his life in New Haven. After a decade he was relieved of mathematical instruction and concentrated on astronomy and meteorology. Olmsted published a series of textbooks and popular expositions of astronomy and meteorology (Dexter 1912, 6:593) that "found a ready sale, and proved a considerable source of income."

In the world of preprofessional science, Olmsted could enter into a scientific career with little in the way of credentials. To the old time-college president (Rudolph 1962: chaps. 4–5), piety was valued above intellect. Yet Olmsted learned his craft and became proficient in several fields. His early scientific publications were in chemistry and geology. Later his interest shifted to meteorology and astronomy. He published a total of seventy-three books and papers (many were short data notes in the *American Journal of Science*). Of these publications, the largest single group (23) dealt with astronomical subjects. Seven were on various aspects of meteorology. Twenty-one discussed topics relating to geology, chemistry, physics, and other aspects of natural philosophy. Religion and educational reform (11 papers) remained a concern of Olmsted's through his adult life.

Boundaries between the sciences were fluid in the first half of the nineteenth century. Meteorology and the study of meteors, the aurora and the zodiacal light all seemed to merge with astronomy. Chemistry, mineralogy, and geology were not clearly separated. Like most of his contemporaries, Olmsted moved across these ill-defined boundaries with ease, conducting research and publishing the results, first in one field and then another. These tendencies were reflected in the undifferentiated nature of undergraduate science teaching (Guralnick 1979). Courses were conglomerates that touched on various aspects of natural philosophy from astronomy to zoology.

In summing up Olmsted's career, his biographer did not cite any major contributions to science. In truth, Olmsted made none. Rather the biography closes (Dexter 1912, 6:594) by noting "Professor Olmsted was a courteous gentleman of the old school in his manners, and an earnest Christian. He was a teacher by nature." Of astronomers active in the first cohort, at least four members of the elite and three from the rank and file studied with Olmsted.

Charles Augustus Young entered into a scientific career in 1856, when he assumed the professorship of mathematics, natural philosophy and astronomy at Western Reserve College in Ohio. Young succeeded to his father's chair of astronomy and physics at Dartmouth College in 1866. Graduating from Dartmouth at eighteen, Young went immediately with his father to visit European observatories and instrument makers. He then secured a teaching position at the Phillips Academy and studied theology at Andover Seminary. Like Olmsted, Young was diverted from a clerical career by an invitation to teach science and mathematics. The Dartmouth graduate was, however, much better qualified than Olmsted. Lacking adequate research facilities, Young spent his time in northern Ohio teaching and reading. Only after he returned to Hanover did he secure adequate instrumentation.

At Dartmouth, Young established himself as the leading American researcher in solar physics. He was the first to photograph solar prominences and discovered the flash spectrum at the 1870 eclipse. In 1877 he was called to Princeton as professor of astronomy and director of the Halsted Observatory. Failing health ended Young's observing career and he devoted himself to the composition of popular works. His astronomy textbooks were among the most widely used at the end of the nineteenth century. Young was in great demand on the lecture circuit. National and international recognition came to Young for his work in solar spectroscopy. He was elected to the National Academy of Sciences in 1872. He was also a member of the American Philosophical Society, the American Academy of Arts and Sciences and a foreign associate of the Royal Astronomical Society. The French Academy of Sciences awarded him the Janssen Medal in 1891.

Young's lifetime publication count is 137 books and papers. With single-minded devotion they all focus on aspects of astronomy or solar physics. Unlike Olmsted's generation, Young and his peers found themselves in an environment in which specialization was both encouraged and rewarded. While he began his career teaching general science and mathematics as well as astronomy, Young focused his interests and narrowed his teaching responsibilities at Dartmouth and at Princeton concerned himself exclusively with astronomy. Like Olmsted, Young wrote textbooks, but there were significant differences. The Yale professor published texts and popular works on astronomy, meteorology, and general science. Young, however, concentrated his efforts on astronomy and his books were seen (Frost 1910: 103–4) as authoritative summaries of current knowledge written by one of the leading research scientists of the day.

In 1859 Simon Newcomb, who by the end of the century would become "America's best-known scientist" (Norberg 1978: 209), was a member of the staff of the Nautical Almanac Office in Cambridge. Born in Nova Scotia to a family of very moderate circumstances, Newcomb had little formal education. He must have been a precocious child, for he began to read history and mathematics at an early age. Newcomb fled to the United States at eighteen after a failed apprenticeship. The young man taught in country schools near Washington, D.C., but spent much time in the library of the Smithsonian Institution, reading mathematics. Here he was befriended by Smithsonian secretary Joseph Henry (1797–1878), one of the leaders of American science in the antebellum period. Henry's patronage led to an appointment at the Nautical Almanac in 1857. Newcomb, almost entirely self-taught up to this point, enrolled in the Lawrence Scientific School at Harvard where he studied mathematics and astronomy with Benjamin Peirce. Newcomb received the bachelor of science degree in 1858, a few months after his first paper appeared in the *Astronomical Journal.*

By 1859 the Canadian immigrant had made the acquaintance of most of the leading astronomers and mathematicians in Cambridge and impressed them with his abilities. With patrons like Benjamin A. Gould and Admiral Davis of the Nautical Almanac Office, he was in position to begin actively managing his career. In 1861, through the help of Gould and Davis, Newcomb was appointed to the staff of the Naval Observatory where he remained until 1877, when he became superintendent of the Nautical Almanac Office.

Newcomb's great contribution was to analyze the movements of the moon and planets and produce new mathematical models of their motions. This program also involved a reexamination of the fundamental astronomical constants. These investigations occupied Newcomb for most of his professional life, earning him major domestic (elected to the National Academy of Science in 1869) and foreign honors (corresponding member of the French Academy of Sciences in 1874). By the 1890s Newcomb was the acknowledged leader of American astronomy and one of the spokesmen for American science in the international scientific community (Moyer 1992).

Simon Newcomb published more than any other astronomer. A total of 512 books and papers can be credited to his active pen. These range from papers on celestial mechanics and mathematical astronomy through discussions of contemporary economic issues to a utopian novel and autobiography. While Newcomb lectured for a few years at the Johns Hopkins University, he was primarily a researcher and did not establish a formal school or leave behind students and disciples.

The lives of Olmsted, Young, and Newcomb illuminate the structure of the American astronomical community in 1859 and suggest a number of themes that will be discussed in later chapters. At one extreme stands Olmsted, the generalist and popularizer, who spent his life teaching and whose research contributions were modest. Young stands toward the middle, his biography illustrating movement toward a more tightly concentrated career. Beginning as a generalist, Young became increasingly specialized in his teaching, while his research concentrated on the emerging field of solar physics. Newcomb did only a little teaching; his energies were given almost exclusively to research. While the available evidence is incomplete, it appears that Olmsted and Young made major career moves on the basis of local contacts at the institutions where they were educated (Yale and Dartmouth) while Newcomb depended on the patronage of major figures in American astronomy, as well as initial recognition by Henry at the Smithsonian. The highest earned degree held by Olmsted, Young, and Newcomb was the bachelor's. To be sure, all had one or more honorary doctorates.

These three lives suggest that a research career provided access to honors and awards more readily than did one involving a mix of teaching and research. The importance of patrons should also be noted in considering the development of careers. An early interest in science may be a significant factor in success. Newcomb made up his mind to study mathematical astronomy at an early age. Young and Olmsted, however, were diverted from preparation for the ministry into science. Young and Newcomb sought institutions that provided maximum opportunities to carry out research, but Newcomb achieved a favorable location much sooner than Young. Newcomb found government research institutions responsive to his career needs. As his interests grew and changed and the vision of his life's work became clear, Newcomb moved from the NAO to the Naval Observatory and then back to the NAO as its chief. Finally, Newcomb and Young represent the two major career options that would mark American astronomy after 1859: the one continuing traditions of celestial mechanics and astrometry and the other following new directions as the field of astrophysics developed.

For American astronomy, 1859 marked an *annus mirabilis*. The old order was passing. The culture of preprofessional science represented by men like Denison Olmsted was coming to an end. The future belonged to younger scientists who, like Young and Newcomb, would be heavily involved in research. While the Civil War closed some observatories and interrupted the work of others, after 1865 it would be the Youngs and Newcombs who would take the lead in articulating new goals and developing new institutions for the American astronomical community. The researches of Bunsen and Kirchhoff, first announced in 1859, would open the way for the development of astrophysics. By the 1890s, astrophysics would challenge celestial mechanics and astrometry for hegemony in the American astronomical community.

3 ✦

THE NEW ASTRONOMY: IDENTITY AND CONFLICT

With poetic license, anthropologist Loren Eiseley (1958: 333) suggested that astrophysicists use the spectroscope to "dip a ladle into the roaring furnace of the sun and stars." The researches of Robert W. Bunsen (1834–99) and Gustav R. Kirchhoff (1824–87) made it possible to analyze the chemical composition of astronomical objects and, in time, spectroscopy would provide information on their movements and distance, temperature, magnetic field, and rotation, as well.

Agnes Mary Clerke (1842–1907), the perceptive historian of nineteenth-century astronomy, who was close to many of the people and events she chronicled, grasped the significance of the new science of astrophysics. She reported (Clerke 1908: 142) that disciplinary lines were blurring. "The astronomer has become . . . a physicist; while the physicist is bound to be something of an astronomer." Clerke's understanding went beyond the development of a new research agenda. She indicated some of the social and psychological factors that made astrophysics an attractive research field. Calling the emergence of astrophysics "a new birth of knowledge," she contrasted its challenge with the mature puzzle-solving science of astronomy (celestial mechanics and astrometry). Clerke suggested that astronomy, because of "its very perfection, had ceased to be interesting"; a science "whose tale of discoveries was told," in which future growth would consist of "minute technical improvements, not of novel and stirring disclosures." In the new field of astrophysics, however, Clerke detected all the "audacities, the inconsistencies, the imperfections, [and] the possibilities of youth." Small wonder that astrophysics quickly attracted a band of dedicated workers who proved its worth. These founders would, in time, be replaced by a much larger group of followers.

This chapter examines the development of astrophysics in the United States in the period from 1860 through the first decade of the twentieth century. Where appropriate, comparisons will be made with European experience. Three points must be made at the outset. It is virtually impossible to argue that astrophysics has deep historical roots in the same way that astrometry or celestial mechanics can lay claim to such a heritage. There is little in ancient, medieval, or early modern science that can be described as ancestral to astrophysics. As the name implies, astrophysics owes its beginnings to developments in the elder science of physics, especially experimental spectroscopy. Only after technical and theoretical developments in physics could the science of astrophysics be imagined.[1] If this argument is correct, it is impossible to contend that a preexisting science was transformed by the emergence of astrophysics. The *uniqueness* of astrophysics is central to the argument in this chapter.

Two other points relate specifically to the American scene. Unlike astrometry and celestial mechanics, astrophysics received virtually no federal support before World War II. This stands in sharp contrast to France and Germany or even England, where government patronage supported various forms of astrophysical work. The need to secure private funding had important consequences for the growth of astrophysics. These ranged from hostility toward well-supported traditional research establishments such as the Naval Observatory and the Nautical Almanac Office to the need for leading astrophysicists to be effective fund raisers and administrators, who knew how to secure maximum research output for minimum dollar investment. It also meant that astrophysicists would be among the suitors at the doors of the great philanthropic foundations that emerged in early twentieth century America.

The final point concerns the relationship between astrophysics and the development of the research university in post–Civil War America. Astrophysics must not be seen as a university-based science. Its founders were only tangentially connected with developing research universities, and the second generation of followers were by no means always established in university positions that entailed access to graduate students.

These historical and cultural conditions pose challenges for those who would explain the development of astrophysics in the United States. Their models must be culture-specific, sensitive to historical circumstances that characterized the American experience during the second half of the nineteenth century.

1. Clerke (1908:125–42) deals only with early nineteenth-century developments in solar astronomy and the history of spectrum analysis during the same period. Meadows (1970) follows a similar course. Both authorities limit their discussion of the pre-1859 history of these topics to the period after 1800. Neither can find deep roots for the subject prior to the nineteenth century. Hufbauer (1991:1–41) provides an overview of solar astronomy prior to 1810. See also Meadows (1984a) on the rise of the new astronomy. For an opposing point of view, see DeVorkin (1996).

It does not seem fruitful to explain the rise of astrophysics as either a revolution involving the transformation of an existing science (Cohen 1985) or as the growth of a new specialty within an established discipline (Edge and Mulkay 1976; Chubin 1976). How, then, should the problem be treated? Clerke's characterization of astrophysics as "a new birth of knowledge" is suggestive. Historians and sociologists of science have come to recognize that the creation of new knowledge involves both cognitive and social dimensions (Shapin 1982). Perhaps, the most fruitful approach is to ask two closely related questions: How was the research agenda for astrophysics formulated and how was the social role of astrophysicist created?

The elegant experiments of Bunsen and Kirchhoff and the theoretical interpretation of the results by Kirchhoff marked an epoch in the history of physical science (McGucken 1969). Once the power of spectroscopy was demonstrated, it became an important research tool in many fields. Kirchhoff went on to develop the concept of black-body radiation and lay the groundwork for advances in thermodynamics. Armed with Kirchhoff's laws and a spectroscope, one could identify an element or compound through its spectroscopic signature. Spectrum lines are unique to the substance that produces them. These procedures work in the physical or chemical laboratory and for the astronomer observing the sun and stars (Hearnshaw 1986).

Theory tells scientists what to look for, how to interpret the observations, and suggests crucial experiments by offering predictions. But, in and of itself, theory is not enough. Scientists must have instrumentation capable of providing observational data demanded by theory. Only if such research technologies are available can scientists gather relevant data and carry out critical experiments.

Photography was the key research technology needed to capitalize on the theoretical innovations of Kirchhoff. Visual spectroscopy quickly reached the limits of accuracy. The photographic plate, even in its early forms, was far superior to the human eye. Unlike the eye, the photographic plate stores light, thus revealing fainter objects than the eye can discern. Further, the photographic plate is sensitive at wave lengths to which the human eye is all but impervious. For example, most nineteenth-century photographic materials were sensitive in the blue-violet region of the spectrum, while the eye reaches maximum sensitivity in the yellow-green. Late in the century, important advances were made in both red-sensitive materials and orthochromatic plates (Lankford 1984a, 1987c). Further, photography provides a permanent record. Plates can be examined by more than one scientist.

Photography demanded new forms of the telescope. Since most photographic materials and the human eye were sensitive to different portions of the spectrum, telescopes designed for visual observations were not suitable for photography. The first phase involved designing objectives that made it possible to use refracting telescopes for photographic work. The second phase witnessed the development of reflecting telescopes, whose optical systems transmitted a far wider portion of the spectrum than refractors. Reflectors also collected more light than did refractors, whose lenses tended to absorb or scatter a significant portion of incoming photons (Van Helden 1984a).

The development of the spectroscope was a second important area. Astronomers sought to maximize the instrument's mechanical stability and its ability to transmit light. Increased resolution was also a major goal. At first prisms were used. Late in the century the ruled grating gained acceptance. A large objective prism permitted the astronomer armed with a photographic refractor to record simultaneously the spectra of many stars (Struve and Zebergs 1962; Van Helden 1984a).

Still another form of instrumentation had to be devised for use in conjunction with photographic spectroscopy. High-dispersion spectra contained a wealth of information, but devices were needed to measure the position of spectrum lines with precision. In time, instruments would also be needed to measure the intensity and width of lines.

In addition to theory and instrumentation, two important social elements must be added to the list of prerequisites. These include a relatively coherent group of founders who are, in turn, succeeded by a second generation of followers (Lankford 1987b). The founders of astrophysics in the United States (here defined as that group of scientists active in astrophysical research between 1859 and 1879) numbered only thirteen (Table 3.1). The founders struggled to articulate the research agenda for astrophysics. They defined problems worthy of investigation and suggested the best technical means for carrying out research. Their agenda might be described as an intellectual hybrid since it involved methods and theories borrowed from other sciences and pressed into service in a new context.

The founders produced important new data on the sun and applied spectroscopy to the brighter stars. Photographs of star fields, solar eclipses, and the moon were appreciated by the lay public and scientists alike. These activities served two purposes. They confirmed the value of the new field to the founders themselves, providing a sense of accomplishment as they pushed back the frontiers of knowledge. Further, important discoveries, as they were acknowledged and then incorporated into the canon of scientific knowledge, served to legitimate astrophysics.

Several of the founders went out of their way to make new findings available to interested laypersons. Samuel P. Langley (1834–1906), a solar physicist at the Allegheny Observatory and later secretary of the Smithsonian Institution, and Charles A. Young, first at Dartmouth and then

Princeton, published extensive popular discussions dealing with solar physics. Lewis M. Rutherfurd (1816–92) shared his work on optics and photography through lectures and correspondence. Henry Draper (1837–82) published the first detailed monograph in English on the construction of reflecting telescopes. Popularizers not only helped to legitimate the new science, they also brought its results to the attention of wealthy donors who might provide financial support for astrophysics.

In addition to scientific activities associated with creating the research agenda for the new science, founders had other responsibilities. Most important among these was the creation of the social role of astrophysicist. This was accomplished in part by publicizing successful programs of astrophysical research and in part by appropriating the role of physicist. The founding generation saw the role of astronomer as one of relatively low status. They felt that physicists had a great deal more status. The process of role hybridization was complex and is best understood against the background of the cognitive and social structure of traditional astronomy.

Another social activity performed by the founding generation involved recruiting followers. Sociologists and historians of science frequently assume this function is most easily carried out in the context of graduate education in a research university. This model, however, does not apply to American astrophysics. The founders did relatively little teaching. Recruitment had to proceed in other ways.

Through their popular writings, the founders demonstrated the power and research potential of astrophysics to recruits. George Ellery Hale (1868–1938), a leader among second-generation astrophysicists, was captivated as a teenager by the writings of Young and Langley (Wright 1966: 39–47, 61–63). Langley's *New Astronomy* (1892) reached a wide audience, first as a series of magazine articles and later in book form. Young's *The Sun* (1881) went through several editions.

Further, the founders served as mentors. Again, Hale provides a case in point. Young at Princeton assisted in the education of the young solar physicist but was not associated with Hale's formal course of studies at the Massachusetts Institute of Technology. James Keeler (1857–1900), second director of the Lick Observatory, while educated at Johns Hopkins, was effectively mentored in astrophysics by Langley (Osterbrock 1984a: 15–35).

It fell to the second generation to carry on the work of the founders. They consolidated and deepened the founders' work in solar physics and expanded spectroscopic research to the stars. In the hands of men like James Keeler at the Lick, astrophysical research expanded to include the study of nebulae as well. The second generation also created many of the institutions and professional structures of astrophysics. These activities included establishing journals and creating scientific organizations. After

1880, both founders and followers were involved in conflict with workers in celestial mechanics and astrometry. These were generational conflicts as well as disputes over the appropriate content of astronomical science. It also fell to some of the followers to begin the process of formally educating the third generation of astrophysicists.

ASTRONOMY AS A NORMAL PUZZLE-SOLVING SCIENCE

In the discussion that follows, astronomy (celestial mechanics and astrometry) is considered as a normal puzzle-solving science. It is important to understand the condition of astronomy in order to see how it contrasted with astrophysics. A normal science provides comparatively few opportunities for discovery and rewards come slowly. Concern for the next decimal place suggests that the general shape of the answer is already known. Only the fine points need to be worked out.

In characterizing astronomy as a normal science, I do not wish to detract from its many strengths. In the United States, Simon Newcomb at the Naval Observatory (after 1877 director of the Nautical Almanac Office) and George W. Hill (1838–1914), Newcomb's associate at the NAO, spent their careers perfecting mathematical models of the motions of the moon and planets. This work carried on the great European tradition of celestial mechanics exemplified by Laplace, Hansen and LeVerrier. The models of Newcomb and Hill remained standard until the middle of the twentieth century.

Astrometry developed rapidly in the early nineteenth century. In the 1860s the *Astronomisches Gesellschaft* initiated a survey of the northern sky in which all stars to the ninth magnitude would be carefully observed with transit circles and their positions determined with a high degree of accuracy. Both the Harvard and Dudley Observatories took part in this project. The U. S. Naval Observatory produced a series of catalogues providing the places of stars with ever-increasing precision. By the end of the 1880s, Lewis Boss (1846–1912), director of the Dudley Observatory at Albany, was moving that institution to the forefront in astrometric work. As catalogues of precision grew in number and accuracy, astronomers could use them to detect the movements (proper motion) of the stars in space. Astronomers were also measuring the parallax (distance) of the nearest stars. These achievements made American astronomy competitive with European science. Men like Newcomb, Hill, and Boss were received as equals by the leaders of English and continental astronomy.

Thomas Kuhn (1970: 35–36) suggests that "the most striking feature" of normal science is how little it aims to produce "major novelties, conceptual or phenomenal." Kuhn describes the research process in a normal science as "achieving the anticipated in a new way." This approach "requires the solution of all sorts of complex instrumental, conceptual, and mathematical puzzles." On Kuhn's model, "The man who succeeds [in normal science] proves himself an expert puzzle-solver, and the challenge of the puzzle is an important part of what usually drives him on." In the second half of the nineteenth century, scientists and laypersons were aware of astronomy as a normal science. As Agnes Clerke suggested (1908: 142), because of "its very perfection" astronomy "had ceased to be interesting."

Seth Carlo Chandler (1864–1913), second editor of *The Astronomical Journal* and a major contributor to solving the problem of variation of latitude, clearly recognized the status of astronomy as a normal science. In an unpublished manuscript, "Glimpses at the Future of Astronomy," Chandler sought to justify the old astronomy.[2] In his introductory remarks Chandler discussed the nature of a perfect science. He observed (n.d.:1) that while many attributed perfection to astronomy, they were wrong. A perfect science would have done its work and be little more than "a monument erected over the grave of a terminated investigation." To Chandler, only those who are ignorant of or do not appreciate "the magnitude of the problems confronting astronomy . . . [problems so difficult they] must be worked out on a broader stage, not merely of the next century but of the next millennium" could label the science as perfect.

The Boston astronomer characterized science as alternating between observation (the collection of data) and testing theory against data. At each step in this cyclic relationship, the means of collecting observational data were improved. Here again, Chandler (n.d.: 4) was speaking of certain characteristics of normal science; the quest for greater precision and knowledge of the next decimal place.

The tone that permeates the lecture is at once defensive, hostile to astrophysics, and self-righteous. Chandler, virtually alone in his generation, was the only astronomer elected to the National Academy of Sciences (1888) who was not provided with a biographical memoir. A protégé of Benjamin A. Gould, Chandler was a difficult man who did not shrink from controversy. He often mixed personal criticism in scientific disputes and was well connected with the Boston press, which he did not hesitate to call into service, planting stories that put his opponents in a bad light. As the home secretary of the National Academy remarked, when trying find an astronomer willing to write a biographical memoir

2. The manuscript is in the archives of the Dudley Observatory. It is unsigned and undated. Internal evidence suggests it was intended as a public lecture and composed toward the end of the 1890s. I attribute authorship on the basis of the manuscript's location among materials relating to the transfer of editorial responsibility for the *Astronomical Journal* from Chandler to Lewis Boss, director of the Dudley Observatory, in 1909.

on Chandler, "I understand that the case of Chandler is rather peculiarly difficult because of certain enmities that occurred in his lifetime." In the end, the home secretary dropped the idea of memorializing the Yankee curmudgeon.[3]

Chandler was quite explicit concerning the paradigm that organized astronomical research. "Astronomy merely has to do with the positions and motions of heavenly bodies, and its sole aim it to be able to predict these for any future time." According to Chandler (n.d.: 2), "this sums up the whole science, and we must not wander from it." Sir George Biddle Airy (1801–92), the great Victorian astronomer royal, provided an authoritative formulation of the paradigm (Maunder 1900: 266–67). "What astronomy is expected to accomplish is evidently at all times the same." Astronomy lays "down rules by which the movements of the celestial bodies, as they appear to us upon the earth, can be computed." Then, in ringing tones, the seventh astronomer royal, went on to pronounce the sentence of excommunication. "All else which we may learn respecting these bodies . . . possesses no proper astronomical interest." The only legitimate concern of the astronomer was "to learn so perfectly the motions of the celestial bodies that for any specified time an accurate computation of these can be given." To Sir George, "that was, and is, the problem which astronomy has to solve."

Chandler was willing to concede one point to those who saw astronomy as a perfect science. He believed, "speaking in a relative way, that the astronomy of the solar system is now under control of mathematical analysis and that we may confidently anticipate the successful explanation of some minor anomalies that yet remain." Chandler could not resist contrasting the knowledge of the solar system developed by celestial mechanicians and astrophysicists. He characterized what astrophysics knew about the sun and planets as "purely speculative."[4]

One of Chandler's motives was his desire to counter the belief that astronomy "had ceased to be interesting" and that any "farther advance must be in the line of minute technical improvements" rather than "novel and stirring disclosures" as was the case with the new science of astrophysics (Clerke 1908: 142). Chandler (n.d.:1) characterized astrophysics as an "*inexact* science" and made a veiled reference to conflict between astronomers and astrophysicists in the 1890s, when he alluded to "a certain superficiality that could be perceived in the conduct of research by

3. Home Secretary of the National Academy of Sciences to Armin O. Leuschner, 4 January 1917. BLUCB, Astronomy Department Papers. For biographical information on Chandler see Marsden (1971) and Jones and Boyd (1971:194–95, 331).

4. Chandler (n.d., p. 5). DOA. The list of "minor anomalies" was headed by problems with the orbit of Mercury that were not solved until Einstein's general theory of relativity.

some of its votaries." Chandler (n.d.:2) told the audience that his aim was to show "how vast is the domain which astronomy has yet to cover" and "how puerile is the notion that its methods have exhausted the subjects for their application, and that it has reached its term as a perfect science, or is in any respect devitalized."

In spite of Chandler's defense, it appeared increasingly difficult to counter the impression that astronomy was a mature puzzle-solving normal science in which few exciting discoveries remained to be made and which, consequently, held out little hope of advancement for an ambitious young investigator. The history of the Royal Astronomical Society of London (Dreyer 1923: 81) substantiates this interpretation. Citing the Victorian mathematician and science critic Augustus DeMorgan (1806–71) with apparent approval, the RAS historian pointed out that "Astronomers had rather given over expecting anything very great in the future: they were inclined to think that nothing was left except to give the existing methods and results additional fullness and accuracy, facility, and neatness." Sampson (1923: 85–86) suggested that as astronomy approached Chandler's state of perfection and became a mature normal science "there was a comparative paucity of great things, accompanied by a constant and gradual improvement of routine."

In the end, Chandler (n.d.: 8) cut the ground from under his own feet when he insisted that "no life yet in being can hope to see any considerable fruit plucked from the tree of knowledge as a tangible and definite result of the labor that is now so industriously bestowed on the accumulation by observation of the facts which will solve these great problems. The astronomer of today, in devoting his life to making these precious records, well knows that he is working for the benefit of coming generations." In the best puritan manner, Chandler characterized the work of astronomers as "a labor of self-sacrifice" and suggested that for them "satisfaction" must come from imagining the future value of their work. Chandler sought to soften the austere nature of these rewards by contrasting the work of astronomers with that of astrophysicists. The latter engage in "idle scientific speculations . . . born of no achievement of knowledge." The work of astrophysicists "will disappear like smoke in the air," and its "authors will lie in forgotten graves."

This heroic vision of astronomy as a normal science stretched far into the future. Data collectors worked like drones, but their labor would bring fourth new knowledge only in distant generations. While this might gain them "niches of honor in the astronomical temples of the future" it provided little satisfaction in the present or hope for advancement within the discipline (Chandler, n.d.:8).

In modern science, status is generally determined by the ability of a field to produce new knowledge and make fresh discoveries that capture the attention of scientists and laypersons alike. By the third quarter of the nineteenth century, the science of astronomy, as described by Chandler, was losing rather than gaining in status. A thousand-year research program did little to excite the general public or attract ambitious young recruits.

The centerpiece of Chandler's argument (n.d.: 6–7) involved extending the methods and techniques of astrometry to the study of distant stars. The motions of these stars in space, although small and very difficult to measure, held out great promise for understanding the structure of the sidereal universe. Chandler urged fellow astronomers to devote whole lifetimes to collecting data on the proper motion of stars, in the hope that at some distant period this vast collection of data would lead to powerful new models.

It is surprising that Chandler did not put the study of stellar motion into its contemporary context. This was one of the few areas in which astronomers and astrophysicists shared common ground. Using the transit circle, astronomers compiled data that led to the determination of stellar proper motions. Using the spectroscope, workers in astrophysics measured stellar motion in the line of sight (radial velocity). When combined, these data provided important new perspectives on the sidereal universe. Further, an international project for photographically mapping the sky had been underway since the late 1880s. Yet Chandler made no reference to the work of the *Carte de Ciel.* Finally, he appeared unaware of important new evidence that suggested a relationship between the spectral class of a star and its motion in space. These omissions suggest the price astronomers like Chandler paid in resisting astrophysics and locking themselves into a narrow conception of the goals and methods of astronomical research.

Another way of looking at the old astronomy as a normal puzzle-solving science involves a close examination of astrometric research. The fundamental instrument for the nineteenth-century astronomer was the transit circle. The instrument is fixed in the plane of the meridian and sweeps an arc from horizon to horizon. It is mounted on massive piers and every one of its mechanical and optical elements is fabricated with strict attention to minute detail. With this instrument the astronomer times the passage of a celestial object across the meridian (the imaginary north/south line bisecting the dome of the sky that intersects the horizon precisely at the north and south compass points).

The view an astronomer has through a transit instrument is illustrated in figure 3.1. Illuminated wires divide the field and the observer times the transit of an object across each wire. These observations provide information from which the right ascension (celestial longitude) can be computed. The declination (latitude) of the object was read off especially con-

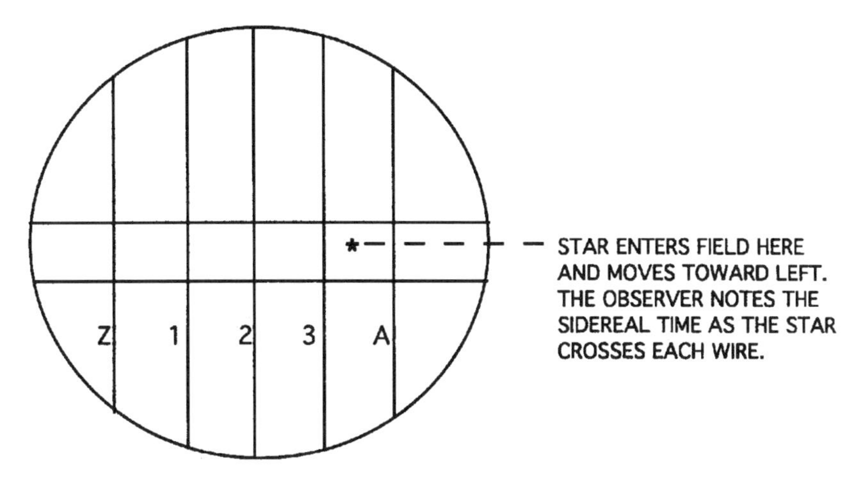

Figure 3.1. View through the eyepiece of a transit circle.

structed circles attached to the instrument. These circles were graduated with great precision and read with microscopes (Gill 1891: 335–39; Lankford 1987a).

As Sir David Gill (1843–1914), Astronomer Royal at the Cape of Good Hope and confidant of leading American astronomers, wrote of the meridian transit (Forbes 1916: 222), "However perfect an instrument may be (and it is the astronomer's business to see that it is perfect), it is the astronomer's further business to look upon it with complete and utter mistrust." This attitude forms a fundamental part of the mind-set of the astronomer. Much time and energy were lavished on studying instrumental errors. Every mark on the graduated circle had to be checked for accuracy. When the circles on the great Olcott transit instrument at the Dudley Observatory were regraduated, Lewis Boss and his staff spent a whole summer, working eight-hour days, checking each of the thousands of divisions under the microscope.

Transit circle observations were affected by many factors. Temperature and barometric pressure were measured several times during each observing session. The massive piers that held an instrument were subject to movement and, though minute, these were studied and taken into consideration when reducing observations. The telescope was held between pivots and the Ys on which the axis rested were subject to wear and various forms of displacement. The telescope was often checked to make sure it was aligned in the plane of the meridian. Nor were observers free from personal errors. Some measured the transit of a bright star too early, others too late. The color of a star also influenced the observer's perception of its passage across the wires. Indeed, there was even a personal error depending on whether an observer's feet were facing the north or south compass points.

In fact, workers in astrometry spent more time analyzing their own personal reaction-times and testing instruments than in making observations. To be sure, observing programs were demanding, often beginning before sundown for observations of bright stars in the twilight and frequently extended past dawn. But still, observations were accumulated in a matter of months. The reduction of observations (the application of all known corrections relating to the instrument and the observer as well adjusting them for refraction, nutation and other strictly astronomical factors) often took years. For example, the Dudley Observatory sent an expedition to Argentina to observe southern stars in 1908 and it finished observational work in 1911. However, the published catalogue did not appear until 1928. This lag was not due to inattention on the part of the staff in Albany. The institution was well funded and a large corps of female computers spent their days reducing the observations. The women did computations on large printed forms using six-place log tables. At every step, the computations were performed in duplicate. Then the material had to be copied into catalogue form and sent to the printer. The proof sheets of numerical data had to be proofread with great care. It is no wonder that the Dudley catalogues took years to complete.

To workers in astrometry, the universe was circumscribed by the view presented in the eyepiece of the transit circle and the printed catalogue of star positions. In catalogues, stars appeared as numbers representing the designation of the object, its right ascension, declination, and magnitude. Other columns might compare this position with previous catalogues and indicate proper motion, radial velocity and parallax. Astronomers invested a great deal of time and energy on the determination of stellar positions, but catalogue values for the brightness of stars were usually only an educated guess, sometimes off by as much as a full magnitude.

Researchers in astrometry saw the stars as abstract points whose positions were to be determined with great precision. The results were represented as numbers. Theirs was a numerical universe in which the physical or chemical characteristics of the stars were of no consequence. Because catalogues represented the sky at a given epoch, most viewed the heavens (at least until the publication of the next catalogue) as unchanging. This view represented the universe as numbers with the stars forever frozen at the epoch to which the data were reduced.[5]

Leaders of astrometric research in the second half of the nineteenth

5. Astronomers like Lewis Boss were deeply concerned with the problem of proper motion as revealed by a comparative analysis of star catalogues. But even here, much time and energy went into reanalyzing earlier observations to make sure motions were real rather than artifactual.

century in America (Boss, Gould, Chandler) were compulsive and oriented to fine detail. They drove their subordinates and themselves unmercifully. The goal was more data and greater accuracy. For each, the ratio of interpretative publications to data papers was small (B. Boss 1920).

The mind-set of workers in astrometry is illustrated by the attitude of Lewis Boss (1836–1912) toward the large refracting telescope acquired by the Dudley Observatory when it was moved to a new location at the end of the nineteenth century. Boss and his staff spent countless hours supervising workmen who moved the Olcott transit circle and then testing the remounted instrument for errors working to tolerances of .00025mm. (about a hundred thousandth of an inch); but they examined the optics of the new refracting telescope in the most casual way. It was with the transit circle that the primary work of the Dudley Observatory would continue to be done, not the large refractor. That instrument would be used to measure the positions of comets and asteroids and perhaps double stars, work that was peripheral to Boss's conception of astrometry: the collection of data from which to determine the positions of the stars and the presentation of those data in numerical form as printed catalogues (L. Boss 1895:172).

The transit circle mentality illustrates the status of astronomy as a normal puzzle-solving science in the second half of the nineteenth century. This is what young, would-be scientists saw when they looked over the field. Astronomy attracted those who were formally educated in mathematics; for others, with ambition but little in the way of college or university training, it was a forbidding prospect.

Entry into the normal puzzle-solving science of astronomy involved clearly defined standards. Of those who chose to work in astrometry or celestial mechanics in the years between 1860 and the end of the century, 59 percent had one or more college or university degrees. Of this group, 19 percent attained a bachelor's degree and 17 percent the master's. Twenty-three percent held the doctorate. Given the high level of mathematical proficiency demanded in the field, there is nothing surprising in these data. Proficiency was gained most readily in an academic context.

Mathematical training involved mastery of the works of the great European scholars. The emphasis was on assimilation into a well-established tradition. George W. Hill, Simon Newcomb's collaborator at the Nautical Almanac Office, recalled (Brown 1916: 275) that his teacher, Theodore Strong at Rutgers, always advised students to return to Leonhard Euler (1707–83) for analytical theorems. For the would-be astronomer, a college education involved studying the great masters from Newton to Laplace. Only after this task was successfully accomplished was the student ready to proceed with an astronomical career. This pattern stands in sharp contrast to the educational attainments of the founders of astrophysics.

It is a remarkable twist of fate that while concentrating on methods, American astrometry lost sight of larger goals. For the leaders of American astrometry, Lewis Boss in the second cohort and Frank Schlesinger after 1900, the collection of astrometric data became an end in itself. Innovative European astronomers like Hugo von Seeliger (1829–1924) in Germany and Jacobus Kapteyn (1851–1922) in Holland were, by the 1890s, turning their attention to statistical cosmology (Paul 1993). The aim of this field of astronomical inquiry was to develop an understanding of the distribution of stars and nebulae as a function of such parameters as luminosity and spectral type. Information on stellar parallaxes and proper motions, luminosity, radial velocities, and spectral type were key to these inquiries.

At the Dudley Observatory, Lewis Boss concentrated on astrometric catalogues that would provide a firm empirical foundation for the determination of stellar proper motions. He published only a handful of papers on solar motion: the direction in which the sun and its planets are moving with respect to the stars. Schlesinger, first at Allegheny and then Yale, devoted his attention to the mass production of stellar parallaxes using photography. He developed new forms of instrumentation and an efficient systems for the reduction of data so that assistants could produce parallaxes rapidly. Schlesinger's bibliography lists no publications at all in the field of statistical cosmology. Apparently, he left it to others to utilize the data he so eagerly and efficiently collected.[6]

That American astrometry focused on the collection of data rather than their interpretation, helps explain why many intending astronomers opted for astrophysics, where discovery was the order of the day. It would seem that, at least in the hands of Americans, astrometry field never rose above the level of mere data collection.

There is a remarkable irony here, worthy of deeper investigation. It was George Ellery Hale, the astrophysicist, who made Kapteyn a research associate at Mount Wilson and who directed that the new 60-inch telescope and Mount Wilson staff members be used to supply data for Kapteyn's selected areas program (Wright 1966: 235–36). There is more involved here than access to large telescopes in order to study the characteristics of faint stars. Perhaps, by the beginning of the twentieth century, the new astronomy was more sympathetic to developments in statistical cosmology than were practitioners in astrometry, the very field from which interest and sympathy might have been expected.

6. Comments on publications by Schlesinger and Boss are based on an examination of their bibliographies (B. Boss 1920; Brouwer 1945) included in the biographical memoirs published by the NAS. There is only one passing reference to Schlesinger in Robert Paul's detailed study (1993) of statistical cosmology and none to Boss.

ASTROPHYSICS: THE FIRST GENERATION

The classic autobiographical statement of a pioneer astrophysicist was written by Sir William Huggins (1824–1910), whose observatory in suburban London was long the center of astronomical spectroscopy in the United Kingdom and who had many friends among American astrophysicists. Like many others in their autobiographical recollections, Huggins is guilty of rewriting history (Becker 1993: 1: chaps. 1–2) to suit his own purposes. At one point, however, his recollections do have the ring of truth. Huggins (1900: 450) looked back on the early development of astronomical spectroscopy as "a time [of] expectation and of scientific exaltation . . . almost without parallel." As he recalled, "nearly every observation revealed a new fact, and almost every night's work was red-lettered by some discovery." This stands in sharp contrast to Chandler's astrometric research program.

The founders of astrophysics in America must have differed from those who went into astronomy in at least one critical respect. They sought areas of scientific activity in which they would have the chance to make discoveries. Mature science offers little opportunity for those who believe that discovery is the most basic reward a scientific career could provide. As the English mathematician and philosopher Bertrand Russell remarked, "It is difficult to imagine anything less interesting [than a mature science] or more different from the passionate delight" of discovery.[7]

While it is doubtless true that those who entered astrophysics or astronomy were ambitious, ambition must have manifested itself in different ways. Of the thirteen individuals listed in table 3.1 each made one or more significant discoveries (observational, theoretical, or related to instrumentation) in astrophysics or an allied science. This suggests ambitions far different from that manifested by entrants into the mature science of astronomy. Only a long time later, after observational data had been reduced and analyzed, would its meaning be revealed.

Discovery is central to the reward system of science. Rates of individual advancement vary according to the number and kind of discoveries made by scientists. Astrophysics offered a much wider field for discovery than did the mature science of astronomy. Quantitative evidence confirms these suppositions. Until the 1880s, astrophysicists were elected to the National Academy of Sciences at a younger professional age than were workers in astrometry or celestial mechanics. From the late 1880s to the

7. Quoted in Galison (1983: 42–43). For a summary of the literature on the psychology of scientists, see Fisch (1977) and the essays in Jackson and Rushton, eds. (1987). On Huggins, see Becker (1992) and her (1993) Johns Hopkins University dissertation on Sir William and Lady Huggins and the development of astrophysics in Britain.

end of the century, the professional age at election to the NAS was about equal. In an era before the academy was organized by sections that were responsible for nominating individuals, this reflected the judgment of the American scientific elite, represented by the membership of the Academy voting on all nominees.

Between the year of Kirchhoff and Bunsen's great discovery and the end of the 1870s astrophysics in America grew slowly. In 1860, four of the cohort that entered in or before 1859 were included in the ranks of astrophysics (table 3.1). Their number was augmented by two additions (Langley and Draper) in the 1860s. During the same period, astronomy added twenty-four new recruits. In the 1870s, seven individuals cast their lot with astrophysics as compared to forty-seven in astronomy. By 1879 the founders of astrophysics were a group of thirteen individuals whose median age was thirty-nine. At the center of the tiny group stood men like Charles A. Young at Princeton, Henry Draper with his magnificently equipped private observatory at Hastings-on-Hudson, Samuel P. Langley at Allegheny, and Edward C. Pickering (1846–1919) at Harvard. They were assisted by physicists of considerable stature: Henry A. Rowland (1848–1901) at Johns Hopkins, Albert A. Michelson (1852–1931), soon to move from Clark University to the University of Chicago, and optical theorist Charles Hastings (1848–1932) at Yale. Toward the edge stood one of America's early eclipse-chasers, David Todd (1855–1939), who would soon move from the Nautical Almanac Office to the directorship of the Amherst College Observatory and Edgar Larkin (1847–1924), who worked almost exclusively on the visual observation of nebulae.

American astrophysics started small. A handful of dedicated scientists laid the foundations for a new field. They made major discoveries in solar and later stellar spectroscopy, worked to legitimate the new science and to attract recruits. The levels of educational attainment of the founders were comparatively high. Three had no more than a secondary education while six attained the bachelor's degree. One had a master's and one the Ph.D. Two were M.D.'s. The fields in which the founders were educated and/or continued to work include chemistry, physics, and medicine as well as engineering and mathematics. Three made important contributions to photographic instrumentation (Bond, Rutherfurd, and Henry Draper). Six worked in solar physics (Rowland, Young, Langley, Todd, Very, and Rutherfurd). Stellar spectroscopy claimed the attention of two (Henry Draper and Pickering). Pickering was an innovator, first in visual photometry and then the application of photography in the measurement of stellar magnitudes. Rutherfurd was a pioneer in photographing the solar spectrum. Bond and Larkin used visual methods to discover and catalogue nebulae. Many of the early astrophysicists depended on Hastings for technical advice on the construction of special optical equipment.

Table 3.1 The Founders of Astrophysics in the United States

	Life Dates	Educational Attainment	Research Fields
Entered astronomy in or before 1859			
Bond, George	1825–65	B.A.	comets, nebulae, astrophotography
Draper, John W.	1811–82	M.D.	chemistry, astrophotography
Rutherfurd, Lewis M.	1816–92	B.A.	solar spectroscopy, astrophotography, instrumentation
Young, Charles A.	1834–1908	M.A.	solar physics
Entered 1860–1879			
Draper, Henry	1837–82	M.D.	stellar spectroscopy, astrophotography, instrumentation
Langley, Samuel	1834–1906	High School Only	solar physics, instrumentation
Hastings, Chas.	1848–1932	Ph.D.	physics, optics
Todd, David	1855–1939	B.A.	solar eclipses
Pickering, Edw. C.	1846–1919	B.A.	stellar spectroscopy, photometry, astrophotography, instrumentation
Michelson, A. A.	1852–1931	B.A.	physics, light
Very, Frank	1852–1927	High School Only	solar physics
Larkin, Edgar L.	1847–1924	High School Only	nebulae
Rowland, Henry A.	1848–1901	B.A.	solar physics, spectroscopy, instrumentation

Nine of the founders were elected to the National Academy of Sciences, America's premier honorific body. Frank Very (1852–1927) was awarded a star in the first edition of *American Men of Science* (1906), bringing the number of elite scientists in the group to ten. That 77 percent of the founders achieved elite status is significant. It means that despite their pursuit of scientific activities that were out of the astronomical mainstream, they were not denied access to the reward system of American science. Further, high status meant visibility, thus facilitating recruiting and fund-raising activities.

One of the most important uses of demographic data is to differentiate

Table 3.2 Cohort Two (1860–99) by Astronomy Type and Level of Education (N=303)

Highest Degree	Old Astronomy	New Astronomy	Mixed Astronomy
High School Only	69 (40.4%)	47 (63%)	4 (31%)
Bachelor's	32 (19%)	15 (20%)	2 (15.4%)
Master's	29 (17%)	6 (8%)	2 (15.4%)
Doctorate	40 (23.4%)	6 (8%)	5 (38.5%)
Medical	1 (0.6%)	1 (1.3%)	
Total *(Data missing for 44 individuals.)*	171	75	13

between populations, in this case between those in the second cohort (1860–99) who entered astrophysics and those who chose careers in astronomy. As table 3.2 indicates, the two groups differed significantly in levels of educational attainment. A much larger portion (63 percent) of astrophysicists as compared to astronomers (40 percent), had only a secondary school education. The two populations were comparable at the level of baccalaureate degree, but twice the percentage of astronomers attained the master's. At the Ph.D. level these differences show clearly. Almost three times (23.4 percent) as many astronomers held the doctorate as did astrophysicists (8 percent).

Career data suggests another important difference. Table 3.3 shows the institutional potential of the first job of astronomers and astrophysicists who were members of the second cohort. Institutional potential is a measure of an institution in terms of its ability to foster a research career (for a full discussion see chapter 6). At the top of the scale are private research institutions like the Dudley or Mount Wilson observatories. This category also includes the private research establishments of Draper and Rutherfurd. The scale moves downward from there. Private universities are seen as offering more potential than public institutions while government research establishments are often involved in mission-oriented research and may be highly bureaucratized. Conflict related to political patronage also hampered government scientists. Private or public colleges provided little in the way of equipment and even less time and financial support for research activities.

The distribution of these two populations by first job provides interesting contrasts. By far the largest single group (70 percent) of astrophysicists began their careers in private universities. This compares to 29 percent of those who worked in astronomy. For astronomers, the largest

Table 3.3 Cohort Two (1860–99) by Institutional Potential of First Position and Astronomy Type (N=303)

IP First Position	Old Astronomy	New Astronomy	Mixed Astronomy
(5) Private Research Institution	26 (16.1%)	5 (7%)	0
(4) Private Research University	47 (29.2%)	51 (70%)	6 (46.2%)
(3) Public Research University	13 (8.1%)	6 (8.2%)	3 (23.1%)
(2) Government Research Institution	57 (35.4%)	4 (5.5%)	2 (15.4%)
(1) College	19 (11.8%)	7 (9.6%)	2 (15.4%)
Total	**161**	**73**	**13**
(Data missing for 56 individuals.)			

single group (35.4 percent) started their professional lives at a government research establishment such as the Nautical Almanac Office, the Naval Observatory, or with one of the exploring or mapping expeditions that needed the services of an astronomer to determine geographic positions. Neither population seemed to find many employment opportunities in public universities, and about the same percentage were employed in private or public collegiate institutions. Twice the percentage of astronomers worked in private research observatories as compared with astrophysicists.

During the twenty years from 1880 to the end of the century, sixty-seven individuals joined the ranks of astrophysics. These individuals followed in the footsteps of the founders who had entered the American astronomical community before 1879. Of these, twelve were women, including Williamina P. Fleming (1857–1911), first curator of astronomical photographs at the Harvard College Observatory; the first woman to hold a Corporation appointment at Harvard; Annie J. Cannon (1863–1941), who supervised the Henry Draper catalogue of stellar spectra at Harvard; and Henrietta S. Leavitt (1868–1921), who worked in spectroscopy and photometry at HCO. Male recruits included James E. Keeler in 1881, William Wallace Campbell (1862–1938), who entered in 1888, George Ellery Hale in 1890, Heber D. Curtis (1872–1942), who joined the ranks in 1896, and Charles St. John (1857–1935) in 1898. These individuals would be among the leaders of early twentieth-century astrophysics. During the same period, astronomy attracted ninety-nine new recruits of whom fourteen were women.

In the first edition of her classic *Popular History of Astronomy During the Nineteenth Century*, Agnes Clerke (1887: 463) commented on the interdisciplinary nature of astrophysics. "The science of the heavenly bodies has . . . become a branch of terrestrial physics, or rather a higher kind of integration of all their results." From the beginning, what was noticeable for participants and knowledgeable lay persons alike, was the lack of clear lines of demarcation between the new science of astrophysics and the established field of physics. Astrophysics started life as an intellectual hybrid, appropriating many of the tools, techniques, and theories of physics and applying them to situations far removed from the physical laboratory. For astrophysicists, the sun and stars became laboratories where conditions (temperature, pressure, mass, rate of rotation) obtained that no physicist could approximate in terrestrial experiments. Essential to the process of defining astrophysics was its interdisciplinary nature.

The editors of the *Supplement* to the *Oxford English Dictionary* (Sampson and Weiner 1989, 1:735) report that astrophysics and astrophysicist entered the language in 1869. What the *OED* fails to note is that for many years, the word was printed with a hyphen: *astro-physics*. This was true late in the century when the *Astro-physical Journal* (1895) began appearing and the founders of the Astronomical and Astro-Physical Society of America kept the convention when the society was launched (1899). This usage vividly suggests the hybrid nature of the new science.

In 1879, Allegheny Observatory director, Samuel P. Langley, reviewed "The Recent Progress of Solar Physics" in a vice-presidential address to the Mathematics and Astronomy Section of the AAAS. Langley began with the distinction between an interest in the motions and positions of solar system objects or the stars and the new approach that focused on astronomical physics and chemistry. Langley pointed to the high degree of precision achieved by astronomers, characterizing accurate prediction and observation as the most striking triumph of the old astronomy. Indeed, Langley saluted workers in astrometry and celestial mechanics with a poetic tribute.[8]

> The little Vernier on whose slender lines
> The midnight taper trembles as it shines,
> Tells through the mist where dazzled Mercury burns,
> And marks the spot where Uranus returns.

However, the Allegheny astrophysicist was perplexed. How could it be that observers "who had learned this excellent lesson of precision, had learned no other?" Why, Langley wondered (1879: 2), have astronomers remained "indifferent to a great question to which the old methods did not apply?" The question, of course, was the physical nature of the sun. "We are called into existence by a great central fire, the sun, by which we continue to exist from one hour to another. What is it? What is the heat which it pours into space, and with whose cessation we shall cease? How long will it feed our lives?"

8. Langley (1879: 1). I have not been able to identify the verses. Only poets would be able to locate Mercury at midnight!

If we put this into a social-psychological framework, Langley was discussing differences in imagination and perception. The old astronomy viewed the heavens through the eyepiece of a transit circle and conceived of the stars as numbers in printed catalogues. To the celestial mechanician, the sun and planets were to be treated as mass points and discussed according to the laws of Newtonian mechanics. The old astronomy rested on a mathematical rather than a physical perception of the universe. Its equations dealt with the three-body problem and its practitioners worried about errors of observation rather than the source of solar energy.

In a series of popular articles published in *The Century Magazine* in 1884 and revised as a book in 1892, Langley elaborated on ideas first suggested in his AAAS address of 1879. He developed an elaborate analogy between practitioners of the old astronomy and the astronomer-priests who built Stonehenge. While not religious shrines, the "great national observatories, like Greenwich or Washington, are the perfected development of that kind of astronomy of which the builders of Stonehenge represent the infancy." Langley (1884: 712) then pressed his advantage. "Those primitive men could know where the sun would rise on a certain day, and make their observation of its place . . . without knowing anything of its physical nature." Emboldened by his flight of rhetoric, Langley suggested "the astronomer of today can still use his [transit] circle for the special purpose of fixing the sun's place in the heavens, without any more knowledge of that body's chemical constitution than had the men who built Stonehenge." Lest any reader mistake the nature of this characterization, Langley concluded, both ancient priest and nineteenth-century astronomer "aim at the common end, not of learning what the sun is made of, but of where it will be at a certain moment; for the prime object of astronomy, until very lately indeed, has still been to say *where* any heavenly body is, and not *what* it is."

But the situation was changing. "A new branch of astronomy has arisen, which studies sun, moon and stars for what they are in themselves, and in relation to ourselves." Ever the polemicist, Langley (1884: 713) reported that "This new branch of inquiry is sometimes called Celestial Physics, sometimes Solar Physics, and is sometimes . . . referred to as the New Astronomy." The polemical intent is clear. In late Victorian America, characterizing anything as "New" was sure to attract attention. The promise of interesting *new* material stands in sharp contrast to the abstract, remote, and arcane subject matter of the *old* astronomy. Middle-class readers were ill prepared to grasp the subtleties of celestial mechanics with its formidable equations or astrometry with its precise methods of observations and myriad corrections. Nor would page after page of numerical information on the position and motion of stars speak to the imagination of most readers. Langley's richly illustrated exposition of the new astronomy was a masterful defense of astrophysics and by indirection, an indictment of the old astronomy.

Langley (1879: 10) often suggested that solar physics, through its study of solar radiation and sun spots, might lead to a better understanding of changing weather patterns on earth. Indeed, in time it might be possible to develop long-range weather forecasting and thus avert flood and famine. Here Langley was presenting the new astronomy as a practical science, much as the old astronomy was once justified because it provided information useful for navigation and time keeping.

In the preface to *The New Astronomy*, Langley took the debate one step further. He pointed out that the new astronomy was not funded by the federal government and had been ignored by philanthropists. In 1892 this latter assertion could not be defended, as the history of Pickering's administration at Harvard amply demonstrates. In any event, Langley (1892: i) directed his book to the "educated public on whose support" scientists are "so often dependent for the means of extending the boundaries of knowledge." Through the generosity of the government and private benefactions, the old astronomy found itself "munificently endowed" while the new astronomy "so fruitful in results of interest and importance, struggles almost unaided." Langley continued in a backhanded manner: "We are all glad to know that Urana, who was in the beginning but a poor Chaldean shepherdess, has long since become well-to-do. . . . It is far less [well] known than it should be that she has a younger sister . . . bearing every mark of her celestial birth, but all unendowed and portionless." The Smithsonian secretary concluded "It is for the reader's interest in the latter that this book is a plea."

THE SCIENCE OF ASTROPHYSICS

In October 1897 astronomers and physicists gathered at Williams Bay, Wisconsin, to dedicate the Yerkes Observatory. With its forty-inch refracting telescope, Yerkes was by far the most impressive astrophysical research establishment yet constructed in America. The creation of George Ellery Hale, the Yerkes Observatory was attached to the new University of Chicago. Several of the speakers at the dedication took the opportunity to make thoughtful statements concerning the nature and scope of astrophysics. Among these addresses, none was more insightful than that of Lick Observatory Director James Keeler.

Keeler spoke on the relations between astrophysics and the other physical sciences, but his deepest concern was with the identity of the new science. Keeler (1897: 271) began his remarks with a discussion of the constituent domains of physical science. Unlike political divisions, these domains could not be separated "with mathematical precision." The sciences "pass into one another by imperceptible gradations" because the unity of nature opposes "rigid systems of classification." Keeler then asked the audience to visualize the location of astrophysics in the "broad ground" between astronomy and physics. This broad ground Keeler considered to be "so extensive and fertile as to justify the development of a new science for its special cultivation." Keeler hoped that, in time, the new science of astrophysics would help bring physics and traditional astronomy closer together. He also believed that astrophysics would enter into an active exchange of knowledge with its neighbors.

Keeler made every effort to be irenic rather than confrontational. After all, the redoubtable Simon Newcomb, doyen of the old astronomy, was sitting only a few feet from the young astrophysicist at the Yerkes dedication. A substantial portion of the address was devoted to legitimating the new astronomy. Keeler began with a discussion of taste in selecting research problems and went on to point out areas of disagreement not only between the old and new astronomy, but within the ranks of traditionalists as well. Keeler defined the rapid rate of discovery in astrophysics as normal for a new science. The Lick director acknowledged the public appeal of astrophysics as opposed to the old astronomy, but suggested that heightened public awareness would benefit both sciences.

In his address Keeler made a major concession to the old astronomy. Citing the "high order" of "mental discipline" developed by the study of astronomy, Keeler (1897: 277), who had concentrated in physics and mathematics as an undergraduate at the Johns Hopkins University, conceded that "training in the methods of the older astronomy should be regarded as indispensable preparation for astrophysical work." For good or ill, this was the first public articulation of the division of labor that would mark the organization of astronomical teaching and research at the University of California and the University of Chicago. In effect, educational activities would be divided, with the old astronomy controlling virtually all of the undergraduate offerings and much of the graduate program on campus while the remote observatory (Mount Hamilton or Williams Bay) concentrated on astrophysical research and provided advanced students an opportunity for hands-on experience and the opportunity to do thesis research. To be sure, candidates for advanced degrees in astrophysics took courses in physics and chemistry, but also had to endure hours of classwork in celestial mechanics and survive marathon sessions devoted to computing orbits for newly discovered solar system objects.

Keeler also acknowledged significant differences in the perceptual foundations of the old and new astronomy. The Lick director used photographs of the Milky Way taken by Edward Emerson Barnard (1857–1923) to make his case. These remarkable plates suggested physical relationships between bright and faint stars and wide spread nebulosity. Keeler (1897: 278) argued that Barnard's plates indicated that the real diameters of the stars must vary greatly and that their distribution in space was more complex than previous discussions implied. Although workers in astrometry might want to treat Barnard's plates mathematically, the approach would be "to no purpose" because the first step in interpreting such numerical data would involve "the construction of just such a chart as the photograph places . . . in our hands." To astronomers endowed with physical imagination, Barnard's plates revealed a wealth of information by simple inspection. For the student of astrometry, the plates would be objects to measure, reduce and analyze in numerical terms.

Keeler saved his strongest arguments for the closing section of his address. Admitting that there are many areas (especially the astrophysics of the solar system) in which the new astronomy had made few contributions, he pointed to the unexpected universe revealed by the spectroscope. This new universe pushed back the frontiers of science and revealed unsuspected wonders. It was a universe which the old astronomy had in no way anticipated. The spectroscope revealed undreamed of objects, from highly luminous Wolf-Rayet stars to spectroscopic binaries. It led to the discovery of the element helium and confirmed the existence of the complex Balmer series of hydrogen lines, so important in the later Bohr model of the atom. Motion in the line of sight (radial velocity) was detected by the spectroscope, thus adding a new dimension to the study of the sidereal universe. Hale's innovations in instrumentation permitted a whole new way of looking at the sun at different levels of its atmosphere. Both the sun and stars became laboratories in which physicists and astrophysicists could study the behavior of matter under conditions they could not duplicate on earth. Finally, the young science of astrophysics invited reflection on the problem of stellar evolution. This unexpected universe proved an endless source for fresh discoveries. The frontiers of astrophysics lay open and inviting to young investigators. And it was a science that provided good copy for the mass circulation urban press as well as monthly magazines.

Of course, there were some who would never agree. Writing in 1878, the irascible Asaph Hall (1829–1907), discoverer of the moons of Mars, speculated that "when the novel and entertaining observations with the spectroscope have received their natural abatement and been assigned their proper place" astrophysics would prove to be a fad rather than a science (Turner: 1902: 9). Perhaps to Hall's disappointment, by the end of the century, the new universe revealed by astrophysical research was being rapidly incorporated into the corpus of scientific knowledge.

THE SOCIAL ROLE OF ASTROPHYSICIST

As sociologists, Joseph Ben-David and Randall Collins suggest in a seminal paper dealing with social factors in the origins of a new science, ideas alone are not sufficient to trigger "take-off into sustained growth in a new field; a new role must be created as well."[9] The development of the social role of astrophysicist is of great importance for understanding the emergence of the new science after mid-century.

As a normal puzzle-solving science, astronomy held out few opportunities for ambitious young investigators in a hurry to make their reputations through dramatic discoveries. In the eyes of aggressive young scientists, astronomy did not have the status of a field like physics where the work of Bunsen and Kirchhoff was simply one among many exciting advances (Jungnickel and McCormmach 1986). Physics was seen as an active, creative, high-status field producing ideas and techniques that could be applied to problems located in the broad ground between astronomy and physics that Keeler described in his Yerkes address.

In rejecting the normal science of astronomy as a career choice, these young people also rejected the social role of astronomer. When the founders of astrophysics appropriated tools and techniques from physics, they also borrowed some of that field's status. As recruitment figures suggest, before 1900 most entrants still identified with astronomy and opted for the older, established field rather than astrophysics. But, especially after 1880, a growing number perceived the potential of astrophysics.

George Ellery Hale, a leader of the second generation of American astrophysicists, characterized the roles of astronomer and astrophysicist. Hale (1908: 9) described the astronomer as one who, "soon after the setting of the Sun, retires to a lofty tower, from whose summit he gazes at the heavens throughout the long watches of the night." Following Langley's lead, Hale suggested that the astronomer is a "lineal descendent of the seers and soothsayers of the Chaldeans" who "dwells apart, finding little of interest in the ordinary concerns of the world, so occupied are his thoughts with celestial mysteries." This vivid polemic suggests a social role that is irrelevant not only to the development of science but to the progress of society as well. Americans of the 1890s, no less than their early Victorian parents, placed great stock in being up-to-date. The highest accolade for both scientists and businessmen was "progressive." Those who inhabited lofty towers far from the "ordinary concerns of the world" were hardly worthy of the world's attention.

In sharp contrast to Hale's caricature of workers in the old astronomy, the astrophysicist used the telescope in a routine manner to expose photographic plates, worrying about dew on the lenses and the steady motion

9. Ben-David and Collins (1966:451). I have not applied the Ben-David and Collins model in the same way they use it for analyzing the rise of psychology in nineteenth-century Germany. The cultural and scientific contexts are quite different and the dynamics do not follow similar patterns. What is important is the concept of limited career opportunities within a discipline and the perception of status differentials between disciplines.

of the mechanism that moved the instrument as it followed objects across the night sky. Hale (1908: 11–12) said of the astrophysicist, "His most interesting work is done, and most of his discoveries are made, when the plates have been developed, and are subjected to long study and measurement under the microscope." In addition, the astrophysicist must spend "much of his time . . . in the laboratory, imitating, with the means placed at his disposal by the physicist and chemist, the various conditions of temperature and pressure encountered in the stars, and watching the behavior of metals and gases in these uncommon environments." With each new research problem, the astrophysicist had to design and construct appropriate instrumentation. "Kept thus in touch with the newest phases of physical and chemical investigation, the countless applications of electricity, the methods of modern engineering, and the practical details of workshop practice, his interest in these things of the world is likely to be quite as broad as that of the average man."

For Hale, the social role of astrophysicist involved two salient dimensions. Those who aspired to the new role approached astronomical research *as physicists*, using the conceptual and theoretical tools of that science, all of which rested on the cultivation of a physical imagination.[10] Further, they presented themselves as experimentalists. In contrast to astronomers, astrophysicists defined themselves as *practical* individuals, skilled in the fabrication of research instrumentation and dedicated to using those instruments to make important discoveries about the sun and stars. In essence, astrophysicists were as practical, progressive and hardworking as any other American.

OF BOUNDARIES AND COMMON GROUND

As Keeler suggested, boundaries between various domains of natural science cannot be distinguished with the clarity of political divisions. For purposes of analysis, however, lines of demarcation must be established. One way of doing this is to indicate what two domains have in common. Agnes Clerke took this approach in her discussion of *Problems in Astrophysics*. Clerke (1903: 286) argued that "between the old and new astronomy lies a region claimed by both, yet belonging by exclusive right to neither." This region involved the study of stellar motion in the line of sight: the radial velocity of stars. As Clerke reminds us, these observations can only

10. It was European physics that Hale and other second- generation astrophysicists held up as the ideal. In the 1890s, physics in the U.S. (Kevles 1978: chaps. 3–4) had not reached European levels of excellence. Hale, in particular, was well acquainted with European physics and physicists and became a major source of information for younger, less well read or traveled Americans.

be spectroscopically determined. When the observations are combined with data on the proper motion of stars, astronomers gain a deeper understanding of the structure of the sidereal universe. However, the process of collecting data on stellar radial velocities proved as prone to instrumental error as transit-circle observations. Roughly speaking, the same kind of approach had to be taken to radial-velocity work as to making and reducing observations with the transit circle. Clerke (1903: 287) suggested that it was this increasing concern for precision that broke down "the barrier between celestial mechanics and celestial physics." Astronomers came to take more interest in spectroscopic work and astrophysicists had to develop a much greater understanding of what it meant to make precise observations and then to eliminate all possible errors when reducing the data. In this field of research, "sidereal physics merges into sidereal mechanics." Yet, as Clerke continued (1903: 5), there must be a line of demarcation between the two sciences. Clearly, the shared common ground is the study of stellar motions. By the late 1890s, the common ground also included statistical cosmology. But this field was almost exclusively a European endeavor (Paul 1993). Away from that point the cognitive terrain becomes clearly differentiated.

Lewis Boss, a leading figure in astrometric work, early saw the possibilities of this shared ground. Writing to Edward S. Holden (1846–1914), first director of the Lick Observatory, Boss drew a sharp distinction between "stargazing" and real scientific work. The Dudley Observatory director conceded that "for the first season or two stargazing [observing the moon and planets and making accurate drawings of their surface features] with the big glass [the Lick 36-inch refractor] and *experiments* with photographic processes seem to me very appropriate."[11] But, Boss continued, "What I am mainly anxious about is the measurement of stellar motion in the line of sight." Boss believed that "the biggest and most epoch-making haul is there. And the recent experiments of [the German astronomer Hermann] Vogel [1841–1907] seem to increase our hopes vastly." Boss went on to describe to Holden his grand design for the study of stellar proper motions, a project that would occupy Boss for the rest of his career.[12]

CONFLICT

In the area of stellar proper motions, the old and new astronomy seemed to work together in harmony. This pocket of cooperation, however, was

11. Lewis Boss to Edward S. Holden, 3 October 1888, p. 3. SALO, Holden Papers. Emphasis in the original.

12. Boss to Holden, 3 October 1888, p. 4. SALO, Holden Papers.

the exception rather than the rule. Elsewhere, astronomy either ignored astrophysics as irrelevant, attacked the new science as physics rather than astronomy, or, in a few cases, denied the validity of its methods and findings. This latter situation involved areas in which astronomy and astrophysics used different methods to approach the same research problems.

By far the most comprehensive attack on astrophysics came from the pen of Naval Observatory astronomer John R. Eastman (1836–1914). In his address as retiring vice president of Section A (Mathematics and Astronomy) of the AAAS, Eastman deliberately rejected the traditional hortatory review of recent progress or bland historical disquisition. Instead, he (1892:18) chose to deal with substantive issues: "The importance of one branch of astronomical work . . . which, owing to the present tendency towards specialization, is likely to suffer from serious neglect." Eastman's remarks differed from those of Chandler in that he carefully compared the old and new astronomy in addition to mounting a defense of the old.

Eastman could not resist the temptation to begin with the Chaldean astronomer (by now a literary cliché in the debate between old and new astronomy). But his use of the symbol was very different from Langley's. For Eastman (1892: 17), the ancestral priest-astronomer "on the plains of Chaldea" had an advantage over nineteenth-century astronomers. The science had not yet "reached that critical epoch when he must choose between the 'old' and the 'new' astronomy." Eastman went on to lament what he termed specialization. The trend was pernicious. "The range of study and investigation has spread beyond the efficient grasp of any individual and specialists are rising up in all directions." In using the word specialization, Eastman was referring to the emergence of astrophysics whose paradigm was wholly different from that of astronomy as defined by Chandler or Sir George Biddle Airy. Eastman (1892: 18–19) asserted: "Physics has so expanded in all directions and so adapted itself to its new surroundings, that we find it, in one department at least, casting aside its former title and masquerading under the name of Astronomy." Eastman bitterly resented this new form of astronomical research, arguing that investigators in astrophysics selected problems that promise quick returns. "A few hours' labor with the spectroscope or camera may be spread attractively over several printed pages."

Eastman (1892: 19) questioned the validity of astrophysical research. "From the true astronomical point of view, all these questions are at least secondary to the fundamental problems of finding the true position of the solar system in the stellar universe and determining the relative positions and motions of those stars . . . that compose that universe." The Naval Observatory astronomer seemed to suggest that the old astronomy formed the foundation on which all research, including astrophysical investigations, must rest. Like Chandler, Eastman (1892: 20–22, 25) defined the old astronomy as labor-intensive, involving long years of data-collecting by highly skilled observers and the careful and painstaking reduction of those observations.

The most novel aspect of Eastman's critique (1892: 28) involved an economic interpretation of the rise of the new astronomy. Philanthropists who supported scientific activities demanded a price. Patrons looked for "immediate and novel results," or "material benefits" for the donor, or, failing that, an increase in "popular reputation." Since it is hard to see how research in astronomy or astrophysics could lead to significant material benefits, an enhanced public image must be understood as the most important reward for benefactors. Eastman expressed concern that this situation led to "the evolution of a certain type of astronomer, and also of a corresponding type of astronomical patron."

Even as Eastman was addressing Section A, one of its members, George Ellery Hale, was about to begin his second adventure in observatory building. Hale, clearly an example of the type of astronomer Eastman abhorred, learned at this meeting of the AAAS of a pair of 40-inch lens blanks in the vault of the telescope-maker Alvan Clark. Hale rushed back to Chicago to mobilize William Rainey Harper, president of the new University of Chicago and one of the greatest modern academic fund-raisers. Together they would search out the kind of patron that Eastman deplored, in this case Chicago tycoon, Charles T. Yerkes. To be sure, Eastman could not have known what was afoot at the time he delivered his address, but his timing was uncanny. In decrying what he considered base motives behind endowing astrophysical research institutions, it was as if Eastman (1892: 31) could see what soon would be constructed on the shores of Lake Geneva in Wisconsin. "Novelty, combined with a desire for architectural display and an absurd ambition to secure the largest telescope and the greatest variety of astronomical instruments," were not, to Eastman, acceptable reasons for establishing a new observatory. Yet this was the course on which Hale was about to embark.

Eastman (1892: 31) felt the circumstances surrounding private benefactions for the new astronomy were intolerable. "It is the duty of astronomers to see to it that, for their own reputation and for the present and the ultimate welfare of their science, the true purpose of astronomical study and research and the grounds for the existence and support of observatories should be frankly given and courageously maintained." The only honest justification for astronomical research is that it "stimulates the highest form of intellectual activity . . . and increases the sum of human knowledge." Personal aggrandizement or public image had no part to play in the support of science. Of course, it is difficult to imagine Seth Carlo Chandler opening the purse of a wealthy patron with his descriptions of the old astronomy.

Eastman (1892: 32) then restated the paradigm of astronomy as a normal puzzle-solving science. Astronomy deals with the figure and motions of the earth, the movement of the planets, and the motions and distances of the stars. In a major concession to astrophysics, Eastman included solar physics in the paradigm. To my knowledge, this is the first time that any astronomer amended the paradigm to include a portion of astrophysics. How sincere Eastman was in making this gesture is not clear. It is clear that in the closing paragraph of the address, he demanded that astronomers maintain standards "well above the popular fancies of the hour" and devote themselves to "fundamental work that shall not only satisfy the rigorous demands of the present time, but shall make the last decade of the nineteenth century an important epoch in the real progress of astronomy." By this point, however, the tide was running against the old astronomy. More and more donors chose to support astrophysical research.

Benjamin A. Gould and his Boston circle, especially Seth Carlo Chandler and John Ritchie, Jr. (b.1853), are of special importance for understanding conflict between the old and new astronomy. They represent both a specific approach to research in photometry and the study of variable stars and deep personal antagonism toward the director of the Harvard College Observatory, Edward C. Pickering and all that Pickering stood for as one of the founders of astrophysics in America. Further, through Ritchie, they could mobilize journalistic resources. John Ritchie Jr. was an early specialist in science journalism. He wrote for the Boston newspapers as well as editing his own publication, *The Science Observer*. Ritchie often followed a yellow journalism approach in his science reporting. Readers in the 1890s frequently had their science served up in the same way that mass circulation dailies presented domestic politics or international affairs.[13]

Gould lent his personal and professional stature to these attacks, so it is important to know something about the man. The best analysis of Gould is by historian Mary Ann James (1987). As James points out, there was a deep element of instability in Gould's personality that marriage and a successful career seemed to hold in check. In 1885 he retired to Boston, a widower, after fifteen years as the director of the Cordoba Observatory. Asaph Hall, retired from the Naval Observatory and lecturing on celestial mechanics at Harvard, knew Gould well. Hall characterized Gould as "a crabbed old Puritan, who says sharp things, but he is fair in conduct, and sometimes his independence is refreshing." After Gould's death, Hall recalled that the older man "had a strong dislike of Pickering's astron-

13. See for example, Lankford (1979). The presentation of science in the late nineteenth-century print media and the political uses made of science reporting by scientists and patrons of science would make an interesting and valuable monograph. For an overview of science in the American media, 1910–1950, see LaFollette (1990).

omy."[14] Indeed, so strong was this antipathy that it prevented Gould from calling on the Astronomer Royal, Sir William Christie (1845–1922), when Christie visited Boston. The Astronomer Royal was Pickering's guest at the Observatory.[15]

Gould used the *AJ* to encourage research in astronomy and sometimes to attack the work of astrophysicists, especially at Harvard. He increasingly took a pessimistic view of the future of astronomy. As he wrote to Lewis Boss at Dudley, "But for you and Chandler, I should almost lose faith in the progress of astronomy in our country." Gould could not tolerate the drama of frequent astrophysical discoveries and the quest for new and bigger instruments. Nor did he approve of the search for patrons. "Biggest things on earth, 'hustling' struggles for (constantly short-lived) notoriety, employment of physical energies in trumpet blowing instead of study and useful research,—these seem to have taken possession of the field."[16] Gould's point of view was shared by Chandler and Ritchie.

John Ritchie was a devious character. Writing to the first director of the Lick, Edward S. Holden, Ritchie indicated that he was "much interested' in the conflict between Holden and several members of the Lick staff. Ritchie attempted to ingratiate himself with Holden. "I am not altogether without influence in shaping the opinion of our local press, and while I don't wish to pry into your affairs, if the time comes that you can use me, and your cause is just, I hope that you will count on me."[17] Ritchie clearly wanted information and was willing to become deeply involved in the complex clash of personalities and research agendas that was making life on Mount Hamilton chaotic.

Two years later, Ritchie was again fishing in troubled waters. William H. Pickering (1858–1938), younger brother of the Harvard Observatory director and a leader in the field of lunar and planetary research, announced that at least one of the four bright satellites of Jupiter was egg-shaped rather than spherical. This report brought a storm of criticism (Jones and Boyd 1971: 307–9). Lick astronomers were especially harsh in their treatment of the younger Pickering. Hoping to inflame relations between Harvard and Lick, Ritchie indicated to Holden his opposition "to Harvard getting any more money until they start in with better work." Ritchie asked for more information concerning Lick observations of the Jovian satellites.[18]

14. Asaph Hall to Asaph Hall, Jr., 19 November 1896. LC, Hall Family Papers.

15. Hall to Hall, Jr., 29 November 1896, pp. 2–3. LC, Hall Family Papers.

16. Benjamin A. Gould to Simon Newcomb, 26 October 1896. DOA, Gould letter book II, p. 51.

17. John Ritchie to Edward S. Holden, 19 October 1892, pp. 1–2. SALO, Holden Papers.

18. Ritchie to Holden, 14 June 1894. SALO, Holden Papers.

Word of Ritchie's attempt to involve the Lick director in his crusade against the Harvard College Observatory was transmitted directly to Pickering by Holden. In response, the Harvard director asked Holden to "remember that Mr. Ritchie . . . is anxious to prevent, as he has stated in print, money being given to aid modern astronomical methods [astrophysics] and would probably not be sorry if he could involve the two principal American observatories devoted to this kind of work, in a controversy."[19] The Harvard director deplored any scientific debate that shifted from facts to personalities.

Gould and his protégé, Chandler, both had experience with visual photometry, using methods developed earlier in the nineteenth century by Friedrich W. A. Argelander (1799–1875) at the Bonn Observatory. Gould claimed Argelander as one of his mentors and defended the master's methods. Chandler was equally insistent that these methods could not be surpassed. When Pickering (Jones and Boyd 1971: 184ff) allocated a large share of the Harvard Observatory budget to photometry, first using the new meridian photometer and then photographic techniques, Gould and his circle took great umbrage. The situation was made more complex because Pickering became involved in a dispute over visual photometry with English and German astronomers (Jones and Boyd 1971: 341ff).

In a field like photometry, there were contending research methods and forms of instrumentation. These ranged from Argelander's technique of visually estimating differences between stars of known and unknown magnitudes to various devices using artificial stars as well as photography. Under the best of circumstances, it was a research domain surrounded by a good deal of conflict. Only after World War I did the international astronomical community agree on fundamental photometric standards.[20]

Pickering's photometer used the pole star (incorrectly assumed to be of constant magnitude) as a comparison and measured how much brighter or fainter an object was than *Polaris*. Edward S. Holden (1883: 173), then at the University of Wisconsin, praised the Harvard photometric studies. Pickering reduced his data to the magnitude scale suggested by the English astronomer Norman Pogson (d. 1891). On this scale each stellar magnitude differs from the next by a fixed ratio. In effect, Pickering placed the study of stellar magnitudes on a firm quantitative basis. Holden believed that "the work of Professor Pickering must inevitably, from its intrinsic merit, at once replace all existing scales of magnitude."

At the 1886 meeting of the AAAS, Chandler (Anon. *The Observatory*

19. Edward C. Pickering to Edward S. Holden, 15 June 1894, pp. 2–3. SALO, Holden Papers.

20. Lankford (1984a: 4–9) provides an introduction to the history of astronomical photometry. For an overview, see Hearnshaw (1996).

1886: 367) launched an attack on Harvard photometry. Gould's associate argued that instrumental photometry had thus far proved a failure." It had not "developed a more uniform scale of magnitude than Argelander's, nor had the accuracy of individual determinations been increased," but were "far more uncertain than the old differential estimates." Chandler made sure his paper appeared in the major international journals. He also compiled a series of variable-star catalogues which were published in the *Astronomical Journal.*

In his first catalogue of variable stars, Chandler had harsh words for observers who used methods other than those devised by the sainted Argelander. Chandler described his work compiling a catalogue of variable stars as a "discouraging and thankless labor." He alleged the literature was filled with reports of "pseudo-variables." Thus no star was included unless its variability received independent *visual* confirmation. Chandler used amateur astronomers to check suspected variables.[21]

By 1893, the year in which Chandler published the second catalogue of variable stars, photographic photometry at Harvard was well under way, but the results were not at all to his liking. Compared to observations secured with a transit circle or a refractor equipped with graduated circles, the reported positions of Harvard variables taken from photographic plates were relatively rough approximations. Chandler (1893: 90) deplored this and urged Harvard astronomers to "raise their standard of precision."

The problem of locating variables according to the standards of precision of the old astronomy was not, however, Chandler's major criticism. He (1893: 91) distrusted photography as a research tool. "In view of the profuse announcements of variability during the past few years [coming primarily from HCO], the very moderate increase in the present Catalogue over the previous one may excite surprise among those who do not know the facility with which rubbish collects about the subject." He held up the visual work of Argelander as the standard that all workers should follow. It is also interesting to note that in addition to the preparation of ephemerides, stars accepted into the canon were treated to extensive mathematical analysis in order to find the laws of variability. This is the approach of the mathematical astronomer, at work on problems that were essentially physical in nature.[22]

21. Chandler (1888a: 92). In the same year, Chandler sought to apply the Argelander method to the determination of star colors using red and blue filters with a 6-inch refractor. Chandler (1888b: 137–40).

22. Gould himself illustrated the limited physical understanding shown by workers in old astronomy. He wondered if some of the very short period variables discovered by Chandler might be rapidly rotating stars "only half of which is luminous to its full extent." Benjamin A. Gould to Herbert H. Turner, 3 November 1895. DOA, Gould letter book I.

The most thoughtful reply to Chandler appeared in an unlikely place. During the course of an address on women in astronomy, presented at a congress held in conjunction with the 1893 Chicago World's Fair, Williamina Paton Fleming, Pickering's trusted associate at Harvard, responded to Chandler. Without ever mentioning names (she hardly needed to, the audience would immediately recognize the principals), Fleming (1893: 687) suggested that "One must not always cling to the earliest method of accomplishing anything" or "assume that because it was the earliest . . . it must consequently be the best, and also the only way." Fleming, like Hale at a later date, sought to identify astrophysics as a progressive science. "While the old time astronomer clings tenaciously to his telescope for visual observations, astronomical photography is leaving him far behind and almost out of the field in many investigations." Fleming turned the knife when she reported an example of the difficulties experienced by visual observers. Referring to a variable discovered photographically in the constellation *Delphinium*, Fleming recounted the experience of two "skilled visual observers," who attempted to observe the star. One concluded that it was not variable with a constant magnitude of nine and the other that it was not a variable, but of magnitude eleven. In the end, it turned out that they were observing different stars, neither of which was the variable. Fleming suggested a moral to the story. "No such error could have occurred from the comparison of the photographic charts."

Despite Fleming's comments, Chandler continued to snipe at photographic methods in supplements and revisions of his catalogue. While distrusting the accuracy of Harvard reports "especially in the matter of positions and identifications," Chandler (1895: 81) did soften his attacks somewhat. Pickering assured him that all discoveries were confirmed by independent examination of the plates and Chandler agreed to include Harvard variables, assuming that "confirmation necessarily includes all that the word implies, namely right identification and correctness of position, as well as the fact of variability." In 1901, Chandler, burdened with the editorship of the *AJ* and more interested in research on variation of latitude, relinquished the compilation of catalogues of variables to a committee appointed by the *Astronomisches Gesellschaft.* The *AG* committee was willing to accept variables discovered photographically at Harvard.

By the beginning of the twentieth century, open conflict between the old and new astronomy tended to fade from public view. But senior workers in the old astronomy did not forget. As Boss wrote to Simon Newcomb in 1905, "In the interest of harmony Dr. Chandler gave up his variable star work; and that was a great pity, for Chandler is a master in that line. No one can handle it with the promptness, judgment, or skill that he has displayed." But there were practical consequences as well. Speaking as editor of the *AJ*, Boss lamented the fact that Chandler's decision, "cut off some material for the *Astronomical Journal*."[23] The decline in submissions to the *AJ* provides one of the first clear signs that astrophysics was moving ahead in competition with astrometry and celestial mechanics.

Nor did those associated with the new astronomy forget the conflict. Robert W. Wood (1868–1955), a Johns Hopkins physicist and colleague of astrophysicists working on spectroscopy and experimental optics, wrote the third director of the Lick Observatory, William Wallace Campbell (1862–1938), thanking him for a volume of the Lick Observatory *Publications*. Wood reported that at first he was afraid the large package contained a Naval Observatory Catalogue giving the positions "of four million or so stars, which volumes I use daily for supporting apparatus" in the laboratory.[24] By the end of the first decade of the new century, much of the tension was gone, but the American astronomical community would remain conscious of the division between those who worked in astrophysics and those who devoted their energies to celestial mechanics and astrometry.

In the 1890s, second-generation astrophysicists set about creating important new institutions for the American astronomical community. One such institution was the *Astrophysical Journal*. The *ApJ* facilitated the publication of research in the new astronomy. After 1895, the old and new astronomy each had a major publications vehicle. Further, at the end of the century, leaders of the two camps cooperated in founding a national professional organization, the Astronomical and Astrophysical Society of America (AASA). The society would retain this name (indicating the self-conscious coexistence of the old and new astronomy) until 1914. The net result of these activities was to reduce conflict within the astronomical community.

However, the creation of the AASA entailed some tension. Simon Newcomb and George Ellery Hale clashed over naming the society. Newcomb pointedly remarked to Hale (Berendzen 1974: 37) that he hoped "astrophysicists will consider their science as a continuation of the ancient and honorable science of astronomy and allow the new body to be called the American Astronomical Society." This was not acceptable to Hale. "I feel that the time has fully come for according astrophysics a distinct place among the sciences." Hale could not agree with the proposition implicit in Newcomb's request. "If, as you remark, we were all working in so ancient and honorable a science as astronomy, everyone would undoubtedly be willing to adopt the simple and effective name 'American Astronomical Society.' *But this does not seem to be the case, for it would be stretching*

23. Lewis Boss to Simon Newcomb, 27 December 1905. LC, Newcomb Papers.

24. Robert W. Wood to William Wallace Campbell, n.d. (sometime in 1909). SALO, Campbell Papers.

our definition of astronomy farther than it would go" to include the activities of both physicists and astrophysicists working on a wide variety of problems, most of which were remote from the concerns of celestial mechanics and astrometry. At length, Newcomb capitulated and the society was christened according to Hale's wishes, but Hale also compromised. He originally wanted a constitution that provided for the presidency to alternate between representatives of the old and new astronomy. This provision was not included in the final draft of the constitution.

THE HALE SYNTHESIS

George Ellery Hale was an institution builder first and a research scientist second. He was a man in a hurry, pursuing grand schemes that ranged from fathering observatories (Kenwood, Yerkes, Mount Wilson and Mount Palomar) to reforming the National Academy of Sciences and creating the California Institute of Technology. Yet, among second generation astrophysicists, it was Hale who provided the most comprehensive statement of goals for the field. The Hale synthesis took shape over a decade. It was first articulated in his address at the 1897 dedication of the Yerkes Observatory. The founding of the Mount Wilson Solar Observatory provided the opportunity to reformulate and elaborate his ideas. Hale's 1908 book, *Stellar Evolution* marked the end of the process.

At the heart of the Hale (1908: 5) synthesis lies the belief that physics is central to understanding the natural world. "An investigator who has been confined to the traditional methods of a department of science where physics has . . . played little part, may . . . find in physical methods a powerful means of advancing his subject." He (1897: 310–11) saw one of the primary missions of the Yerkes Observatory as "strengthening the good will and common interest which . . . draw astronomers and physicists into closer touch." On the Hale (1905b: 42) model, astrophysical research establishments would come "to resemble a physical laboratory more closely than an observatory." Hale refused to be confined by traditional disciplinary boundaries. He (1908: 6) believed that "astrophysics has become . . . almost an experimental science, in which some of the fundamental problems of physics and chemistry may find their solution."

But astrophysicists should not wait until their colleagues in terrestrial physics laboratories found time to assist them. Hale's (1897: 313) ideal was one in which astrophysicists turn from the telescope to a "well-equipped laboratory" where "experiments . . . could be performed at any time." Hale (1897: 311) did not wish to maintain clearly marked boundaries between physics and astronomy. In a rare display of public humor, he defined the telescope "as an instrument for forming an image of a celestial object on the slit of a spectroscope." These beliefs guided Hale in creating the *Astrophysical Journal*, whose subtitle was *An International Review of Spectroscopy and Astronomical Physics*. Both astronomers and physicists were on its editorial board and in the *ApJ* readers were "likely to find a paper on radiation in a magnetic field in close proximity to an account of nebular photography or a discussion of stellar motion in the line of sight." In Hale's view, physics and astrophysics would blend into one activity.

Astrophysics, like astronomy, was a data-driven science. Astrophysicists (Osterbrock 1985) are caught up in a never ending "quest for more photons." Instruments are central to the development of astrophysics. Before Pickering took over as head of the Harvard College Observatory, most observatory directors had been content to add one major new instrument or technical innovation during their tenure. Under Pickering, however, innovation became institutionalized. Photometers, astrographs, spectrographs, and specialized telescopes followed one another in quick succession.

New astrophysical research problems demanded specialized instrumentation for their solution. The astrophysicist had to learn optical and mechanical engineering. Hale (1908: 13) aptly limned this aspect of the ideal astrophysicist, who was pictured in the laboratory, "surrounded by lenses and prisms, gratings and mirrors," and "other elementary apparatus of a science that subsists on light." In these circumstances, the astrophysicist "cannot fail to entertain the alluring thought that the intelligent recognition of some well-known principle of optics might suffice to construct, from these very elements, new instruments of enormous power." Hale spoke from experience. As an undergraduate, he hit upon the design of the spectroheliograph, an instrument based on well-known principles of optics that revolutionized research in solar physics.

In his Yerkes address, Hale spent considerable energy justifying the great 40-inch refractor, largest working instrument of its kind. He spoke of the light-gathering power of the telescope and described how it could detect close double stars using magnifications in excess of 3,000 diameters. The great refractor was also well suited for the study of faint nebulae. During the day, the instrument was dedicated to solar research.

As he came to the end of his address, Hale invited the attention of the audience to the mechanical and optical shops of the Yerkes Observatory. Hitherto, observatories depended on optical firms like the Clarks or Brashear to construct instrumentation. Relying on external sources, however skilled, led to delays and inevitable confusion over fine points of design and construction. Hale (1897: 319–20) proudly noted that "with the facilities here . . . it is possible to construct the various pieces of special apparatus which are constantly in demand, particularly in astrophysical and physical work." But the talents of Yerkes opticians and instrument makers extended far beyond the construction of auxiliary apparatus. Hale urged his auditors to inspect work in progress including the 24-inch reflector, large heliostat, and 60-inch mirror blank. The Yerkes operation was a telescope-manufacturing establishment in its own right.

Hale learned many lessons in building Kenwood and Yerkes. These experiences were the proving ground for Hale's skill as a fund-raiser and administrator as well as his ability to design new instrumentation. The centrality of instrumentation is most clearly demonstrated in Hale's discussions of the Mount Wilson Observatory. At Yerkes, he was limited by the great refractor itself. The instrument had astrophysical capabilities, especially in the fields of solar and stellar spectroscopy, but it was not specifically designed for research in the new astronomy. Hale had seized the main chance in 1892, when he learned of the existence of the lens blanks. The development of Mount Wilson would proceed along very different lines.

From the beginning, Hale considered the Mount Wilson Solar Observatory a research institution whose instruments would be carefully designed to meet the needs of specific astrophysical investigations. Hale (1908: 121) outlined the research agenda for Mount Wilson as follows: "(1) Solar investigations, to contribute toward our knowledge of the physical and chemical condition and nature of the Sun (a) as a typical star and (b) as the central body of the solar system; (2) photographic and spectroscopic studies of stars and nebulae, bearing directly upon the physical nature of these bodies, with special reference to their development; (3) laboratory investigations, for the interpretation of solar and stellar phenomena." Instrumentation followed from this research agenda.

Solar physics required a large image of the sun and heavy, securely mounted instruments for various lines of spectroscopic investigation. Hale and his associates developed a tower telescope in which the solar image was reflected from a large coelostat to a bank of instruments almost two hundred feet below. Later forms of this instrument had even longer focal lengths (and consequently larger solar images), and spectroscopes and spectroheliographs would be located underground, in constant-temperature laboratories. The 60-inch reflector was designed with a coudé focus that permitted the light of a bright star to be fed by a series of small mirrors to a fixed high-dispersion spectrograph located in a constant-temperature room. The great reflector could also be used to photograph faint nebulae and other extragalactic objects (Hale 1905a: 42–8).

At Yerkes, Hale proposed a mixed agenda. He had to develop programs that made maximum use of a large refracting telescope. The agenda (Hale 1897: 319) included investigations in both the new astronomy (solar and stellar spectroscopy) as well as traditional problems including the study of stellar distances, double stars, faint planetary satellites, and the surface markings of the planets themselves. At Mount Wilson, no such compromise was necessary. Instrumentation followed from the research agenda. Telescopes and spectrographs were designed in accordance with the demands of astrophysical investigations.

In simplest form, Hale (1905: 42) summed up this agenda as the "study of the constitution of the Sun and the problem of stellar evolution." In his discussion of Mount Wilson, much more space was devoted to describing new instrumentation than to outlining the research goals of the institution. This prefigures an important aspect of twentieth-century astrophysics. Instruments would come to embody the research agenda of a particular scientist or observatory.

The Hale synthesis also reflected his interest in evolution. Hale's successor as director at Mount Wilson, Walter S. Adams (1876–1956), recalled (1941: 185) that reading Darwin's *Origin of Species* "affected [Hale] profoundly and gave him a lasting desire to apply evolutionary principles" to astronomy. References to evolution often occurred in Hale's writings but never received extended treatment. He invoked evolution as a metaphor, suggesting that it was the "single greatest problem" for all students of natural science and the social sciences as well. For Hale (1908: 3), the evolutionary approach to science begins with "the origin of the stars in the nebulae" and culminates "in those difficult and complex sciences that endeavor to account for . . . the laws which control a society composed of human beings." In his own way, Hale was a popularizer of evolutionary thinking applied to both science and society.

Readers who thought Hale's 1908 volume was, in fact, an essay on stellar evolution must have been disappointed. While the topic was often mentioned, there was no attempt to develop it in theoretical or observational detail. Hale remained skeptical of the most comprehensive theory of stellar evolution at the time, put forward by his contemporary, Sir Norman Lockyer (1836–1920). In 1908 Henry Norris Russell (1877–1957) at Princeton and Einar Hertzsprung (1873–1967), a young Danish astronomer, were deep into research that would lead to the color-luminosity diagram and a conception of stellar evolution grounded in physical theory. By the end of the first decade of the new century, however, Hale's research career had, in fact, ended and he did not play an active part in this important area of investigation. For Hale, the study of stellar evolution remained a powerful metaphor indicating the direction, but not the content, of astrophysical research.

What is most obviously missing from the Hale synthesis is any reference to contemporary physical theory, especially the work of Max Planck (1858–1947) or Albert Einstein (1879–1955); nor did Hale make reference to experimental investigations in atomic physics. This was cutting-edge science, but noting these omissions reminds us that Hale grounded his research program in late-nineteenth-century experimental spectroscopy. In the twentieth century, as physics provided new ways of looking at atomic structure, the Hale synthesis became outdated. Frequently, third-generation American astrophysicists were little more than data collectors, whose research programs lacked the guiding hand of contemporary physical theory. It is a remarkable irony that, in the end, astrophysics and astrometry both reached the same point: data collection as an end in itself. Before the early 1930s, only a handful of individuals in the American astrophysical community were conversant with research in atomic physics, quantum mechanics, or relativistic physics.[25]

By the eve of World War I the old and new astronomy had learned to live in relative harmony, and the astrophysicists of the second generation were pursuing goals broadly outlined by the Hale synthesis. In many ways, the new astronomy had taken a commanding position in the American astronomical community. Subsequent chapters explore various aspects of the conflict that had divided the two groups and trace the ways in which astrophysics gained hegemony in the American astronomical community.

25. David DeVorkin of the National Air and Space Museum at the Smithsonian Institution is at work on a biography of Henry Norris Russell. When complete, this study will be a major addition to our knowledge of the history of twentieth-century astrophysics.

4 ✦

THE EDUCATION OF ASTRONOMERS

We assume today that careers in science depend on acquiring the proper credentials. The education of scientists begins with mathematics and science courses at the secondary level and goes on to include rigorous undergraduate training followed by a specialized graduate program and then a postdoctoral fellowship. This pattern obtained after World War II, but it was not the case earlier, at least for American astronomy. Before 1940, the Ph.D. had not yet become the sine qua non; individuals with no more than a high school certificate or, at best, an undergraduate degree were not excluded from successful careers in astronomy.

How a scientific community defined educational requirements for entrants and developed educational institutions tells us much about its values. When, where, and under whose supervision did undergraduate and graduate programs in astronomy develop? How did these programs reflect divisions between the old and new astronomy? Was access to power related to educational attainment? Were only those with advanced degrees recipients of major honors and awards? Was there a necessary connection between those who produced the best science and those who had the highest levels of educational attainment? How did the specialist research communities and the larger generic community share in the process of credentialling? Answers to these questions will help us understand the values of the American astronomical community and the ways in which the community functioned across the lifetimes of three cohorts of scientists.

DEMOGRAPHIC CONSIDERATIONS

The data discussed in this chapter are dawn from the collective biography. This section examines levels of educational attainment for each of the three cohorts. This approach permits a comparative discussion of changing credentialling patterns. In turn, this information helps explain changes

Table 4.1 Level of Education by Cohort

Highest Level	1859 or Before	1860–1899	1900–1940
High School *(N = 523, data missing for 26 individuals)*	20 (30.3%)	130 (43.1%) Δx=+555%	347 (43.3%) Δx=+164.9%
A.B. *(N = 182, data missing for 3 individuals)*	35 (53%)	59 (19.4%) Δx=+69%	85 (11%) Δx=+44%
M.A. *(N = 156)*	7 (11%)	50 (16.4%) Δx=+614.3%	99 (12.3%) Δx=+98%
Ph.D. *(N = 339, data missing for 4 individuals)*	2 (3%)	63 (21%) Δx=+3050%	270 (34%) Δx=+328.6%
Medical Degree *(N = 4)*	2 (3%)	1 (0.3%)	1 (0.1%)
Total *(Data missing for 34 individuals.)*	**66**	**303**	**802**

in the way the community viewed educational requirements for its members between 1859 and 1940.

Table 4.1 presents information on educational attainment by cohort. For many members of the American astronomical community, secondary education was the highest level of attainment. The most dramatic growth in this category occurred between the first and second cohorts. While the percentage of the community whose education stopped at the secondary level remained virtually constant between the second and third cohorts, the numbers continued to increase.

Most individuals in this category were short-term workers, women and men who served as computers or assistants and then left astronomy for other occupations or marriage. However, lack of academic credentials did not necessarily restrict access to research-grade instrumentation or the reward system. Maria Mitchell was trained in observational astronomy by her father and read celestial mechanics while working at the Nantucket Athenaeum. Lack of a college degree did not prevent Mitchell from becoming the first professor of astronomy at Vassar and director of its observatory. With no more than a secondary education, Sherburne W. Burnham (1838–1914), gained an international reputation as America's premier double star observer. And Samuel P. Langley, pioneer solar physicist and founder of the Smithsonian Astrophysical Observatory, was not hampered by lack of an undergraduate degree. Even in the twentieth century, want of collegiate training did not prevent Milton Humason (1891–1972) from becoming a respected expert in the field of extragalactic nebulae, collaborating with Edwin P. Hubble (1889–1953) at Mount Wilson.

These data suggest that before World War II, the American astronomical community did not enforce strict standards of certification. In each cohort it was possible to rise to positions of distinction and power without a collegiate education. Nor was access to the reward system tied to education. The lack of postsecondary education did not prevent Langley from achieving election to the National Academy of Sciences (NAS) or Humason the award of a star in the 1933 edition of *American Men of Science (AMS)*.

In discussing the history of American astronomy, it would be incorrect to label individuals like Langley or Humason amateurs because they lacked college degrees. Such thinking assumes that current standards for certification were in use at an earlier epoch. This was not the case. Professional status was *not* necessarily linked to formal education.

For the second cohort (1860–1899), 24 percent of those whose education stopped with high school were women. This figure rose to 62 percent for the group that entered between 1900 and 1940. The figures for males without collegiate training stood at 76 percent for the second cohort but fell to just over a third for the cohort that entered after 1900. While the number of men with only a secondary education declined between the second and third cohorts, they did not disappear. Even in the twentieth century both men and women with minimum educational credentials found employment in the American astronomical community.

The second category includes those whose education stopped with the baccalaureate degree. This group numbers 182 individuals. While their percentage fell across three cohorts, the number of those with only an undergraduate degree increased. Table 4.2 indicates that the academic origins of more than half (54.3 percent) of this population can be traced to ten institutions. Harvard and Yale led the way with Michigan third. Given the importance of the Lawrence Scientific School at Harvard (1848) and the Yale Sheffield Scientific School (1854), it is perhaps not surprising to find these institutions at the head of the list. For much of the nineteenth century, Benjamin Pierce (1809–80) presided over mathematical astronomy at Harvard. His successor teaching in undergraduate astronomy was Robert Wheeler Willson (1853–1921), who established the Student's Astronomical Laboratory in 1903 (Stetson 1928). Yale was well served by a succession of able undergraduate teachers including Denison Olmsted and Elias Loomis (1811–89). Hubert A. Newton (1830–96) was probably the most distinguished of the Yale astronomers, winning international recognition for his researches in meteoric astronomy.

Michigan is the sole representative of public education in this class. Undergraduate instruction in astronomy began at Ann Arbor in 1843. After the arrival in 1854 of German-born Franz Brünnow (1821–91), astronomy received special treatment from the administration. Brünnow was succeeded by James C. Watson (1838–80). When Watson moved to

Table 4.2 Institutional Origins of Those for Whom the Highest Level of Education Was an Undergraduate Degree: The Top Ten Institutions (N=182)

Rank Order	Institution	Graduates
1	Harvard	18
2	Yale	15
3	Michigan	13
4	Radcliffe Vassar	11 each
5	Mt. Holyoke	8
	California	7
	Lehigh	6
6	Cornell Smith	5 each

NOTE: The top ten institutions produced 54.3% of those for whom an undergraduate degree was the highest level of educational attainment. A total of seventy-three institutions account for the degrees of this population.

Madison in 1879, astronomy at the University of Michigan languished until the early twentieth century (Rufus 1951).

Radcliffe and Vassar stand tied for fourth place. Radcliffe A.B.s passed through the Willson laboratory or were products of the Harvard department of astronomy, which dates from the mid-1920s. At Vassar the tradition of undergraduate teaching established by Maria Mitchell was carried on by her successor, Mary Whitney (1847–1921) and, in the third generation, by Whitney's student, Caroline E. Furness (1869–1936).

Three institutions vie with one another as we descend the list. Mount Holyoke, California, and Lehigh University combined to produce twenty-one individuals whose highest level of educational attainment was an undergraduate degree. After Vassar, Mount Holyoke probably had the strongest astronomy department among the New England women's colleges. The subject was first taught by Elizabeth Bardwell (1831–99) and her successors included Anne S. Young (1871–1961). By virtue of its leadership in Ph.D. production, it is not surprising to find California on the list; its graduate program rested on a strong undergraduate major. Lehigh served as a feeder for the Dudley Observatory. Lehigh astronomy was under the supervision of Charles L. Doolittle (1843–1912) until he moved to the University of Pennsylvania in the 1890s. Cornell and Smith round out the top ten producers. The Cornell program was very limited before World War II. Under the leadership of Mary Byrd (1849–1934) and Harriet Bigelow (1870–1934), astronomy at Smith ranked third among the women's colleges before 1940.

In each cohort, a number of elite astronomers had only an undergraduate degree. Simon Newcomb, long-time director of the Nautical Almanac Office, was not hindered in his professional career because he had nothing more than an undergraduate degree from Harvard. The first three directors of the Lick Observatory, Edward S. Holden, James E. Keeler, and William Wallace Campbell had no formal credentials beyond the baccalaureate. George Ellery Hale, founder of the Yerkes and Mount Wilson observatories studied briefly in Germany but never earned more than an undergraduate degree from the Massachusetts Institute of Technology. Lack of credentials beyond an undergraduate degree did not prevent any of these individuals from being elected to the NAS; nor, at a later date, did the same level of educational attainment block election for Naval Observatory astronomer Gerald M. Clemence (1908–74) or Michigan solar astronomer Robert A. McMath (1891–1962).

The master's degree is the most problematic level of educational attainment discussed in this chapter. The problem lies in the impossibility of identifying those cases in which the degree was earned. Through at least the third quarter of the nineteenth century, many institutions awarded the master's to baccalaureate students after a specific number of years had elapsed (generally between two and five) on condition that the individual was of good moral character and professionally successful. It was not unheard of for graduates to apply to their alma mater for this degree. Only later did the master's become a step on the way to the Ph.D.

For 156 individuals the master's represented the highest level of educational attainment. As a percentage of the astronomical community this group did not fluctuate much, but the numbers increased in each cohort. Table 4.3 indicates the leading producers of master's degrees for those who ended their education at that point. Eleven institutions produced 45 percent of the degrees. The rest were scattered among sixty-eight institutions. Chicago and Radcliffe rank first and second. The Chicago department was strong in celestial mechanics while students working in astrophysics trained at the Yerkes Observatory. Three institutions tied for third place. The appearance of California and Harvard is not surprising given the role they played in doctoral education in American astronomy. At first glance, George Washington University in Washington, D.C., seems out of place, but many of its M.A.'s were on the staff of the Naval Observatory or Nautical Almanac Office. Dartmouth stands fourth on the list. Dartmouth astronomy was reorganized by Charles A. Young and carried on by Edwin B. Frost and his successors. Princeton and Swarthmore contest fifth place, producing five master's each. The Princeton tradition can be traced to Stephen Alexander (1806–76) and his successor, Charles A. Young. In the twentieth century Princeton astronomy was led by Henry Norris Russell. At Swarthmore, astronomy and mathematics were closely linked. Susan Cunningham (b.1842) and John A. Miller (1859–1946)

Table 4.3 Institutional Origins of Those for Whom the Highest Level of Education Was the Master's Degree: The Top Eleven Institutions (N=156)

Rank Order	Institution	Graduates
1	Chicago	10
2	Radcliffe	9
3	California George Washington Harvard	8 each
4	Dartmouth	6
5	Princeton Swarthmore	5 each
6	Indiana Kansas Michigan	4 each

NOTE: The top eleven institutions produced 46% of those for whom the master's degree was the highest level of educational attainment.

created a program that would remain strong and dynamic. Three public institutions vie for sixth place. While the Michigan department grew in stature after its reorganization in 1905, astronomy at Indiana and Kansas lacked cutting-edge research instrumentation and effective leadership.

A number of distinguished astronomers were in this group. The second director of the Yerkes Observatory, Edwin B. Frost, and the second director of Mount Wilson, Walter S. Adams, had no more than a master's degrees. Charles G. Abott, second director of the Smithsonian Astrophysical Observatory and later Secretary of the Smithsonian Institution, achieved success in solar physics with a bachelor's and master's from the Massachusetts Institute of Technology. These three individuals became members of the National Academy of Sciences. With a master's degree from George Washington University, Eleanor Lamson (1875–1932) became the chief of the computing division at the Naval Observatory.[1]

Perhaps the most dramatic change in educational attainment was the rise in the number who earned the doctorate. Only 3 percent of the cohort that entered through 1859 held the Ph.D. The numbers and percentages increased sharply for the second and third cohorts. Table 4.4 shows the leading Ph.D.-producing institutions.

Nine institutions produced 68 percent of the doctorates who entered astronomy. These nine institutions will be discussed at length in this chap-

1. For an overview of graduate education in the United States, see Veysey (1965: 121–79). For a detailed discussion of selected institutions, see Geiger (1986). By far the most important recent discussion of graduate education in the twentieth century is Kohler (1990).

Table 4.4 The Nine Leading Ph.D.-Producing Institutions, 1859–1940

Rank	Institution (Number of Ph.D.'s)	Percentage of Leading Nine	Percentage of Total Ph.D. Production
Leaders	California (64)	28%	19%
	Chicago (36)	16%	11%
Rising Stars	Harvard (Includes Radcliffe Ph.D.'s) and Michigan (27 Each)	12%	16%
	Princeton (23)	10%	7%
Failing Departments	Yale (15)	7%	4% each
	Hopkins (14)	6%	
	Columbia (13)	6%	
	Virginia (12)	5%	

NOTE: The top nine institutions produced 68.1% of all Ph.D.'s who entered the American astronomical community between 1859 and 1940. The total number of Ph.D.-producing institutions was fifty eight.

ter. Forty-nine other institutions account for the remainder. The nine leading Ph.D.-producing institutions are grouped into three classes. The first group (California and Chicago) were the leading producers of astronomy Ph.D.'s. The University of California at Berkeley held pride of place, producing just under 19 of all Ph.D.'s who entered astronomy. It was followed by the University of Chicago, whose share of Ph.D.'s stood at just under 11 percent.

Below the leaders, two other levels can be distinguished. The second, designated rising stars, includes three institutions. Harvard (including Radcliffe Ph.D.'s) and Michigan tie with twenty-seven doctorates each, followed by Princeton with twenty-three. A third group of institutions account for the lower end of the production scale. For reasons discussed later in the chapter, they are classified as declining departments. This group produced almost 16 percent of the Ph.D.'s who entered astronomy. Three are private universities while the fourth, Virginia, hardly fits the traditional state university model because of its patrician traditions and distinctly Southern perspective.

It is useful to examine levels of educational attainment from other perspectives as well. Table 4.5 provides information on Ph.D. production by type of astronomy. Fifty-two percent of the Ph.D.'s who entered astronomy can be classified in astrophysics while just over a quarter were in celestial mechanics or astrometry. The category of mixed astronomy (individuals who did research in both areas) numbers only 8 percent.

Given its corner on Ph.D. production, it is not surprising to find that California led in both research areas, but its share was larger for docto-

Table 4.5 Ph.D. Production of the Top Nine Institutions by Type of Astronomy

Astrophysics N=120 (52%)	Celestial Mechanics and Astrometry N=62 (17%)	Mixed Astronomy N=18 (8%)
California 36 (30%)	California 13 (21%)	California 9 (50%)
Harvard 21 (18%)	Columbia 10 (16%)	Chicago 3 (17%)
Chicago 18 (15%)	Yale 9 (15%)	Columbia, Harvard, Hopkins, Michigan, Virginia, Yale } 1 each (33%)
Princeton 17 (14%)	Virginia, Chicago } 8 each (26%)	Princeton 0
Michigan 16 (13%)	Michigan 7 (11%)	
Hopkins 7 (6%)	Princeton 4 (6%)	
Yale 3 (3%)	Hopkins 2 (3%)	
Virginia 2 (2%)	Harvard 1 (2%)	
Columbia 0		

(Data missing for 31 cases.)

rates in astrophysics. Harvard followed California in the production of workers in the new astronomy. Chicago stood third on the list with Princeton and Michigan close behind. John Hopkins, Yale, and Virginia brought up the rear in granting Ph.D.'s to workers in astrophysics.

For astrometry and celestial mechanics, Columbia followed California with Yale, Virginia, Chicago, and Michigan close behind. Princeton, Hopkins, and Harvard were at the end of the list. Ph.D.'s whose careers fell into both categories produce much less clear-cut patterns. Again, California led with Chicago a distant second. With the exception of Princeton, the other institutions cluster with one Ph.D. each. Before 1900, the vast majority of doctorates were in the old astronomy. Between 1900 and the coming of World War II, doctorates in astrophysics (117) far outstripped those in astrometry and celestial mechanics (48).

Table 4.6 reports dissertation fields of Ph.D.'s entering astronomy between 1859 and 1940. In all cohorts, astronomy Ph.D.'s led the way. For the 1860–1899 cohort, Ph.D.'s in mathematics stood second and physics doctorates third. After 1900, Ph.D.'s in physics increased significantly to take second place behind astronomy with mathematics third. Ph.D.'s in mathematics generally were engaged in work in old astronomy, especially celestial mechanics, while physics doctors most often worked in some

Table 4.6 Dissertation Fields of Ph.D.'s Entering Astronomy

Ph.D. Field	Cohort One 1859 or Before	Cohort Two 1860–1899	Cohort Three 1900–1940
Astronomy	2	29	194
Physics	0	5	28
Mathematics	0	11	17
Engineering	0	2	1
(Data missing for 50 cases.)			

SOURCE: *Dissertation Abstracts.*

branch of astrophysics. Chicago and Johns Hopkins led the production of Ph.D.'s in mathematics who entered careers in astronomy while Hopkins and Princeton produced more physicists who turned to astronomy.

The undergraduate origins of astronomy Ph.D.'s throws light on some interesting practices in the American astronomical community. For the first and second cohorts there are no clear-cut patterns. The numbers are relatively small and the scatter large. After 1900, however, patterns do emerge. They are especially interesting when the third cohort is broken down by status. The undergraduate origins of female Ph.D.'s are discussed in chapter 9.

The undergraduate origins of elite Ph.D.'s who entered after 1900 reveal unexpected scatter. These findings differ from the conclusions of scholars dealing with larger populations that include several fields of science (Visher 1947: 273). California led the way, producing six bachelors who went on to take doctorates and achieve elite status. Six other institutions (Brown, Indiana, Leiden in Holland, Minnesota, Princeton, and Virginia) produced two each. The rest were scattered among the thirty-one institutions. It appears that the identification and initial nurturing of those who went on to become elite astronomers took place at a number of institutions rather than in a few.

An examination of the undergraduate origins of rank-and-file doctorates reveals a very different situation. Here there are clearly defined patterns for the cohort that entered astronomy after 1900. California and Harvard led the way with eleven individuals each. Chicago and Michigan followed with six apiece. Princeton produced five bachelors who went on to the doctorate. Six institutions produced four undergraduates each who went on for the doctorate. The last group is made up of nine institutions that graduated three bachelors apiece. The rest were scattered among fifty-two institutions. For whatever reasons, an undergraduate education at an institution whose astronomy Ph.D.'s frequently attained elite status did not necessarily ensure a similar career path for undergraduate majors.

The contrasting patterns in the undergraduate origins of elite and rank-and-file doctorates underscore significant differences in the experiences of these two populations. This finding is important. It points to differing patterns in the education of the elite and rank-and-file. These differences suggest that the populations are products of historical development, not artifacts of analysis.

Forty-four American astronomers studied in Europe. Twelve earned European degrees; thirty-two studied abroad for varying lengths of time, but returned without degrees. The twelve Americans who earned European doctorates are distributed as follows: one in the cohort that entered through 1859, seven in the cohort that entered astronomy between 1860 and 1899, and four in the post-1900 cohort. Benjamin A. Gould's Göttingen degree was the first. During the second cohort, the legendary Thomas Jefferson Jackson See (1866–1962) earned a doctorate at Berlin. So did his distinguished contemporary Armin O. Leuschner (1868–1952), who presided over the California astronomy department from its inception through the 1930s. Princeton's Raymond Dugan (1876–1940), earned a doctorate at Heidelberg under Max Wolf (1863–1932). The bulk (7) of these degrees came from German universities.

Until the third cohort, European degrees tended to be in celestial mechanics and astrometry rather than astrophysics. The four post-1900 degrees were equally divided between the old and the new astronomy. It is not clear that a European doctorate provided significant career advantages. Five (41.7 percent) of this group achieved elite status; five were rank-and-file members of the astronomical community and two were women.

Thirty-two individuals studied in Europe, but did not take advanced degrees. Of this group 62 percent (20) would achieve elite status. Twenty-eight percent achieved rank-and-file status and just under 10 percent (three individuals) were women. Those whose careers were in astrophysics (43.8 percent) outnumber workers in astrometry and celestial mechanics (37.5 percent) by a relatively small margin. The bulk of this group (56.2 percent) is found in the second cohort. James Gilliss, father of the Naval Observatory, is the sole representative for the pre-1859 cohort. Harvard's Robert W. Willson and the longtime Cincinnati Observatory director, Jermain G. Porter (1852–1933) are examples from the second cohort. Thirteen individuals studied abroad after 1900, but did not remain to take degrees.

Career differences between those who took European doctorates and those who only studied in Europe demand explanation. It would seem that a European Ph.D. provided no clear advantage in the contest for honors and awards. More members of the group that studied in Europe without taking degrees achieved elite status (62 percent, N=20) than did those who returned with a doctorate (41.7 percent, N=5). It may be that those who spent time working on graduate degrees in Europe cut themselves off from important mentoring relations available to those who did graduate work in America.

One other group merits consideration. A total of 104 non-U.S. born individuals entered the American astronomical community. Of these, sixty-two held earned doctorates. This population divides almost equally between those who took their degrees before coming to the United States (30) and those who earned them from American institutions. The fact that a number of non-U.S. born scientists obtained American doctorates reminds us that wandering scholars moved in more than one direction. Indeed, Europeans and Canadians taking American doctorates exceed the number of Americans who obtained European degrees by a considerable margin.

The distribution of Ph.D.'s earned before immigrating to the United States breaks down as follows. Berlin (5) and Utrecht (4) lead the field with Copenhagen, Göttingen, and Lieden each contributing three doctorates. Groningen followed with two. The rest were scattered among ten institutions. Thus six universities produced two-thirds of the European Ph.D.'s who later became active in American astronomy. This group of scientists ranges from Franz Brünnow (Ph.D., Berlin), who directed astronomy at Ann Arbor in the 1850s, through Kaj Strand, (b.1907), a Copenhagen doctorate, whose first American appointment was at Swarthmore on the eve of World War II.

Seventy-eight percent of the Ph.D.'s earned by those born outside the U.S. came from seven institutions. California (7) and Harvard (5) led the way. Chicago, Michigan, and Syracuse tied with three each while Columbia and Princeton brought up the rear with two each. This group includes Russian-born astronomers Otto Struve (1897–1963), Nicholas Bobrovnikoff (1896–1988) and Alexander Vyssotsky (1888–1974). The United Kingdom was represented by Cecilia Payne-Gaposchkin (1900–79) and Canada by Alfred S. Mitchell (1874–1960) and Peter Millman (b.1906).

ADVICE ON ASTRONOMY AS A CAREER

Among the notes and papers of astronomers are letters to students interested in astronomy as a career. These provide a measure of the expectations astronomers had for new recruits. Writing to Edward S. Holden, director of the Lick Observatory, California-born Frederick H. Seares (1873–1964) asked for a comparison of the undergraduate astronomy programs at Harvard and Berkeley. Aware that the California program was then part of the Engineering Department, Seares wondered if it would be wise to supplement a California degree with postgraduate work at the

Lick Observatory.[2] In 1891, Seares did not feel the need to do more than earn an undergraduate degree and, perhaps, serve an apprenticeship on Mount Hamilton. In the end, the young man took an undergraduate degree at Berkeley and went abroad to study celestial mechanics in Paris under Jules Henri Poincaré (1854–1912). After a first job at the University of Missouri, Seares was called to Mount Wilson, retiring as assistant director in 1940.

Although Holden's reply to Seares apparently does not survive, we know his ideas on the education of astronomers from other correspondence. Responding to a Chicago high school student, the Lick director made a number of suggestions. Central to Holden's thinking was the importance of a broad education. While urging the young man to develop his skills in both mathematics and physics, Holden suggested "it is very useful to be a good mechanic and a good photographer," and stressed the importance of robust health, urging physical exercise such as football.[3] Holden encouraged the student to complete a scientific course in high school, and then to apply to Wisconsin, Michigan or, perhaps, Northwestern University. He also advised reading popular books on astronomy. But, Holden insisted, "Go slow and be thorough." It is important to notice that at no point does Holden discuss the doctorate. This suggests that Holden, whose postsecondary education consisted of two bachelor's degrees (one from Washington University and one from West Point), did not feel graduate work was imperative.

In 1913 William Wallace Campbell, third director of the Lick Observatory, composed a short essay, "To the Boy Who Wishes to Become an Astronomer." Director Campbell first advised the young man to take every opportunity "to improve your knowledge of elementary astronomy." Then Campbell went on to admonish: do not "neglect your other studies in order to become a precocious boy-astronomer. You need a *broad* education more than you need *early* knowledge of astronomy." Central to his conception of a broad education was the ability "to speak and write your own language correctly" as well as standing "at the head of your class in arithmetic."

In selecting a college or university, Campbell argued for one with a working observatory, not an institution where astronomy was part of the mathematics department. The undergraduate's primary job, however, was to become "as good a mathematician and as good a physicist as the time will permit." Chemistry, geology, and meteorology were also recommended as important subjects. The prospective astronomer must be able

2. Frederick H. Seares to Edward S. Holden, 16 March 1891. SALO, Holden Papers.

3. Edward S. Holden to Ernest J. Wilzin, 13 May 1892. Copy in UCA, Astronomy Department Papers.

to read French and German. Campbell did not ignore other aspects of general education, stressing the importance of both cultural history and the history of science and technology.

Summer vacations should be spent acquiring practical skills. Campbell recommended at least two summers as apprentice in a machine shop. "Nothing could be more useful than a knowledge of how instruments are made" because "the successful astronomer . . . makes the general designs of his own instruments . . . and he must know what designs are good and what designs are bad." In addition, the prospective scientist should consider spending a summer with a surveying crew or as a volunteer at an observatory. Campbell broke off the discussion with completion of an undergraduate degree. "Your instructor would know how to advise, and you would know how to decide, as to post graduate studies in astronomy and one or two of the closely related sciences."[4] Apparently, Campbell, who directed numerous doctoral dissertations, felt that decisions concerning graduate work had to be made on an individual basis. In any event, he did not take the opportunity to make the case for graduate study.

During the summer following his sophomore year at the Southern Branch of the University of California, mathematics major Robert S. Richardson (b. 1902), later a staff member at the Mount Wilson Observatory, sought advice from Vesto M. Slipher (1875–1969), Director of the Lowell Observatory. Richardson indicated a long-standing interest in astronomy, but friends suggested a more practical major followed by a career in business. Richardson indicated awareness of the necessity of a Ph.D. in order to gain entry into the profession but had a number of questions for Slipher concerning the possibilities for employment. He also wanted to know about salaries and whether he should specialize in astrophysics or celestial mechanics.[5] Slipher reported that the supply of astronomers did not meet the demand and "it ought not to be a difficult experience to find a position." If a large income was the young man's goal, Slipher doubted whether Richardson should choose astronomy. Slipher urged the undergraduate to contact the Berkeley astronomy department about graduate work. He declined to discuss the choice between the old and new astronomy, leaving the decision to be made when Richardson knew more about the fields and his own interests.[6]

By the end of the 1920s a few astronomers were thinking about postdoctoral training. Kenneth P. Williams (1887–1958), a celestial mechani-

4. William Wallace Campbell, "To the Boy Who Wishes to Become an Astronomer." SALO, Campbell Papers. The typescript, less than two and a half pages in length, was apparently a draft.

5. Robert S. Richardson to Vesto M. Slipher, 19 July 1924. LOA, V. M. Slipher Papers.

6. Slipher to Richardson, 21 July 1925. LOA, V. M. Slipher Papers.

cian at Indiana University, discussed the ideal candidate for appointment at Bloomington. The individual would have training in observational astronomy at the Lowell Observatory as well as a year at Princeton with Henry Norris Russell mastering astrophysical theory. It was assumed, of course, that the candidate had taken the doctorate at Indiana and been kept on.[7] The National Research Council was the major sponsor for postdoctoral fellowships in the years between the two World Wars. Only a few went to astronomers.

UNDERGRADUATE EDUCATION IN ASTRONOMY

In his influential study of the education of nineteenth-century American astronomers, Marc Rothenberg concluded that by 1875 a number of major changes had occurred. "Professors of Astronomy with research experience had taken the place of college teachers of the physical sciences." Students had access to instruments in working observatories rather than portable telescopes taken out on the college lawn. But, most important, Rothenberg argues (1974: 245), was a change in attitude. Astronomy was no longer "conceived of as merely a building block in the development of character." A growing number of public and private universities "recognized that astronomy could and should be taught with a research orientation; [and that] the objective of such an education was to produce new astronomers."

By the last quarter of the nineteenth century, astronomy education was moving in new directions. Unfortunately, there are no studies comparable to Rothenberg's for the later period. Information on the undergraduate experiences of astronomers is extremely meager. A few well documented examples must serve to illustrate patterns. Several themes tie these case studies together. Perhaps the most important is increasing specialization.

If we know relatively little about the structure and development of undergraduate programs in astronomy at American colleges and universities after the 1870s, even less is known about psychological factors involved in electing to major in astronomy.[8] While Harlow Shapley (1969: 17), tongue in cheek, tried to convince readers of his autobiography that he stumbled into an undergraduate major in astronomy because the subject was listed first in the university catalogue, the process of selecting a major in science is much more complex.

7. Kenneth P. Williams to Vesto M. Slipher, 3 July 1929. LOA, Slipher Papers.

8. The exception to these generalizations is the women's colleges discussed in chapter 9. Here we know a great deal about the structure of the programs and the reasons why individuals became astronomy majors. For an introduction to the psychology of science and scientists, see Fisch (1977) and Jackson and Rushton, eds. (1987).

Among elite astronomers, seventy-seven (40.7 percent) developed an interest in astronomy during high school or earlier. Comparable data are not available for the rank-and-file or women. The young George Ellery Hale illustrates this point (Wright 1966: 31–47). He early developed a preoccupation with microscopy and then physics, chemistry, and astronomy. The child of a successful Chicago industrialist, Hale soon had his own well-equipped laboratory and observatory.

Antecedent to an early interest in astronomy are certain traits of mind that may predispose an individual to select science as a vocation. An autobiographical sketch composed in the 1940s by the second director of the Mount Wilson Observatory, Walter S. Adams, is especially revealing. Adams offers a rare and candid glimpse of the tastes, preferences and predispositions that incline an individual toward a career in science.

Adams, a dour and parsimonious Yankee, was one of the great spectroscopists of the first half of the twentieth century (Joy 1958). His research set new standards in the field. Investigations by Adams and his associates led to a deeper understanding of stellar temperatures, the use of spectra to develop information on stellar parallaxes and greater knowledge of the interstellar medium. Adams often presented himself as a data collector, but never shrank from interpreting observational material, using physical theory to inform the analysis.

In trying to explain why he was attracted to science during his undergraduate years at Dartmouth, the future Mount Wilson director ("Autobiographical Notes": 4) suggested that he was driven by an "innate preference for exact subjects. . . . Those in which the fundamental processes of reasoning and application are relatively unchanged, as in mathematics, astronomy, physics and chemistry." Adams contrasted these with other academic disciplines. "I had much less interest in social sciences, the bases of which are subject to constant change." Nor was he attracted to "modern languages which are always in a state of flux, varying with usage, grammar, and even spelling." Enumerating other areas of intellectual endeavor he found frustrating, Adams (4–5) pointed to "history when it is taught as a series of great social movements, which are subject to many different interpretations instead of a direct record of past events."

Of nonscientific fields to which he was exposed, Adams (3, 5) admitted an affinity for Latin and Greek, "languages in which the construction of a sentence somewhat resembles a mathematical formula." Indeed, Adams confessed that he could well have gone on to become a professor of Greek. He remarked that "this preference for the concrete and definite subject in which problems have a single answer may be subject to criticism of narrowness and lack of imagination" (5). Adams, however, defended himself with classic Yankee understatement. Such a mind-set "does have a tendency to lead to exactness of thought and reasoning."

It is probable that many intending astronomers entered their undergraduate careers with a well-developed interest in the subject, predisposed to concentrate in the exact sciences and mathematics rather than the social sciences or humanities. Early specialization and a tendency to dismiss nonscientific approaches to knowledge may be part of the psychological baggage would-be astronomers brought to campus as freshmen.

Undergraduate lecture notes preserved in the papers of Ormond Stone, first director of the McCormick Observatory at the University of Virginia, indicate that at least at the old University of Chicago, where Stone took an undergraduate degree in 1870, there was still some balance between humanistic and scientific studies.[9] Form an examination of his commonplace books and class notebooks, it appears that Stone had a reading knowledge of German, French, and Italian. More extensive are the notes from a course in practical astronomy taught by Truman H. Safford (1839–1901), whose varied career included positions at the Harvard College Observatory, the Nautical Almanac Office, and as professor of astronomy at Williams College. Safford was a specialist in transit-circle work and produced important star catalogues. Safford was demanding, and Stone, a junior, examined many star catalogues and read papers by European astronomers as well as works in English. Clearly, by the end of the year, Stone was familiar with the transit circle and the reduction and analysis of astrometric data.[10]

Two early twentieth-century case studies provide interesting contrasts. John C. Duncan (1882–1962) graduated in astronomy from Indiana University in 1905 while Edison Pettit (1890–1962) took an undergraduate science degree at the Nebraska State Normal School at Peru in 1911. Duncan went on for a Ph.D. at California (1909) and Pettit at Chicago (1920). Duncan spent most of his career as a teacher at Wellesley College while Pettit became an expert in solar physics at the Mount Wilson Observatory.

Writing to William Wallace Campbell concerning a fellowship to work on a Ph.D. at Berkeley, Duncan provided a concise summary of his undergraduate work at Indiana where John A. Miller and Wilber A. Cogshall (b. 1874) constituted the astronomy faculty. As an undergraduate, Duncan took a total of fifty-two hours in astronomy including descriptive astronomy, celestial mechanics, theory of orbits, least squares and practical observatory work with the 12-inch refractor, transit circle and photographic camera as well as a 15-inch reflector. Thirty hours of physics were included in Duncan's undergraduate program. The courses in spectroscopy made use of a 4-inch diffraction grating ruled in the laboratory of Henry Rowland at Johns Hopkins. Twenty-eight hours of mathematics

9. These materials are in Box 8 of the Ormond Stone Papers, UVA.

10. These notes are in Box 7 of the Stone Papers, UVA.

through differential equations, together with an introductory chemistry course, rounded out Duncan's studies in science. In addition, he took extensive work in French and German as well as courses in English literature and composition. After graduation, Duncan spent a year as a fellow at the Lowell Observatory. Indiana University awarded him a master's degree on the strength of this experience. By the spring of 1907, Duncan was ready to move on to work for the doctorate.[11]

Later that same year, Edison Pettit also wrote to Campbell, applying for a position on the Lick staff. Pettit indicated an interest in graduate work, but it is not clear that he wanted a doctorate or even understood its significance. He entered the Peru State Normal School directly out of eighth grade. In 1903 he was appointed assistant in physics and later served in the same position at the school's observatory. While Duncan was studying celestial mechanics using the latest texts by University of Chicago professor Forest Ray Moulton (1872–1952) and François Tisserand (1854–96), of the Paris Observatory, Pettit was reading a book by Elias Loomis dating from the first half of the nineteenth century. At the time he wrote Campbell, his mathematics included only basic calculus. Pettit took a number of courses in physics, but the level of instruction is not clear. Where Duncan provided a listing of credits earned, Pettit's letter was far more general. What is unusual is that Pettit included a selection of problems from his notebooks for which he worked out solutions.[12] Duncan was successful in his application; Pettit was not.

However difficult it is to compare undergraduate training at Indiana University and the Peru State Normal School, several things stand out. Duncan's fifty-two hours in astronomy and the informal readings in astronomy done by Pettit, bespeak intense concentration in science. By the end of their undergraduate years, both Duncan and Pettit were well on the way to becoming specialists.

The undergraduate career of Ernest Cherrington, Jr. (b.1909) provides a final point of comparison. Writing to Robert Grant Aitken (1864–1951), fourth director of the Lick Observatory, Cherrington discussed his interest in astronomy and reviewed his undergraduate education as part an application for the Berkeley Ph.D. program.

Cherrington traced his interest in science to junior high school. He began as a collector (rocks, flowers, fossils) and moved on to read popular works in chemistry and physics which soon led to experiments with homemade equipment. In high school, astronomy attracted Cherrington's interest and he began to learn the constellations, soon mounted a two-inch refractor and then built his own reflector. With this equipment, he felt

11. John C. Duncan to William Wallace Campbell, 4 March 1907. SALO, Campbell Papers.
12. Edison Pettit to William Wallace Campbell, 4 July 1907. SALO, Campbell Papers.

ready for research and took up the study of variable stars. The director of the Perkins Observatory soon permitted Cherrington to use the 10-inch refractor.

By the fall of 1929, Cherrington was a junior at Ohio Wesleyan. The new director of the Perkins Observatory, Harlan T. Stetson (1885–1964), recruited Cherrington to take charge of solar photography and to act as his teaching assistant. He does not list the course of study in detail, but it appears that Cherrington completed the general education requirements by the end of his sophomore year and went on to take all the astronomy courses at Ohio Wesleyan, as well as all the advanced courses in physics. So industrious was Cherrington that he managed to accumulate almost half the thirty hours needed for a M.A. in astronomy during his senior year. In addition, he was elected to Phi Beta Kappa and joined the American Astronomical Society. Cherrington took an undergraduate degree in 1931 and a master's the following June. He then went to Berkeley where he earned the doctorate.[13]

In these examples, there are several points of similarity. Duncan, Pettit, and Cherrington illustrate a significant degree of specialization during the undergraduate career. For each there was considerable hands-on experience at the telescope. In addition, Pettit and Cherrington served as teaching or research assistants during their undergraduate years. With the apparent exception of Pettit, each became proficient in at least one foreign language.

It is difficult to ascertain whether any of these individuals were involved in significant research as an undergraduate. However, research, even at the undergraduate level, quickly became an important factor in science education. This tendency can be found at least as early as the mid-1880s. Henry A. Rowland, the irascible Johns Hopkins physicist who worked in solar spectroscopy, remarked to the young Henry Crew (1859–1953), that his chances for admission to the graduate program "would be very much diminished" if Crew did not submit a written report on some original research he had carried out as an undergraduate at Princeton.[14]

One final element of undergraduate education should be noted. The undergraduate context is one in which a few exceptional individuals acquire a mentor. The mentor not only nurtures undergraduate students, but helps them make key decisions concerning graduate education. Sometimes, this relationship extended far beyond the undergraduate years. Unfortunately, mentoring relationships are not often documented. The case of Frederick H. Seares and Harlow Shapley is an exception. Seares discov-

13. Ernest H. Cherrington Jr., to Robert G. Aitken, 4 November 1931. SALO, Aitken Papers.

14. Henry A. Rowland to Henry Crew, March 1884. NWUA, Crew Papers.

ered Shapley and encouraged him to take part in the work of the Lawes Observatory. Soon, the undergraduate was running the time service and reducing photometric observations. In short order, Shapley moved from the desk to the telescope, taking an active part in research programs. Seares described Shapley was "the type of student who will do exceptionally well" in astronomy and rated his abilities as "far above average." Indeed, the only flaw Seares could see in his young charge was a tendency to work "incessantly—much to continuously for his own good." What amazed Seares was the fact that Shapley not only put in long hours at the desk and the telescope but that "he *thinks* about what he is doing."[15] Seares, never one to mince words, summed up his view of mentoring in a brutal fashion. Shapley "is a student in whom I am greatly interested—one who is worth helping, else I would not take this trouble."[16]

In 1938, Alice H. Farnsworth (1893–1960), professor of astronomy and director of the Williston Observatory at Mount Holyoke College, surveyed undergraduate programs at twenty institutions. Fifteen responded to the questionnaire with thirteen providing comparable data. Some responses were perfunctory while others went into considerable detail. Taken together, these replies provide an overview of undergraduate education in astronomy on the eve of World War II.[17] The data are not always complete and it is not easy to compare institutions as diverse as the University of Michigan and the New Mexico State Teachers College, but some reliable patterns do emerge. Specialization at the undergraduate level is evident, but there were other tendencies as well. Thirteen required practical astronomy courses in which students learned to use instruments and make observations. Seven mandated courses in celestial mechanics and seven demanded their majors become familiar with the rudiments of astrophysics. Only two required students to study the history of astronomy. Surviving notes indicate that these courses were taught with emphasis on chronology and the great men (and sometimes women) who dominated the science. Walter S. Adams would have approved! Mathematics requirements varied. At seven institutions, introductory calculus was required while three demanded advanced work through differential equations. Only

15. Frederick H. Seares to Russell T. Crawford, 8 March 1909. BLUCB. Department of Astronomy Papers.

16. Frederick H. Seares to Armin O. Leuschner, 23 March 1909. BLUCB, Astronomy Department Papers. See also Seares to Leuschner, 29 April 1911 and 3 June 1911. BLUCB, Astronomy Department Papers. For Shapley's version of the story, see Shapley (1969: 17–27).

17. These materials are contained in the "Astronomy Survey" file in the Farnsworth Papers, MTHC. Since the questionnaires are not paginated, it is difficult to provide exact citations. Each questionnaire has the name of the institution and respondent at the top of the first page. There are also a number of pages on which Farnsworth reduced and analyzed the responses.

three institutions mandated that their majors take statistics. Physics requirements had changed little since the beginning of the century. Optics and spectroscopy were still the areas most often required. The introductory text most frequently cited was that of Robert Baker (1883–1964), who directed astronomy at the University of Illinois.

THE MAJOR PH.D.-PRODUCING INSTITUTIONS: AN OVERVIEW

By post-World War II standards, the size of major Ph.D.-producing departments was small. In 1901 the Berkeley resident faculty numbered just two astronomers, Armin O. Leuschner, a Berlin Ph.D. and Russell Tracy Crawford (1876–1958), who had just earned the doctorate under Leuschner's supervision. By 1920 the Berkeley faculty had increased to five. All of the additions were products of the department. Strula Einarsson (1879–1974) joined the staff in 1910. William F. Meyer (b.1880) was hired to teach both astronomy and mathematics in 1919. Charles D. Shane (1895–1983), who also occasionally taught mathematics, was appointed in 1920. For a few years Seth Barnes Nicholson (1891–1963) lectured at Berkeley, but joined the Mount Wilson staff in 1915. In 1940 the Berkeley department still numbered five. Leuschner was officially retired but remained actively involved. Swiss-born, Robert J. Trumpler (1886–1956), whose Ph.D. was from Göttingen, joined the department in 1938 after a long career at Mount Hamilton.

With the exception of Trumpler and Shane, the resident faculty at Berkeley were involved in celestial mechanics and astrometry. Only Leuschner and Trumpler were members of the National Academy. Crawford and Shane earned stars in *AMS* and Leuschner made several unsuccessful attempts to secure Crawford's election to the Academy. Einarsson and Meyer were not researchers, spending most of their time on teaching and service.

The Chicago department was, if anything, an even smaller operation. In 1896 the resident faculty included Kurt Laves (1866–1944), and Thomas Jefferson Jackson See, both Berlin Ph.D.'s, plus Forest Ray Moulton, who would earn a Chicago Ph.D. in 1899. Soon T. J. J. See left Chicago for the Naval Observatory and for many years the campus astronomy faculty numbered just two professors. By 1915 the size of the department had risen to three with the addition of William D. Macmillan (1871–1948). Macmillan was a Chicago Ph.D. (1908). By the mid-1930s, after Laves retired and Moulton left for a career in business, the university realized that the campus astronomy program was in serious difficulty. The reorganized astronomy department sought to develop closer ties between the campus and the observatory at Williams Bay. Dutch-born Gerard Kuiper (1905–73) accepted responsibility for astrophysical instruction on the campus and sought to coordinate a program that involved the observatory staff lecturing at the university. Of the Chicago faculty only Moulton was elected to the National Academy. See, Laves and Macmillan were awarded stars in *AMS*. Kuiper would be elected to the Academy after World War II.

The origins of the Berkeley astronomy department are complex. Until the mid-1890s, astronomy was taught by Frank Soule, a civil engineer, whose primary interest was geodesy. By the mid-1890s, however, there was considerable pressure to develop a formal astronomy program and to define the instructional responsibilities of astronomers on Mount Hamilton.[18] In 1899, the astronomy program was reorganized.[19] Instruction was divided into three areas: astrometry and celestial mechanics, astrophysics, and geodesy. The core of the undergraduate program included ninety-six hours distributed between physics, chemistry, mathematics, and astronomy, plus French and German. Students who elected to study celestial mechanics had to take additional work in mathematics. Those in astrophysics were required to do extra courses in physics. Geodesy students had to take additional credits in both mathematics and civil engineering. Requirements for the doctorate (the 1899 report makes no mention of the master's degree) included a three-year residency, with one academic year to be spent at Berkeley. The remaining two years were to be spent at Mount Hamilton. This requirement could be modified for students whose major interest was geodesy or mathematics rather than observational astronomy.

At Mount Hamilton, doctoral candidates devoted themselves to hands-on activities. They served as research assistants, learning the art of observational astronomy as apprentices to resident staff members. In this capacity, students were expected to master the literature related to the problems under investigation. In addition, the committee in charge of a student's graduate program (Berkeley faculty and Lick astronomers) was to outline a reading program and students were expected to devote some part of every day to library work. Committees evaluated each individual's experience as an observer, making sure that every graduate student knew how to make both visual and photographic observations with the Clark

18. There were also tensions between the campus and Mount Hamilton and these may have been another reason why the administration moved to create a department at Berkeley. For documentation of the conflict, see Armin O. Leuschner, "To the Committee on Internal Administration and on the Lick Observatory of the Board of Regents of the University of California through President Martin Kellogg," 2 December 1895. BLUCB, Astronomy Department Papers. On Holden, see Osterbrock, (1984a).

19. A printed copy of the report as well as a manuscript version are in the Astronomy Department Papers, BLUCB. The printed report is dated 9 November 1899 and the typescript copy 20 October 1899. In the same file is a printed report from Holden supplementing a presentation to the Regents in 1895.

refractor and Crossley reflector. In addition, those doing graduate work in astrometry learned to use the Lick transit circle and to reduce and analyze astrometric data.

Before World War II, the Berkeley department is best understood as the fiefdom of its longtime head, Armin O. Leuschner (Hergert 1978; Osterbrock 1990). Appointed director of the Student's Observatory in 1898 and full professor in 1907, Leuschner used his years as graduate dean (1913–23) to consolidate power and enhance the status of the department. Leuschner's marriage to the daughter of a powerful member of the Board of Regents also helped him develop a solid base of support for astronomy.

By 1913 the Berkeley undergraduate program had grown to the point where teaching assistants were needed. After this time, it was common to begin graduate work as a teaching fellow and then apply for a research fellowship to cover work at Mount Hamilton. Teaching fellows were required to take charge of three or four discussion sections in the introductory astronomy course plus two or three sections of elementary practical astronomy.[20] Those who successfully completed the Ph.D. program found that Leuschner and the Mount Hamilton staff worked diligently to secure them employment.

Given the cognitive, administrative, and geographical division between campus and observatory, conflict between the two was virtually inevitable. Relations were made more difficult by the fact that both Leuschner at Berkeley and William Wallace Campbell, director of the Lick Observatory from 1901 to 1923, were strong-willed, hypersensitive individuals, ready to take offense at the slightest criticism. Campbell ran the observatory and the village on Mount Hamilton like a medieval lord. As Horace Babcock (b. 1912), who earned a California doctorate in 1938 (Babcock 1975–77:14), recalled of his graduate days, the department and observatory "were separate and rather distinct organizations." Babcock suggested that for many graduate students it was difficult to make the transition from taking courses at Berkeley to research work at Mount Hamilton. "You were taking up a different life, with different people." The future director of the Mount Wilson Observatory concluded that "it's very difficult to mix" an academic department "with the business of doing research at a major mountain observatory."

For almost forty years, Berkeley astronomy was dominated by the long shadow of Leuschner. His approach to graduate education was

20. The description of duties is taken from a letter sent by Leuschner to a number of university and college observatory directors and heads of astronomy departments. For example, Armin O. Leuschner to John C. Duncan, 21 August 1913. BLUCB, Astronomy Department Papers.

straightforward (Babcock 1975–77:13). "You're going to learn to be thorough, and to know exactly what you're doing. At the same time you should be getting something in astrophysics." Students acquired thoroughness and precision through the study of celestial mechanics, including Leuschner's famous marathon computing sessions during which orbits were calculated for newly discovered solar system objects. Astrophysics, by implication a subject that lacked rigor, was given a back seat to the discipline of mathematical astronomy. Looking back on the late 1930s, Horace Babcock (1975–77:13) recalled that the department at Berkeley was a "post-mature organization, with old professors who were specialists in celestial mechanics." Graduate students "spent much more time on celestial mechanics than was really going to be very useful." But there were compensations. "We got exposed to a lot of current physics, so it wasn't too bad. And there was some math on the side."

California stands as the exemplar of the leading Ph.D.- producing departments. Its development is well documented. Unfortunately, the history of the Chicago astronomy department is known only in broad outlines. Certain points of similarity do, however, stand out. As at California, there was both a physical and cognitive division between campus and observatory. Astrophysics dominated research activities at Williams Bay, while campus astronomers were interested in celestial mechanics and astrometry. But California and Chicago differed in several important respects. Chicago had no equivalent of Leuschner. There was no campus leader committed to developing the Chicago astronomy department in the same way Leuschner was dedicated to Berkeley. Looking over the history of the Chicago department, the impression is one of drift. Further, astronomy on the Chicago campus was always a smaller operation than at Berkeley. If anything, Yerkes astronomers must have played a greater role in graduate education than did the Lick staff. The University of Chicago produced only about half as many doctorates as Berkeley, but it is instructive to note that the percentages of Ph.D.'s in astrophysics (56.2 percent for Berkeley and 50 percent for Chicago) and in celestial mechanics and astrometry (20.3 percent at Berkeley and 22.2 percent for Chicago) were almost identical.

The first two Yerkes directors, George Ellery Hale and Edwin B. Frost, were more interested in research than in teaching. Otto Struve, who succeeded Frost, became more involved in graduate education than his predecessors. Struve inherited a department and observatory that were in rundown condition. His great accomplishment was the revitalization of both. Because Struve was among the few astrophysicists who grasped post-1920 developments in physics, Yerkes became a center for research in the new astrophysics. Struve also linked the Universities of Texas and Chicago in a cooperative project which created the McDonald Observatory. This provided Chicago astronomers with access to the second largest reflecting telescope in the world.[21]

As Horace Babcock remarked (1975–77: 18), in recalling his association with the Yerkes and McDonald observatories at the end of the 1930s, "any astronomy department that Chicago had was essentially at Yerkes." Babcock characterized the intellectual milieu at Yerkes as more exciting than Lick. "Struve knew everything that was going on, he kept track of everybody . . . and he gave opportunities, too." As editor of the *Astrophysical Journal* and director of the combined Yerkes and McDonald Observatories as well as chairman of the Chicago astronomy department, Struve was, indeed, in a position to know almost everything. This provided the Yerkes staff with important advantages; and, as Babcock indicated, Struve was deeply interested in nurturing scientific talent.

We can examine the astronomy departments at Chicago and Berkeley from one other perspective. How did the honors and awards garnered by their graduates compare? Thirteen California graduates achieved elite status. Eight earned stars in *AMS* and five were elected to the National Academy of Sciences. Nine Chicago graduates attained elite status. Six earned stars in *AMS* and three were elected to the Academy. Given the proportion of Ph.D.'s (about two Berkeley Ph.D.'s for every Chicago doctorate) the figures indicate a rough parity. Using this metric, there appears to be no fundamental difference between the two graduate programs.

A second group of Ph.D.-producing institutions (Michigan, Harvard and Princeton) make up the class designated rising stars. While their total output was less than that of California and Chicago, these three institutions were responsible for approximately one-third of the Ph.D.'s produced by the nine leading graduate institutions or 22 percent of all Ph.D.'s entering the American astronomical community between 1900 and the eve of World War II.

What made these three programs different from Berkeley and Chicago was the fact that their observatories and departments of astronomy were neither physically nor intellectually separated. Research and teaching in astrometry, celestial mechanics, and astrophysics went on in the same department. The observatory was either on campus or a short distance away. Both Harvard and Michigan also had southern stations, which produced data to be used by faculty and students. Integration of the department and observatory provided a number of advantages. Graduate students moved

21. William Rainey Harper to George Ellery Hale, 19 December 1902 and Hale to Harper, 20 December 1902. UCA, Harper Papers. Hutchins (1947: 195). A large portion of the 5 September 1947 issue of *Science* was devoted to papers by current or former Yerkes staff members. Significantly, only the brief introductory remarks by Chancellor Hutchins even touch on the Department of Astronomy. In his autobiography (1933), Frost says more about undergraduate astronomy at Dartmouth than Chicago.

freely between the two, and while cognitive divisions remained, they did not entail conflict between the director of the observatory and head of the department. Indeed, at Michigan, Princeton, and Harvard, the positions of observatory director and department head were combined.

These three departments experienced significant growth, especially after World War I. An increase in undergraduate and graduate enrollments translated into justification for new staff and equipment. This was especially true for Michigan and Harvard. Inbreeding existed, but it was not on the scale of Berkeley and Chicago. In the case of both Michigan and Harvard, expansion also included new instrumentation.

Michigan serves as the exemplar of this class. Brünnow and Watson developed the astronomy department and observatory at Ann Arbor, but after Watson moved to Madison (1879) Michigan astronomy fell upon hard times. Mark W. Harrington (1848–1926), who succeeded Watson, was more interested in meteorology. Indeed, he founded the *American Meteorological Journal* (1884) and left Michigan in 1891 to become chief of the U.S. Weather Bureau. Harrington was followed by Asaph Hall, Jr. (1859–1930), son of the distinguished Naval Observatory astronomer. Asaph Jr. obtained a Yale doctorate but, compared to his father, a scientist skilled at both observational astrometry and celestial mechanics, the son appeared lacking in motivation and scientific ability. The Hall family papers at the Library of Congress include a number of letters that are filled with advice from Asaph, Sr. to his son, on how to conduct himself as director at Michigan and how to make observations with the transit circle. The younger Hall cleaned and repaired the transit circle and 12-inch Fitz refractor and adjusted the clocks, but he published little astrometric research. Hall left in 1905 for an appointment at the Naval Observatory, where he did not find it easy to pass the standard examination required of all who sought senior positions.

With the appointment of William J. Hussey (1862–1926), the Michigan astronomy program was revitalized. Hussey, who earned an engineering degree at Michigan in 1889, joined the Lick Observatory staff in the 1890s and, with Robert Grant Aitken, embarked on a study of double stars for which they shared the Lalande Prize of the French Academy of Sciences in 1906. After taking over at Michigan, Hussey, a man of great energy and vision, secured for Michigan a 36-inch reflector, thus providing instrumentation for astrophysical investigations, and sought (unsuccessfully as it turned out) to develop a cooperative program between the Universities of Michigan and La Plata in Argentina. Hussey worked to secure a large telescope to be located in the southern hemisphere and used primarily for the study of double stars, but he died before the project came to fruition.

In 1905 Hussey was the sole member of the astronomy department at

Michigan. The department soon doubled with the addition of Ralph H. Curtiss (1880–1929), a Berkeley Ph.D. Curtiss, an expert in astrophysics, succeeded Hussey as director of the observatory and chairman of the department in 1926. By the mid-1920s the department included three Michigan graduates: Hussey, Will C. Rufus (1876–1946), Ph.D. 1915, and Richard A. Rossiter (1886–1977), whose doctorate was granted in 1923. Rossiter left to direct the South African station in the early 1930s. The Ann Arbor astronomy department grew to six, including a woman, Hazel Losh (b.1898), who earned a Michigan Ph.D. in 1924. Curtiss was succeeded by Heber D. Curtis (1872–1942), a Virginia Ph.D., and Rossiter was replaced in 1935 by Robley C. Williams (b.1908), a Cornell Ph.D. . Allan Maxwell (b. 1901), a California Ph.D. (1927) and Dean B. McLaughlin (1901–65), who earned a Michigan doctorate in 1927, both joined the Ann Arbor department on completion of their graduate work. By 1925 the Michigan department surpassed Chicago in size and by 1930 had more full-time faculty members than Berkeley.[22]

During the Hussey regime, the first order of business involved setting up an observatory shop to take care of mechanical work. A three-man staff was soon fabricating the mounting and tube for the 36-inch reflector. Next Hussey turned his attention to astrophysics. Ralph Curtiss joined the faculty in 1907, offering courses in astrophysics as well as variable stars and spectroscopic binaries. A year later Hussey began negotiations that would eventuate in the 27-inch refractor at the Lamont-Hussey Station in South Africa. From 1911 to 1917 Hussey served as director of the La Plata University Observatory as well as carrying on the work at Ann Arbor. Financial difficulties at La Plata forced his resignation.

Astronomy at Michigan prospered under Hussey. By the academic year 1922–23 undergraduate enrollment totaled 650 and two years later it was reported that the astronomy department had fifteen times as many students as it did when Hussey arrived (Rufus 1951: 459–60). In the 1920s, Michigan produced a dozen doctorates, and eleven during the depression decade. During this period, a new observatory was constructed for undergraduate instruction and plans made to move the large reflector from Ann Arbor to a dark-sky location in the countryside. By the eve of World War II, astronomy at the University of Michigan rested on solid foundations. With the addition of the McMath-Hulbert Solar Observatory, Michigan became the leading American institution for research in solar physics.

22. I have excluded instructors from consideration where they were apparently not on tenure tracks. During these years there were at least five individuals who served as instructors at Ann Arbor. These included two California and two Michigan Ph.D.'s as well as one from Princeton.

In 1900 Charles A. Young was the only member of the astronomy department at Princeton. Professor of astronomy and director of the Halsted Observatory for more than a quarter century, Young was a popular undergraduate teacher but did not turn Princeton into a center for graduate education. In part, this was due to policies of the trustees. When Young retired, a new administration moved to replace him with a scientist who would expand the astronomy program. The new professor, Edgar O. Lovett (b.1871), held a Virginia Ph.D. (1895), and specialized in astrometry and celestial mechanics. In many ways his appointment was regressive, moving Princeton away from the latest astrophysical research, of which Young had been a part. However, the administration made two other appointments in 1905: Raymond S. Dugan, who earned his doctorate at Heidelberg and was a specialist in the photographic study of asteroids and comets, and Henry Norris Russell, a Princeton Ph.D. (1900) who spent several years in postdoctoral study at Cambridge University. In the year that Young retired, then, Princeton astronomy included four faculty members: Professor Lovett, instructors Dugan and Russell plus William Reed (b.1871), who divided his time between mathematics and astronomy. Russell developed into the leading American astrophysical theorist of the first half of the twentieth century and his impact on U.S. astronomy was immense. Part of his influence took the form of Princeton Ph.D.'s. When Lovett left Princeton for an administrative position at Rice University, Russell succeeded him, becoming professor in 1911 and director of the Halsted Observatory a year later. John Q. Stewart (1894–1972), Princeton Ph.D. (1919), joined the faculty in 1921. In the 1920s the triumvirate of Russell, Dugan, and Stewart published a two-volume revision of Young's classic textbook which continued the tradition of an authoritative compendium of empirical findings and current theory in astronomy and astrophysics.

For two decades after 1915 the Princeton astronomy department numbered three professors. With the addition of Theodore Dunham (1897–1984) in 1934 the number rose to four, but Dunham, (Princeton Ph.D., 1927), soon returned to Mount Wilson, and in 1940 the department again numbered three. After 1900 there was a steady rise in the number of Princeton doctorates who entered astronomy. For the first decade of the new century the number stood at four. It fell by one during the second decade and in the 1920s reached six. During the Depression decade nine Princeton doctorates entered the American astronomical community. Harvard came late to graduate education in astronomy. The Harvard College Observatory was conceived as a research institution and took no part in instructional activities (Jones and Boyd 1971; Gingerich 1990). Astronomy education went on either in the mathematics department or the Student's Astronomical Laboratory. From time to time directors of the observatory accepted volunteers who came to learn the craft of observational astronomy, but these individuals were not involved in degree programs.

The way Harlow Shapley (1969: 95–96) told it, the first Harvard doctorate (Cecilia Payne Gaposchkin, 1925) was almost an accident. Encouraged by the administration, Shapley added staff to assist with graduate instruction. He hired Willem J. Luyten (1899–1995) whose doctorate was from Leiden and then Harry H. Plaskett (1893–1980), son of the director of the Dominion Astrophysical Observatory, an expert in spectroscopy and physical theory. Luyten, brilliant but difficult, soon left for the University of Minnesota and a long career in astrometry. Plaskett did not find Harvard congenial and accepted a call to Oxford. In 1932 Shapley, himself a Princeton doctorate, hired another Russell student, Donald Menzel (1901–76), and two years later recruited Bart J. Bok (1906–83), a Dutch astronomer trained at Groningen. Menzel would become a leading astrophysical theorist and specialist in solar physics, while Bok developed into a leading student of galactic astronomy, specializing in the Milky Way. During the first thirty years of the twentieth century, only two Harvard Ph.D.'s entered the American astronomical community. During the fourth decade, the Harvard department produced a record twenty-two doctorates. No Harvard Ph.D.'s became permanent members of the department before World War II.

By the 1930s, Harvard and Princeton were important centers for graduate work in astrophysics. At Harvard, the observatory and department were integrated under the leadership of Harlow Shapley. Princeton, lacking modern research facilities, concentrated on astrophysical theory rather than observational astronomy. Shapley expanded the Harvard facilities by the addition of a 61-inch reflector mounted at the Oak Ridge Station in rural Massachusetts. Graduate students at Harvard could also draw on the rich collection of astronomical plates secured at the South American station, moved in the 1920s to South Africa. As a consequence of Shapley's close relationship with his Ph.D. director, Russell often visited Cambridge and shared his knowledge of astrophysical theory with graduate students and faculty. Shapley also developed a system of summer schools (DeVorkin 1984b), featuring leading European scientists, who provided the faculty and observatory staff as well as graduate students and visitors with knowledge of contemporary research in astrophysics and physical theory.

How did these three departments compare in terms of honors? Princeton could point to two of its members who were elected to the NAS and three who earned stars in *AMS*. At Harvard, a much younger department, one member was elected to the NAS and four starred in *AMS*. The oldest of the three astronomy departments, Michigan, did not see any of its faculty elected to the Academy between 1900 and the coming of World War II. It did, however, have four staff members who were awarded stars in *AMS*.

How do these departments compare in terms of graduates who won recognition? Princeton led the way with three of its doctorates awarded stars in *AMS* and two elected to the Academy. Harvard had one graduate listed in *AMS*, but no astronomy Ph.D. elected to the Academy before 1940. Michigan's record was similar.

These institutions shared several characteristics. They were young departments, either reconstructed or created after 1900. At each, the primary commitment was astrophysics rather than astrometry and celestial mechanics. Harvard produced no doctorates in the old astronomy after 1920 while Michigan and Princeton produced six and four respectively. Only the Michigan department attained the size of the Berkeley faculty. It is remarkable that at both Princeton and Harvard major graduate programs were mounted with such limited faculty resources. Each department had able leadership, but Harvard and Princeton exhibited greater administrative continuity and their leaders (Shapley and Russell) were astronomers of great visibility, elected to the National Academy relatively early in their careers.

Four departments (Yale, Columbia, Johns Hopkins, and the University of Virginia) are defined as declining astronomy programs. Together they account for 23 percent of all doctorates granted by the nine leading producers or 16 percent of all Ph.D.'s who entered astronomy between 1900 and 1940. Between the wars they graduated a combined total of only ten doctorates, while the leaders (California and Chicago) awarded fifty-nine and the rising stars (Michigan, Harvard and Princeton) sixty-one. During the interwar period, Yale led the way with six of the ten doctorates. Johns Hopkins graduated no Ph.D.'s who entered astronomy between 1920 and World War II, while Columbia produced one and Virginia two. Three of these institutions concentrated in producing Ph.D.'s whose area of specialization was in celestial mechanics or astrometry. Before 1900, this was also true of the Hopkins department.

These programs began granting Ph.D.'s much earlier than Chicago or California, and produced more doctorates in the last quarter of the nineteenth century than did any of the other institutions under discussion. During the 1870s, Yale awarded two astronomy doctorates. In the 1880s Yale, Hopkins, and Columbia each produced one doctorate. The 1890s witnessed a sharp rise in the productivity of this group. Together they graduated a dozen Ph.D.'s, Columbia leading with a total of five, followed by Hopkins, with three, and two each for Virginia and Yale. After 1900 Ph.D. production declined.

The declining departments were rooted in nineteenth-century administrative traditions and models of a university. On balance, this generalization applies equally well to both Johns Hopkins and Yale as well as to Virginia and Columbia. For example, professors did not necessarily see themselves as having responsibility for graduate instruction. Further, before the 1890s, there is an inescapable impression that graduate work in astronomy in American universities developed in a haphazard manner. However important Hopkins became in physics, medicine, or mathematics, its graduate astronomy program was neither actively supported by the administration nor led by professors who desired to build a dynamic unit. It may be that the four declining departments were situated in institutions too deeply tied to academic values of a earlier day. Although a number of Yale faculty became leaders in various branches of American science before World War I, the College remained the dominant power, and the success of the Sheffield School, the Peabody Museum, or the Observatory depended on the administrative and fund-raising skills of their directors.

Virginia remained a teaching rather than research university until after World War II. By the opening of the new century, Hopkins had settled into a pattern that emphasized a few programs, rather than seeking to emulate comprehensive research universities like Berkeley or Chicago, that were committed to excellence across a wide range of disciplines. At Columbia astronomy did not participate in the revival of the physical sciences between World War I and II.[23]

As historian David F. Musto observed, "Yale's achievements in astronomy were preeminent among American colleges during the 1830s, with a high reputation for both the quality of instruments and the creativity of theoretical research."[24] Yale, however, lost its position of leadership to Harvard by mid-century. While astronomy became part of Yale's Sheffield Scientific School in the 1850s, it suffered from a lack of advanced research instrumentation and, more important, shared with the other Yale sciences what Musto calls "the customary devaluation and disinterest in science by the College, the real center of power."[25] Part of the short-lived movement after the Civil War to transform the sciences at Yale involved plans for a major research observatory. With financial help from the New Haven

23. These observations are based on extensive archival research. My thinking has also been influenced by Geiger (1986), especially chapters two and three. Of course, Geiger is working at the macro level and I at the micro. The findings developed using these different approaches are not always in accord.

24. Musto (1968: 7). Musto's paper must be used with caution. It was written before the great explosion of scholarship dealing with American science and his generalizations, especially concerning public support for astronomy, have little foundation. Further, Musto apparently did not make use of archival sources that provide information on events following the dismissal of Waldo. The decline of Yale astronomy continued unabated after the critical events of 1884. On the history of Yale astronomy see Hoffleit (1992).

25. Musto (1968: 13). The Articles of Incorporation (1871) clearly state that the observatory staff shall be selected on the basis of "scientific and moral qualities . . . not by ecclesiastical or political considerations." YUA, Astronomy Department Papers. Part of the devaluation of science at Yale involved appointments made on the basis of piety or politics rather than scholarship. See also Pierson (1952, I: 189).

arms manufacturer O. F. Winchester, a land speculation scheme was set in motion that seemed to promise a princely endowment. A board of directors for the Winchester Observatory was incorporated in 1871, but the panic of 1873 put an end to the dream. In 1880 the land given by Winchester was conveyed to the Yale Corporation and new plans were developed to finance astronomy. The observatory would provide standard time for the state of Connecticut. In addition, Leonard Waldo (b.1853) was appointed to the observatory staff. His assignment was to develop a department that would test watches and thermometers and certify their accuracy. These activities would provide income that might, in time, pay for modern instrumentation.

At twenty-seven, Waldo was an aggressive and self-possessed product of Harvard's Lawrence School, with great dreams for Yale astronomy. Writing to Hubert. A. Newton, director of the observatory and Yale's most eminent astronomer, Waldo described plans for a 26-inch refractor. He also looked forward to the day when research at Yale would go beyond income-generating services.[26] Newton, somewhat shy and deeply committed to research, found it difficult to deal with the hard-driving Waldo. On the research side, Newton was faced with many difficulties. He had secured private funding to pay for a heliometer, the only instrument of its kind ever mounted in the western hemisphere, but lacked the knowledge to erect the instrument and the highly specialized skills to use it. The senior astronomer also had plans for a large refractor, but there were no funds for such an expensive project.[27]

In time Waldo appeared to undercut his own position with the business community, when a large percentage of watches and thermometers were rejected because they were not accurate. This high rejection rate quickly dampened the interest of manufacturers in using the services of the Yale Observatory.[28] Waldo, however, continued to explore ways to increase income. He worked out an agreement with the U.S. Army to station a detachment of signal corpsmen at the observatory to do research in meteorology.

Waldo's arrogation of power and his inability to get along with other

26. Leonard Waldo to Hubert A. Newton, 28 October 1879. YUA, Astronomy Department Papers.

27. Hubert A. Newton to the Secretary of the Observatory Board, 20 March 1882. YUA, Astronomy Department Papers. The heliometer was developed in the early nineteenth century in order to measure angular separation between celestial objects with great accuracy. This instrument was expensive to build and very difficult to use. The Yale heliometer was the only one ever mounted at an American observatory. By this time, the value of the instrument had been called into question by astronomical photography which permitted measurements of this kind to be made quickly and cheaply.

28. Waldo to Newton, 20 July 1880. YUA, Astronomy Department Papers.

staff members led to a crisis in 1884. A subcommittee of the Yale Corporation recommended an aggressive commercial program for the observatory to be directed by Waldo. Newton and the observatory board immediately resigned in protest.[29]

The Yale Corporation quickly moved to support its senior faculty. Waldo was dismissed and the research mission of the Yale Observatory reaffirmed.[30] However, while the corporation might be generous with moral and political support, it did not act to provide increased funding. In 1884, Newton and the observatory board of managers called William Lewis Elkin (1855–1933) to Yale. With a salary provided by a subscription from friends of the observatory, Elkin was to take charge of the heliometer. An American educated in Europe, Elkin served an apprenticeship with Sir David Gill (1843–1914), director of the Royal Observatory at the Cape of Good Hope, where he became skilled in the use of the heliometer (Schlesinger 1940).

For a long time, the Yale Observatory was without leadership. Only in 1896, twelve years after Newton resigned, did the corporation appoint Elkin director. Never physically robust, Elkin was neither an aggressive administrator nor a successful fund-raiser, and Yale astronomy did not flourish under his direction. In 1910, Elkin retired and astronomy at Yale virtually ceased to exist. A decade later, the corporation called Frank Schlesinger, director of the Allegheny Observatory, who set about revitalizing both research and instruction and laid the foundations for the post–World War II eminence of the program at New Haven.

Yale astronomy was small. Only rarely did the size of the department reach three. Loomis died in 1889, leaving a department that included: Newton, William Beebe (1851–1917) and Frederick Chase. All were Yale graduates. In 1900 the astronomy faculty numbered two: Beebe and Chase. A decade later they were joined by Ernest William Brown (1866–1932) who had been a student of John C. Adams at Cambridge. Brown's appointment was in mathematics, but as the leading Anglo-American student of lunar theory he taught courses in celestial mechanics and theoreti-

29. Musto (1968: 15–16). For Newton's letter of resignation see Newton to the Board of Managers of the Yale Observatory, 19 April 1884. Other letters from Newton recount growing tension. See, for example, Newton to Robert W. Willson, 19 December 1883. Willson, who later directed undergraduate astronomy at Harvard, became so angry with Waldo that he left Yale. Waldo's scheme to bring in the Signal Corps is outlined in a letter to Newton dated 23 May 1884. Often Waldo acted if he, rather than Newton, were director of the Yale Observatory. All citations are to materials in the YUA, Astronomy Department Papers.

30. See Newton's draft statement to Corporation subcommittee member Mason Young, 29 February 1884. YUA, Astronomy Department Papers. A clipping from a New Haven newspaper describing the strife at Yale is attached to a letter from Young to Elias Loomis, 3 May 1884. YUA, Loomis Papers.

cal astronomy. After 1920, Brown had an office at the observatory. Beebe and Chase soon left and Yale astronomy languished. By 1930, the size of the Yale operation increased to three with the addition of Jan Schilt (1894–1982), a Dutch-born astronomer who took the Ph.D. at Groningen. When Schilt moved to Columbia, he was replaced by another Hollander, Durk Brouwer (1902–66).

Five members of the Yale department were numbered among the elite. Two were awarded stars in *AMS* (Chase and Brouwer) and Newton, Brown, and Schlesinger were elected to the National Academy of Sciences. After World War II, Brouwer would also be elected to the Academy. Yale doctorates did less well than the faculty. Before 1940, only one Yale Ph.D. who entered the American astronomical community, was elected to the Academy and only one Yale graduate was awarded a star in *AMS*.

The fortunes of astronomy at Columbia University are similar in some respects to Yale's. The size of the department varied in the half century after 1885. Only limited instrumentation was available to Columbia astronomers and the administration apparently did not consider astronomy to be worth a significant investment. In 1885 the department was John K. Rees (1851–1907), a Columbia Ph.D. . In 1895 Harold Jacoby (1865–1932) and Herman Davis (1865–1933), both Columbia Ph.D.'s and specialists in celestial mechanics and astrometry, joined Rees. At the end of the century, George W. Hill, perhaps the greatest celestial mechanician ever produced in the United States, lectured at Columbia but did not enjoy the experience and soon withdrew to the solitude of his New Jersey farm and the pleasures of lunar and planetary theory. In 1905, the size of the department reached four. In addition to Rees and Jacoby, Samuel A. Mitchell (1874–1960) and Charles Lane Poor (1866–1951), both Johns Hopkins doctorates, were appointed. Poor was a celestial mechanician who spent much of his later career attacking Einstein and writing books on navigation for yachtsmen. Mitchell was interested in solar physics but lacked adequate training in physics and access to sophisticated instrumentation necessary to do research in the area. After leaving Columbia he concentrated on astrometry and chased eclipses as a scientific hobby. By 1915, Poor and Jacoby made up the department, and when Jacoby retired in 1929 Charles Lane Poor became the lone astronomer. He was joined by Jan Schilt who moved from Yale. After World War I, the Columbia department was primarily an undergraduate teaching operation. Only one doctorate was granted during the years between the wars.

Four Columbia faculty (Rees, Jacoby, Davis, and Poor) earned stars in *AMS*, but none (unless one counts George W. Hill) were ever elected to the Academy. Three Columbia astronomy doctorates earned stars in *AMS* and one was elected to the Academy.

The McCormick gift of a 26-inch visual refractor in 1884 provided

Virginia with one of the largest telescopes in the world. The university administration, however, made little effort to support research or graduate teaching. The first professor of astronomy and director of the McCormick Observatory was Ormond Stone. He was able to secure private funding for graduate fellowships. It is not surprising that most Virginia dissertations were in celestial mechanics or based on visual research with the 26-inch. Stone had little understanding of developments in photographic spectroscopy or photometry, even though the visual instrument could have been easily adapted for research in these areas. An administration interested in developing a research department might have encouraged Stone and provided resources for modernizing the great refractor and securing new faculty to introduce astrophysics into the graduate program. As it was, only in the 1920s did Stone's successor, Samuel A. Mitchell, offer instruction in astrophysics, almost twenty years after the subject was first taught at Michigan.

Before 1900, Stone was Virginia astronomy. By 1920 the program had grown to three: Mitchell, called from Columbia in 1913 to replace Stone as head of the department and director of the observatory, plus two Virginia Ph.D.'s, Charles P. Olivier (1884–1975) and Harold Alden (1890–1964). Alden would eventually leave to direct the Yale station in South Africa and Olivier moved to head the astronomy program at the University of Pennsylvania. At the beginning of the Depression decade the Virginia astronomy program numbered three. Mitchell hired two foreign-born astronomers, the Russian Alexander Vyssotsky and the Dutch astronomer Peter Van De Kamp. Vyssotsky took his degree at Virginia, while Van de Kamp was unique in having both Dutch and American doctorates. By 1940 the size of the department increased to four with the addition of another Dutch astronomer, Dirk Reuyl (b. 1906), who took the doctorate at Utrecht. Vyssotsky's wife, Emma Williams (b. 1894), earned a Harvard Ph.D. in astronomy and served as an instructor. Between the wars, only two doctorates were awarded in astronomy.

Five of the Virginia astronomy faculty achieved elite status: four (Alden, Olivier, Stone, and Van De Kamp) were awarded stars in *AMS* and Mitchell, belatedly exchanging Canadian for American citizenship, was elected to the Academy in 1933. Five Virginia doctorates achieved elite status, four earning stars in *AMS* and one being elected to the Academy. Thus, using the metric of peer recognition, Virginia graduates outdistanced those of both Yale and Columbia.

Astronomy at The Johns Hopkins University began life as an undernourished stepchild and never developed into a major department. From 1884 to 1893, Simon Newcomb, director of the Nautical Almanac Office, went by train to Baltimore twice a week to lecture on celestial mechanics (Moyer 1992: 79). For a year, William S. Eichelberger (1865–1951), taught astronomy at the Hopkins before moving on, and another Hopkins Ph.D., Charles Lane Poor, joined the faculty in 1891. After Newcomb's departure, Poor was the astronomy department until he left for Columbia in 1900. John Anderson (1876–1959), taught related physics courses between 1908 and 1916, when he left for a research career at Mount Wilson. After Anderson's departure, no one lectured in astronomy or astrophysics at Hopkins until Arthur Adel (b. 1908) spent the academic year 1935–36 at Baltimore as an instructor. He returned, however, to a research position at the Lowell Observatory.

Considering its proximity to Washington, it is curious that the Hopkins administration did not develop a staff of adjunct professors, drawing from senior scientists at the Nautical Almanac Office and Naval Observatory. With such a staff, together with the physics graduate program, Hopkins would then have been roughly similar to Chicago or California, with an astronomy department specializing in the traditional fields of astrometry and celestial mechanics and a physics department that provided both course work and research opportunities in astrophysics, especially solar spectroscopy. The Hopkins situation differed from California or Chicago in that the institution never had more than a small teaching observatory, unsuitable for research. After World War I, no Hopkins Ph.D.'s entered astronomy.

The astronomy faculty at Baltimore received little recognition. Simon Newcomb was its only member numbered among elite astronomers. Graduates, however, achieved elite status in greater numbers than those of any other institution in this group. Two Hopkins graduates were awarded stars in *AMS* and four were elected to the National Academy. This last category includes Henry Crew, who took a degree in physics but began his career as a spectroscopist at the Lick Observatory, later joining the physics department at Northwestern University. Crew remained involved in astronomical spectroscopy throughout his life.

This group of declining departments shared several characteristics. They came early to graduate education, developing idiosyncratic programs that lacked clarity of purpose. With the exception of Johns Hopkins, these four departments were almost exclusively dedicated to the traditional fields of astrometry and celestial mechanics. In order to move aggressively into astrophysics the departments (the exception is Virginia) would have required massive support for new instrumentation. The alternative would have been to develop theoretical astrophysics, but able practitioners were in very short supply. In no case does it appear that university administrators were interested in providing resources for these graduate programs. Finally, with the exception of Schlesinger's administration at Yale, the four departments lacked leadership. There were no Leuschners, Husseys, or Russells. Without institutional support and strong leadership at the departmental level, it is not surprising that the four departments declined.

THREE CASE STUDIES OF GRADUATE EDUCATION

The experiences of three astronomers provide insights into graduate education. These individuals were selected because they left extensive manuscript sources.[31] Samuel A. Mitchell (Abbot 1962), a Canadian by birth, did graduate work at Johns Hopkins, earning the doctorate in 1898. Joel Stebbins (1878–1966), a pioneer in photoelectric photometry, earned the doctorate at Berkeley in 1903 (Whitford 1962). Astrophysical theorist Lawrence Aller (b. 1913) took an M.A. at Harvard in 1938 and was awarded his Ph.D. in 1943. Mitchell worked in both solar physics and astrometry, while Stebbins and Aller were astrophysicists but represented very different traditions. All three were elected to the National Academy, and Stebbins and Mitchell directed observatories. Mitchell entered the American astronomical community just at the end of the second cohort and Stebbins at the beginning of the third. Aller completed his degree during World War II.

Samuel A. Mitchell was born in Kingston, Ontario, in 1874 and prepared for Queen's University at the Kingston Collegiate Institute. Mitchell entered Queen's on a scholarship with advanced standing in mathematics, Latin, French, and German. Astronomy caught Mitchell's attention and soon he was helping out in the observatory, timing transits, regulating the clocks, and observing with a 6-inch refractor. During his senior year, Mitchell was in charge of the observatory.[32]

Little is known about the quality of the undergraduate education Mitchell received at Queen's. Surviving notes suggest that his introductory physics course was organized along classical lines, stressing observation and induction.[33] Mitchell's undergraduate notebooks indicate that he took courses in advanced geometry, trigonometry, and the calculus as well

31. Numerous published sources provide information on the graduate education of astronomers. For example, the education of George Ellery Hale is well documented in Wright (1966: chaps. 1–2) and Wright, Warnow, and Weiner, eds. (1972). James E. Keeler's education is discussed in Osterbrock (1984). A number of *Biographical Memoirs* of the National Academy of Science provide information on the education of NAS astronomers. Among the most informative is Osterbrock and Seidelmann (1987) on Paul Hergert. On the education of Simon Newcomb, see Norberg (1974). Since I do not wish to rehearse readily available materials, I have used manuscripts rather than printed sources for individuals whose graduate work is documented.

32. Samuel A. Mitchell, "Autobiography," pp. 1–4. Typescript in NAS, Deceased Members Files.

33. Notebook labeled "Jr. Physics," entry for 3 October 1890. UVA, Mitchell Papers.

as work in theory of equations. He also studied chemistry and enrolled in courses on English composition and literature.

Arriving in Baltimore in the fall of 1895, Mitchell fell under the spell of Charles Lane Poor, a teacher of "very high quality" possessing a "friendly personality," and decided to switch to astronomy.[34] The young Canadian offered physics as a first minor and mathematics as the second, but took extra course work that gave him sufficient hours to meet the requirements for a doctorate in mathematics, had he wished. During his second year in Baltimore, Mitchell was an assistant in astronomy with responsibility for maintaining the student observatory.

The doctoral program at Hopkins focused on celestial mechanics and theoretical astronomy. Students were expected to develop "a general knowledge of the principal works of the leading astronomers of ancient and modern times," with emphasis on the nineteenth century.[35] This requirement was to be filled by independent reading, especially in the European literature. Students electing a dissertation field in astrophysics were excused from advanced studies in celestial mechanics.

During his first year at Hopkins, Mitchell took a full schedule of courses. The school week began on Monday at 9 A.M. with a lecture on differential equations, and after an hour's break there was another mathematics lecture. Physics lecture ran from noon to one o'clock and lab took up the rest of the afternoon. Tuesday was much the same. Wednesday was a mix of physics and mathematics plus an afternoon lab in astronomical computing. Thursday was a easy day, with two astronomy lectures and one in mathematics. Friday was also light, its three lectures divided between physics, mathematics and astronomy. Saturday morning was devoted exclusively to astronomy. During 1895–96, Mitchell's instructors included Poor in astronomy and John S. Ames (1864–1943) in physics. It is not clear whether Mitchell actually studied with Henry A. Rowland. He did, however, select a problem in astrophysics for dissertation research.

Mitchell decided to experiment with a concave Rowland grating to find out whether it would be as effective as the traditional prism spectroscope for research in stellar spectroscopy. Using the 9.5-inch refractor at the student observatory, Mitchell secured 153 plates, observing for almost two hundred hours over a period of forty-three nights between January and May 1898. The proximity of the observatory to the main line of the Baltimore and Ohio Railroad as well as city lights greatly hindered Mitchell's research. In the end, the dissertation dealt as much with theoretical aspects of the concave grating as it did with applications to stellar spec-

34. Mitchell, "Autobiography," p. 5. NAS, Deceased Members Files.

35. Johns Hopkins University, "Astronomy: Requirements of a Candidate for the Degree of Doctor of Philosophy in Astronomy as a Principal Subject," n.d., p. 3. UVA, Mitchell Papers.

troscopy.[36] Mitchell spent the winter of 1898–99 at the Yerkes Observatory trying to perfect his concave diffraction grating spectroscope, using better instrumentation than was available at Baltimore, but had little success.[37]

Lacking strong direction and guidance, Mitchell selected a dissertation topic of marginal value. It dealt more with methods and techniques than with the solution of a problem that would provide new insights into astrophysics. By the end of the 1890s, enough was known to suggest that concave gratings were of very limited value for astronomy, outside the field of solar physics. There was, however, no one at Hopkins to provide Mitchell with firm direction and prevent him from selecting a topic that involved high risk and limited payoff. Mitchell's graduate education provided an ambiguous professional identity. In terms of his own self-evaluation and the assessment of others, he was neither a celestial mechanician in the Columbia tradition nor an astrophysicist of the Chicago or Berkeley school. His subsequent career reflected the limitations imposed by his graduate training at the Hopkins in the 1890s.

Joel Stebbins was born in Omaha, Nebraska in 1878. He attended the University of Nebraska at Lincoln, earning a B.S. in 1899. Like Mitchell, he remained for a year as an instructor before moving on. Stebbins spent a year at the Washburn Observatory of the University of Wisconsin studying with George C. Comstock (1855–1934), then moved to Berkeley to work for the Ph.D. Stebbins's academic progress was swift and virtually without interruption.

At each phase of his education, Stebbins made important contacts. His undergraduate advisor, Goodwin D. Swezey (b.1851) was a respected and able teacher. George C. Comstock, director of the Washburn Observatory and professor of astronomy at Madison, was one of the leaders of American astrometry. He became a mentor to Stebbins, and the relationship continued even though Stebbins moved to Berkeley to concentrate on astrophysics. At the Lick Observatory, Stebbins wrote his dissertation under William Wallace Campbell, a man on the way to a major position in the American astronomical community. Stebbins remained close to these key individuals and each provided important advice and assistance as his career developed.

36. These comments are based on a notebooks in the Mitchell papers at the University of Virginia. See "Concave Grating Measures" notebook and "Thesis Notebook, 1898." During 1891–92, a Hopkins Ph.D. in physics (1887), Henry Crew, worked as a spectroscopist at the Lick Observatory. His assignment was to develop methods for using a Rowland concave grating for spectroscopic research with the 36-inch refractor. The project was not successful and it is surprising that Mitchell did not know of these negative findings. On concave gratings in astronomical spectroscopy, see Osterbrock (1986).

37. The Yerkes experience is discussed in "Thesis Notebook, 1898." UVA, Mitchell Papers. See also Mitchell, "Autobiography," pp. 7–8. NAS, Deceased Members Files.

Writing to the director of the Lick Observatory concerning the availability of graduate fellowships, Stebbins indicated that he became interested in astronomy at age twelve and "read all the popular works I could obtain and constructed, except for the lenses, two small telescopes." At Lincoln, Stebbins tested out of introductory astronomy and proceeded directly to courses in practical astronomy and elementary astrophysics. Later he studied theoretical astronomy from the most advanced German texts.[38] For an undergraduate thesis, Stebbins prepared a detailed set of charts based on observations made at the university observatory. He remained at Lincoln for a year following graduation, teaching mathematics and acting as assistant in the observatory.

At Madison, Stebbins continued his study of theoretical astronomy under Comstock's supervision and took graduate courses in mathematics and physics. He also had practical instruction in observational astronomy using the 15-inch Clark refractor and the transit circle. He had charge of the time service and spent about sixty hours a month reducing astrometric observations.

In applying for admission to the graduate program, Stebbins indicated to director Campbell, "It is my desire to become a professional astronomer and if the taking of a doctors degree will materially aid me, I shall certainly try to do so."[39] So great was Stebbins' trust in Comstock that he did not secure any other recommendations. "I might get other testimonials from professors here and in Nebraska," but I have "been so closely associated with Professor Comstock that I think he is able to judge better than anyone else."[40] This was no naive decision on Stebbins part. At twenty-three he was an astute judge of character and understood that a favorable recommendation from his Wisconsin mentor would carry great weight.

By July 1901, Stebbins was on Mount Hamilton, making himself at home as a member of the "younger crowd." His assignments included computing for Richard H. Tucker (1859–1952), astronomer in charge of the transit circle as well as for William J. Hussey. Stebbins also assisted Hussey with double-star work when moonlight made it impossible to use the 36-inch Crossley reflector for long-exposure nebular spectroscopy. During interminable hours guiding the Crossley, Stebbins reported he was getting important tips on photography from a senior graduate student, Harold K. Palmer (b.1878). With the departure of Tucker for his annual Maine vacation, Stebbins assumed responsibility for the Lick time service. He also was designated to keep a record of the visual magnitude of Nova *Persei*.[41]

38. Joel Stebbins to William Wallace Campbell, 11 April 1901, p. 2. SALO, Stebbins Papers.
39. Stebbins to Campbell, 11 April 1901, p. 3. SALO, Stebbins Papers.
40. Stebbins to Campbell, 11 April 1901, p. 4. SALO, Stebbins Papers.
41. Joel Stebbins to H.F., 28 July 1901. SALO, Stebbins Papers.

All, however, was not idyllic in this western paradise. Stebbins had neglected to read the fine print. The University of California required doctoral candidates to have the equivalent of three years of high school Latin. Campbell appealed the case to the Graduate Council and discussed it with the Latin Department, but the regulations could not be waived. Somehow, Stebbins would have to achieve proficiency. Stebbins was quite candid in seeking advice from Comstock. It seemed a waste of time to study Latin, but the Ph.D. degree was becoming more important in the American scientific community and without meeting the Latin requirement, Stebbins could not earn a California doctorate.[42]

As Stebbins's first season on Mount Hamilton came to a close, he wrote home describing his experiences. During the winter, graduate fellows went to Berkeley to take courses in the old astronomy as well as in physics and mathematics. During these months, Stebbins thoughtfully reflected on his intellectual development. He stressed the acquisition of technical skills in two areas: photography and spectroscopy. He remarked that Professor Comstock had urged on him the importance of photography for research in astrophysics. Stebbins' knowledge of spectroscopy was, as yet, strictly practical and he looked forward to courses that would provide theoretical foundations. Stebbins concluded his reflections with a very interesting comparison "Lincoln, Madison and Lick are about in a sequence. At Lincoln everything was instruction. At Madison half instruction and half research. Here everything [is] research. The Fellows are supposed to come here as assistants and learn by assisting and by themselves."[43]

During the first winter in Berkeley, Stebbins continued to reflect on his experiences at the Lick. "It is a very active place and now that I am away, I realize how many things I learned while there." He concluded, however, that a permanent position at an isolated mountain observatory was not an inviting prospect. "It is better to be where there are more people. My stay there was in the nature of an outing. I knew when I was coming down. I am afraid it would be very different if I felt I was to be there indefinitely."[44]

Stebbins wasted little time in selecting a thesis topic. He wanted to do research in astrophysics. Deferring to director Campbell, Stebbins wrote to him, "You know much better than I, what would be a promising subject," but indicated an interest in the spectroscopic study of variable stars. In the end, Campbell assigned him the analysis of the spectrum of the long

42. Stebbins to Comstock, 18 August 1901, p. 2. UWMA, Comstock Papers.

43. Stebbins to H.F., 2 January 1902, p. 2. SALO, Stebbins Papers.

44. Stebbins to Comstock, 8 February 1902, pp. 2–3. UWMA, Comstock Papers. Stebbins reported that about two months was the longest any of the junior staff felt they could remain on the mountain without a brief vacation.

period variable Omicron Ceti (*Mira*).[45] Stebbins was the first to observe major changes in its spectra. His thesis research required a series of long exposures in order to record the spectrum of the fading variable. At Mount Hamilton, Stebbins demonstrated his ability as an observer who had no reservations about undertaking a taxing research program.

The educational experiences of Mitchell and Stebbins provide several points of contrast. There were important differences in personality between the two men. Stebbins seems more aggressive, clearly in charge of his career from an early date. He quickly grasped the importance of original research and, under the supervision of Campbell, made a significant contribution to astrophysics. In comparison to Mitchell, Stebbins left graduate school with a clearly defined professional identity. Somehow, Mitchell may not have learned the difference between studying in order to excel in course work and original research, with all that this implies about problem selection and the importance of providing peers with new information. Further, Stebbins early found a powerful mentor in the person of George C. Comstock and developed a good working relationship with Campbell, under whom he carried out his dissertation research.

Mitchell never formed a mentoring relationship with a senior member of the American astronomical community. His letters indicate that he relied for advice on family and members of the Queens University faculty in Canada. Mitchell did not exhibit Stebbins' ability to manage his career. While both men were elected to the Academy, Stebbins moved ahead much faster than Mitchell and his research earned honors and awards earlier and in greater numbers than those which came to the Canadian. Stebbins rose to the status of a major figure within the American astronomical community. He pioneered in the use of photoelectric photometry, producing important results for more than forty years. He also became a leading power broker, working through the American Astronomical Society, the Astronomy Section of the National Academy, and the International Astronomical Union. Mitchell never reached an equivalent status and his research remained at the level of routine data collection.

The third case study is drawn from the 1930s. Lawrence H. Aller (b. 1913) was educated at the University of California at Berkeley (B.A. 1936) and Harvard (M.A. 1938, Ph.D. 1943) where he was elected to the Society of Fellows. Aller was one of the new breed of astrophysicists who sought to apply quantum mechanics in order to understand the physical processes at work in stars and nebulae.

Writing in support of Aller's application for a Harvard fellowship, Lick Director William Hammond Wright recounted the Horatio Alger ca-

45. Stebbins to Campbell, 23 February 1902. See also Stebbins to H.F., 19 May 1902 and 1 June 1902. SALO, Stebbins Papers.

reer of the young scientist. "He came to the University of California with practically no preparation, was admitted as a special student on a limited schedule, and eventually worked himself into full status and finished his undergraduate work in really handsome fashion." Wright concluded that "Mr. Aller has come a long way since his first contact with this university. In his progress he has demonstrated character and ability to the highest order."[46]

After graduating from Berkeley, Aller spent the summer at the Lick Observatory. He reduced and analyzed observations for Arthur Wyse (1909–42) and Nicholas U. Mayall (b. 1906) and prepared finder charts for double-star observers. During the dark of the moon, Aller assisted Mayall in photographing the spectra of star clusters. Using a self-registering microphotometer, Aller prepared tracings measuring intensity of the spectrum lines in a number of stars. He gave special attention to *Deneb* (Alpha *Cygni*), searching for possible relations between the radial velocity of the star and the intensity of major absorption lines in its spectrum. He also used the microphotometer to study novae spectra. In August 1936 a comet provided the subject for a brief spectroscopic investigation. Working with Wyse, he photographed several nebulae with the spectrograph on the Crossley reflector. The object of this investigation was to see if their spectra had changed over time. Using plates borrowed from the Mount Wilson Observatory, Aller employed the microphotometer to make about fifty tracings of the spectrum of *Sirius* (Alpha *Canis Majoris*) as part of a study of the star's atmosphere. Finally, between 20 May and 2 August, Aller prepared seventeen drawings of the planet Mars using the 12-inch refractor.[47]

During one summer Aller gained more experience in observational astronomy and astrophysics than did Mitchell in the whole of his graduate career. His energy even exceeded that of the industrious Joel Stebbins. The record of Aller's activities is altogether remarkable, indicating commitment, superb organization, and willingness to put in long hours at the telescope or microphotometer. As director Wright indicated to the Harvard fellowship committee, Aller was much more than an assistant. The young scientist demonstrated "the qualities that make for success in research."[48]

Aller found at Harvard a very different world from that which he had known in California. "The skies here are miserable compared with those

46. William Hammond Wright to Dean George D. Birkhoff, 3 March 1939. SALO, Aller Papers. The phrase, "in really handsome fashion" coming from the reserved, critical and often bitter Wright is high praise, indeed.

47. Lawrence Aller, "Memorandum to Director Wright on Work Done at the Lick Observatory, 15 May 1937 to 15 August 1937." SALO, Aller Papers.

48. Wright to Birkhoff, 3 March 1939. SALO, Aller papers.

of Mount Hamilton. At present the spectrographic equipment for the 61-inch telescope has not been completed so I have been devoting my time exclusively to teaching and theoretical studies." It was in the area of theoretical astrophysics that Harvard differed most significantly from the Lick. Aller quickly became associated with Donald Menzel and his team, who were applying the theory of radiative transfer to planetary nebulae. Menzel's team made use of the theories developed by Victor A. Ambarzumian (b.1908), a Soviet astrophysicist, and Subrahmanyan Chandrasekhar (1910–95), of the Yerkes Observatory.[49]

Aller was extremely fortunate. He was in the right place at the right time. Menzel joined the Harvard faculty in the early 1930s and quickly developed a seminar in theoretical astrophysics that probably had no equal. When Aller arrived at Harvard, he found a group of graduate students, including James G. Baker (b. 1914) and Leo Goldberg (1913–88), working with Menzel, as well as collaborators from the Massachusetts Institute of Technology and other Harvard departments. Aller was swept into a major research program that led to a series of eighteen papers (1937–45) co-authored by Menzel, his students, and associates. These papers appeared in the *Astrophysical Journal* under the general title "Physical Processes in Gaseous Nebulae" (Menzel 1962). An able and ambitious graduate student could not have found a more challenging and creative milieu.

Aller requested permission to spend time on Mount Hamilton in the summer of 1938 so he could make spectroscopic observations in connection with the theoretical investigations being conducted by the Menzel group. Aller complained that research on gaseous nebulae had been hampered by the need to study for comprehensive examinations. Perhaps he was also discouraged by the "enormous amount of work yet remaining" in the analysis of nebulae spectra and by the fact that "the theory has not yet been developed to the point where it can be compared with the observations."[50] Election to Harvard's prestigious Society of Fellows (1939) may have helped Aller in choosing his research problem. As a junior fellow he had assured support for three years and the freedom to devote himself exclusively to research.

Aller needed the freedom provided by his election as a junior fellow. The analysis of spectrograms using the microphotometer involved countless hours of work, much of which went on after the initial tracings were completed. Aller confided to Wright that "The *enormity* of the task before me is somewhat appalling. You may get some idea of the way in which I have been amusing myself when I explain that I have run well over a thou-

49. Aller to Wright, 20 November 1937. SALO, Aller Papers.

50. Aller to Wright, 11 February 1939. SALO, Aller Papers.

sand feet of [paper] tracings [made with the microphotometer] and that there are many plates yet to be run."[51] Given the amount of work, it is not surprising that it took Aller until 1943 to complete the doctorate. Lawrence Aller went on to a distinguished career in astrophysics. His eminence in the field was recognized by the National Academy of Sciences, to which he was elected nineteen years after earning the doctorate at Harvard.

Aller found purpose and direction in his scientific career much earlier than did Stebbins. The period from about 1905 to 1910 was critical for the young Nebraskan as he developed the new research technology of photoelectric photometry and demonstrated its power in a path breaking study of the eclipsing variable *Algol* (Beta *Persei*). In comparison, Mitchell's research career never came into focus. He spent his life moving between topics in astrometry and solar physics. Another way of measuring differences between these three astronomers is to note that Mitchell began publishing in the year he took the doctorate, Stebbins two years before the award of the Ph.D., and Aller during his senior year at Berkeley.

Aller brought important skills to the Harvard doctoral program. He was already an able observational astronomer, thanks to the tutelage of Wyse and Mayall. He was also well versed in contemporary physical theory and its application to astrophysics, thanks to his teachers at Berkeley and informal discussions at Mount Hamilton. Aller found a mentor in Lick director William H. Wright. He also must have won the esteem of the Harvard astronomy faculty for they supported him for election to the Society of Fellows. These connections helped Aller gain access to the observational materials he needed, as well as providing time to carry out a lengthy analysis of the data. In addition to mentors and sponsors, Aller was extremely fortunate to find at Harvard fellow graduate students like Baker and Goldberg. It is not clear that Mitchell ever credited any of his fellow students during his years at Hopkins while Stebbins referred only once to the lessons he learned from a senior graduate student. The Menzel circle (Goldberg 1977) must have provided a situation in which the members learned a great deal from one another.

These case studies also document changes in graduate education between the 1890s and the 1930s and, at the level of individual biography, illuminate the interplay between opportunity and personality. Mitchell was the product of a doctoral program that did not benefit from the expansion of graduate education in the 1890s. His career was bifurcated between the old and the new astronomy. Stebbins, by comparison, rode the crest of the wave that would soon make the Berkeley department the leading astronomy graduate program in the United States. Stebbins also found powerful mentors and sponsors. Aller's mentor, William Ham-

51. Aller to Wright, 3 July 1939. SALO, Aller Papers.

mond Wright, had the wisdom to send him to Harvard for the doctorate after providing him with superb training in observational astronomy and basic astrophysics. At Harvard, Aller encountered a situation analogous to Berkeley in Stebbins' day: an institution that was moving into a position of leadership, training graduate students in the application of contemporary physical theory to astrophysics. He also found powerful sponsors. All three of these individuals achieved elite status, but their graduate experiences led them into very different careers.

THE FIRST JOB

First jobs go a long way in determining the shape of a scientific career. If it is a position in which a young scientist can grow and develop, the first full-time job will provide a beneficial environment. If the individual is restricted and constrained, forced into research in which she has little interest, then the career may be retarded or even permanently blighted. This section follows Joel Stebbins as he secured his first position subsequent to taking the Ph.D. at Berkeley. Archival sources provide rich documentation of Stebbins' decision-making process. The more general problem, availability of entry-level positions will be discussed in chapter 4.

Stebbins's experiences on Mount Hamilton early led him to conclude he would prefer an appointment in a university. He found life at the Lick Observatory stimulating, but the young Nebraskan was less than pleased with the isolation and consequent tensions that took the form of interpersonal clashes. As he put it in a letter to his family, "If you knew of all the 'scraps' that have occurred here on Mount [Hamilton] you would think that I am a rather peaceful citizen." Stebbins's conception of an ideal social environment was a major factor in selecting a first position.

By the autumn of 1902 it was clear that Stebbins would have to make a choice. Director Campbell was expecting funding from the Carnegie Institution of Washington for postdoctoral fellowships. A Carnegie Fellowship would look good on Stebbins' vita, but he was not tempted. "I would prefer a good university position to something which would be only temporary."[52] The desire for a permanent university position would become an important consideration for Stebbins.

By night Stebbins secured spectrograms of *Mira* and by day found time to write letters inquiring about university positions. Just before Thanksgiving his efforts bore fruit. The president of the University of Illinois showed interest. Illinois had a modest observatory with a 12-inch refractor. There had been no professor of astronomy for several years and the courses were taught by an instructor. Illinois wanted a new man in the

52. Stebbins to H. F., 19 October 1902. SALO, Stebbins Papers.

astronomy post. "We are willing to take a young man, but he would have to have the *stuff* in him to make a professor out of." President Draper continued, "If we could find such a young man we should be willing to afford the time for him to grow up to it." Draper also indicated that the new appointee would probably have to teach courses in the mathematics department. He encouraged Stebbins to submit his application and provide recommendations

To Stebbins, the Illinois position "sounds like a golden opportunity. The new man will eventually be at the head of things."[53] In addition to his desire for a permanent position and the conviction that life on an isolated mountain top was not to his liking, Stebbins listed independence as a third consideration. He desired as much freedom for himself, and control over the astronomy program and observatory, as he could obtain. Illinois seemed to offer the right situation. In addition, it was a state university in transition, moving toward eminence in the physical sciences.

In consultation with Comstock, Stebbins planned his campaign. One of the biggest obstacles was Campbell, who wanted to keep Stebbins at Lick but who had nothing better to offer than a Carnegie fellowship. Stebbins did not wish to leave graduate school with a legacy of animosity between himself and the director. Stebbins turned to Lick astronomer William J. Hussey for help in preparing the Illinois application. Hussey devoted considerable effort to the project, making Stebbins write a first draft, the covering letter, and the responses to the questions on the application form. Then Hussey edited the material and typed the final copy. The senior astronomer advised the younger man "that in some cases . . . the letter is a very important factor. I imagine that a good letter might not get the job, but a poor one might lose everything."[54]

In addition to providing help with the formal application, Hussey must have spent time with Stebbins discussing the management of his career, should he win the Illinois position. Stebbins expressed excitement at the prospects. He imagined a salary in the $900 to $1,200 range, while Hussey suggested that "I would be a full professor by the time I am thirty—which would be exceptionally young."[55] Stebbins told his family that he did not know of a better opening. While indicating that salary was significant, Stebbins returned to the basic theme of independence. The young Nebraskan, who had functioned so successfully in subordinate positions under Professors Swezey and Comstock and director Campbell, longed for the day when he would be independent.

Early in the New Year, Campbell offered Stebbins a Carnegie fellow-

53. Stebbins to Comstock, 20 November 1902, p. 2. UWMA, Comstock Papers.

54. Stebbins to H.F., 23 November 1902, p. 1. SALO, Stebbins Papers.

55. Stebbins to H.F., 23 November 1902, p. 4. SALO, Stebbins Papers.

ship. In his usual parsimonious way, the director was vague about the stipend, mentioning first $1,000 and later $900. Stebbins knew his man and remarked "I guess it will be $1000 all right if I propose to leave." During their discussions, Campbell assured Stebbins that "he would never stand in the way of any member of the staff going where he thought he could do better."[56] Stebbins continued to rely on Hussey as a source of advice. At this important juncture in the young man's career, the senior astronomer assumed a critical role. Hussey reviewed the pros and cons of remaining at Lick versus moving to Illinois. Hussey concluded that Illinois was the better option. "He thinks the chances for promotion will be a great deal better there."[57]

During the early weeks of 1903 director Campbell and President Draper exchanged several letters. In one of these Draper summarized the ideal young faculty member. "If Mr. Stebbins is a man of entirely sound character, good spirit, scholarly tastes and some dignity of character, with a lot of work in him, who could come here and enter an agreeable University community which is already well developed and has excellent prospects before it, I would like to have him come."[58] Apparently, if Campbell recommended Stebbins, the position was his.

On 21 February 1903 Stebbins sat down to write what might have been one of the most important letters of his career. Campbell wanted him to drop work and come to Mount Hamilton at once to discuss a Carnegie fellowship. Stebbins did not want to face the director, whose piercing glance and dominating personality made students and staff members walk softly in the great man's presence. Besides, Stebbins had long since made up his mind. He had been thinking about career options since graduating from college, and "while you may not agree with my choice, I hope that you will see that I have not jumped at my decision." Toward the end of the letter, Stebbins made his position abundantly clear. "I would rather grow up to be a Professor at Illinois, than an Astronomer at Mount Hamilton."[59] Campbell had the good grace to write President Draper immediately; a few weeks later Stebbins learned of his appointment.

Stebbins use of the phrase "grow up to be a Professor" demands explanation. Some might simply explain his language as an obsequious gesture to one of the most iron-willed individuals in the history of twentieth-century American science. But there are alternative readings. A few weeks later, in announcing to Campbell that he had accepted the Illinois posi-

56. Stebbins to H.F., 4 January 1903, pp. 1–2. SALO, Stebbins Papers.

57. Stebbins to H.F., 4 January 1903, pp. 1–3. SALO, Stebbins Papers.

58. President Draper to William Wallace Campbell, 9 January 1903. SALO, Stebbins Papers.

59. Stebbins to Campbell, 21 February 1903, pp. 2–3. SALO, Stebbins Papers.

tion, Stebbins used similar language. "I hope that I may grow up and be able to do something for the Lick Observatory in return for what I have received."[60] Rather than an admission that he was a mere infant scientifically, Stebbins may have had something very different in mind. He was not revealing a lack of ego-strength to his dissertation supervisor, rather he was indicating his conception of the scientific career as a dynamic process. This usage suggests that Stebbins realized he stood at the beginning of a new phase of professional development. Formal education was over and he must now take full responsibility for managing his career. Clearly there were risks in his decision. The first lay in alienating Campbell. Stebbins's career would have been severely damaged had he made an enemy of the powerful Lick director. Further, in leaving one of the finest research observatories in the world, Stebbins put his research career in jeopardy. The skies of the Midwest were not like those of California and the 12-inch refractor was a far cry from the superb instrumentation at Lick. Yet, in the end, the gamble paid off. Luck was on Stebbins's side, and he soon became acquainted with another young Illinois faculty member, F. C. Brown (b. 1881) of the Physics Department, who introduced Stebbins to the light-detecting properties of the selenium cell. In cooperation with Brown, Stebbins was soon developing the selenium cell for astronomical photometry (Whitford 1962: 297). Later in his career, as a visiting astronomer at the Lick Observatory, Stebbins would repay the debt he incurred as a graduate student in ways that neither he nor Campbell could have imagined.

It can be argued that the experience of Joel Stebbins is not representative, but that would be to miss the point. In all probability, the process of finding a first job is one of the unique experiences of a scientific career. In each case, the mix of chance and necessity must differ, and the interplay of personalities produces unique conditions. Stebbins's experience represents some of the constraints and variables involved in the process. These include the desires and goals of the individual and the evaluations and plans of academic advisors. Further, the quality of advice to which an individual has access is important. Considered from this point of view, the experience of Joel Stebbins should be seen as illustrating major elements in the process of securing a first job.

THE QUESTION OF SCHOOLS

Subrahmanyan Chandrasekhar (Wali 1991) was trained by Arthur Stanley Eddington (1882–1944) in theoretical astrophysics at the University of Cambridge and joined the staff of the Yerkes Observatory in the 1930s. From Chandrasekhar's (1977: 61) perspective, American astronomy did

60. Stebbins to Campbell, 16 March 1903. SALO, Stebbins Papers.

not exhibit a tradition of schools (groups of disciples under masters) similar to those that marked the education of European scientists. This view deserves attention. Is it, indeed, the case that there were no schools in American astronomy before World War II? Or, to put the matter another way, is it realistic to expect the American system of graduate education in astronomy to produce schools?

Schools are more often associated with Europe than America. The concept implies formal training and something more: intellectual and, perhaps, personal discipleship. In chemistry and geology, students of Justus von Liebig (1803–73) and Abraham Gottlob Werner (1750–1817) spread the gospel of these masters across Europe and even to America. So did the graduates of Wilhelm Wundt's (1832–1920) experimental psychology laboratory. But where are well-defined schools in astronomy? One can point to the products of J. C. Kapteyn's (1851–1922) astronomical laboratory at Groningen, where students specialized in clearly defined areas and learned important concepts or techniques often developed by the master.[61] But in American astronomy, it is virtually impossible to identify schools in the sense of graduate programs dominated by powerful and sometimes charismatic individuals, whose ideas and methods were so forcefully impressed on students that they became marked as the products of a given master.

The American experience was conditioned by several factors. The nature of graduate education, especially from the 1890s, tended to be open and flexible, offering students a variety of options (astrometry, celestial mechanics, or astrophysics) even in situations where powerful individuals, for example, Leuschner at Berkeley, exerted a great deal of control. From a structural perspective, graduate education was organized in the form of departments that were composed of several members, not according to the European model of institutes dominated by a single professor. Further, American graduate schools introduced diversity through a system that required graduate students to take a portion of their work outside the major department as one or more minor fields. These structural realities tended to dilute the influence and power of any one individual. Finally, the demographic data suggest the doctorate did not become normative in American astronomy until well into the twentieth century. It would be impossible for schools to develop in a context where an undergraduate degree was often the highest level of educational attainment and in which many successful scientists had no more than a secondary education.

There are other factors that militated against the development of

61. E. Robert Paul provides an assessment of the impact of Kapteyn's school on Dutch astronomy in his entry on "Netherlands, Astronomy" in J. Lankford, ed., the *History of Astronomy: An Encyclopedia*. New York, Garland (1996).

schools in the American astronomical community. Many leading astronomers were connected with private institutions such as Mount Wilson, the Harvard College Observatory before the 1920s, or the Dudley Observatory that were committed to research and made no provision for graduate training. Mount Wilson opened its doors to exceptional Berkeley graduate students engaged in doctoral research only at the end of the 1930s. Apparently, the Dudley Observatory never did so. Further, even in the great observatories connected with universities such as Chicago or California, there were scientists whose activities were devoted exclusively to research and whose contact with students was minimal. This was especially true of Yerkes before the Struve administration.

At least one other factor merits consideration. As indicated in chapter 3, astrophysics in America developed outside the university context. At Allegheny, Harvard or the private observatories of pioneers like Draper, Rutherfurd, and later George Ellery Hale, there was no provision for graduate students. These institutions were devoted exclusively to the search for new knowledge.

Professor Chandrasekhar, then, has neglected certain historical and cultural factors. Graduate education in the United States differed significantly from the European experience. Its organization militated against the development of schools. The American astronomical community included a sharp division between institutions devoted exclusively to research and those which divided their energies between research and teaching. True, students went to Berkeley to study celestial mechanics with Leuschner or to Harvard to work in astrophysics with Menzel, but the education they received covered a much wider field than seminars offered by the dissertation supervisor. Graduate education was pluralistic in both an intellectual and structural sense. This pluralism made the development of schools in the European sense a virtual impossibility.

Before 1940 the American astronomical community was open to entrants with a wide range of educational credentials. The Ph.D. became more important after 1900, but it did not become normative for entry into the community. Nor did educational credentials define access to the reward system. Further, astronomers with low levels of educational attainment produced world-class science. These findings point to the pluralism and diversity of the American astronomical community before World War II. Attempts to impose late-twentieth-century standards of educational credentialling on the past lead to a misreading of history.

5 ✦

THE CHANGING SCIENTIFIC CAREER

This chapter provides a quantitative discussion of the scientific career based on three cohorts of astronomers active between 1859 and 1940. Evidence is marshaled that throws light on such questions as the duration of careers, the Ph.D. pool, sources of non-Ph.D.'s, the experiences of three cohorts of scientists as they moved from a first to a second position and then to a third, and decisions relating to leaving and/or returning to the field. Major cognitive developments such as the emergence of astrophysics or changes in the generic community, including the growth of graduate education from the 1890s, helped to shape individual careers.

Sociologist Andrew Abbott (1988: 324) argues that careers are "a strategy invented in the nineteenth century to permit a coherent individual life within a shifting marketplace." While the nineteenth century ideal stressed remaining in a single professional career for a lifetime, recent scholarship (Abbott 1988: 132) indicates there was a great deal of in- and out-migration. Clearly migration was the case in American astronomy. Before 1940, astronomy may be classed as a permeable profession (Abbott 1988:130–31) in which careers were flexible. The flexible career is marked by variation in the age of recruitment and educational requirements, as well as frequent on the job training. Flexible careers in permeable professions might be characterized as entrepreneurial. In the context of a permeable profession, individuals have considerable freedom to manage their careers. By contrast, impermeable careers, such as the law or the military, do not show flexibility in the sense discussed here.

All too often scholars and laypersons alike view careers as a process of moving upward through a series of steps, according to a predetermined schedule. From this perspective (From 1968: 252), careers involve prescribed levels of educational attainment for entry level positions followed by "systematic occupational experience" in which each rank "is considered as technical and social preparation" for the next step. This view represents a *linear progressive model.* It is the product of modern bureaucratic culture. Successful careers move ever forward and upward.

The concept of scientific careers developed in this chapter rests on a *social process model.* Social processes are, by nature, messy and indeterminate, driven by complex interactions between chance and necessity. At the level of individual biography, careers appear to ebb and flow, now in one direction, then in another. Evidence drawn from the collective biography suggests that it would be impossible to employ the linear progressive model as a fruitful explanatory device. The careers of individuals are much more contingent than the linear progressive model supposes. Careers may exhibit retrograde motion. Movement through the ranks can veer back and forth; an entry-level position of high rank may be exchanged for lesser rank at an institution offering greater advantages for research. Conversely, low rank at a leading research institution may be traded for higher rank at an institution that provides limited research opportunities.

Diversity seems to characterize career patterns of astronomers. Before 1940, careers in American astronomy were flexible. Patterns of in- and out-migration, truncated or retrograde careers, and careers that moved forward by leaping over ranks today viewed as necessary way-stations, encompass some of the diverse experiences of these scientists. There are significant career differences between elite scientists, the rank and file, and women.

Considered as a social process, careers can be characterized as a series of critical intersections (Pavalko 1987: 161–65). These intersections involve choices; for example, whether to specialize in the old or new astronomy, the type of institution at which to accept employment, and the selection of patrons or sponsors. Responses of individuals at these critical junctures are not uniform; careers develop in a variety of ways. Where the linear model appears nonproblematic, the social process model provides ample room for choice as well as chance.

RUDIMENTARY AND COMPOUND CAREERS

The careers of American astronomers can be grouped under two broad headings: rudimentary and compound. The rudimentary career is, in comparison to the compound career, relatively uncomplicated. Two individuals have been selected to illustrate the range of rudimentary careers. William Chauvenet is an early example. Chauvenet (1820–70) was born in Milford, Pennsylvania, and entered Yale at sixteen. Following an apprenticeship in geodesy and geophysics with Alexander Dallas Bache (1806–

67) and astronomy under Seares Cook Walker (1805–53), Chauvenet was appointed (1841) to teach mathematics, astronomy, and navigation at the Naval Academy. Four years later he was promoted to professor and director of the observatory at Annapolis. An expert in astrometry and celestial mechanics, Chauvenet did much to make European ideas and methods available to American astronomers. Called to Washington University in St. Louis in 1859, Chauvenet left science when he became chancellor of that institution in 1862. A founding member of the NAS, he was its vice president at his death in 1870.

Harold Alden was born in Chicago in 1890. His career, while rudimentary, illustrates a more complicated pattern of development than Chauvenet's. Alden attended Wheaton College (A.B. 1912) and earned a master's degree at Chicago in 1913. He was awarded a doctorate in astrometry and celestial mechanics at Virginia in 1917 and given a faculty appointment. Alden moved through the ranks, serving as an instructor and then assistant professor until 1924, when he was promoted to associate professor. In 1925 Alden moved to Yale. He left an associate professorship at Virginia for the rank of assistant professor at Yale but had the additional title of astronomer in charge of the Yale South African station. Alden had no teaching responsibilities and could concentrate on research.

Compound careers are more complex; the category includes a much wider range of variation. The four individuals selected for discussion illustrate different patterns. These include the failed career (Asaph Hall, Jr.), movement from research to teaching (Blair), an astronomer (Eichelberger) whose career oscillated for more than a decade before stabilizing, and a scientist (Ross) who left astronomy for industry and then returned.

Asaph Hall, Jr., represents the failed career. Born in 1859, Hall grew up in scientific Washington where his father, an astronomer at the United States Naval Observatory (USNO), won acclaim as discoverer of the moons of Mars. Hall attended Harvard (A.B. 1881) and earned a doctorate at Yale in 1889. After graduating from Harvard, he was an assistant at the USNO for three years and then served four years in the same rank at Yale. Hall returned to the USNO in 1889 as assistant astronomer. In 1892 he moved to Ann Arbor as professor of astronomy and director of the observatory. In this position Hall had the opportunity to rebuild astronomy at Michigan, but psychological and intellectual limitations prevented him from meeting the challenge and it remained for others to modernize the Ann Arbor program. The year 1905 found Hall back at the USNO with the rank of assistant. At age forty-six, Hall occupied the same rank in the same institution at which his career had started almost a quarter century before. In 1908 Hall was promoted to astronomer and accepted an adjunct professorship at George Washington University. When he retired in 1930, Hall had little to show in the way of scientific accomplishments.

The career of Gilbert B. Blair (b.1879) illustrates movement away from research to teaching. Blair attended Tabor College (A.B.1902) and earned a master's degree from Washburn University (1904), where he served for a year as an assistant before moving to the Allegheny Observatory at the same rank. For a year (1906–07) Blair was an assistant at the Lick Observatory. The next year he accepted a professorship of astronomy and physics at Morningside College. In 1909 Blair moved to Oregon College as an assistant professor of astronomy and physics and remained in that rank until 1919 when he accepted a call to the University of Nevada. His rank and duties at Nevada were identical to those at Oregon College. In 1923 Blair was promoted to associate professor and in 1936, at age fifty-seven, attained the rank of full professor. At a time when many observatories found it difficult to attract qualified staff members, Blair spent only two years engaged in full-time research. After 1907 his energies were devoted to the classroom at a series of institutions that could hardly be called distinguished.

William S. Eichelberger served for a year as as an instructor at Johns Hopkins, then moved to the Nautical Almanac Office (NAO) as assistant astronomer in 1889. From 1890 to 1896 he was an instructor at Wesleyan University in Connecticut but returned to Washington in 1896, again as an assistant astronomer at the NAO. In 1898 Eichelberger transferred to the USNO, but moved down a step to the position of computer. In 1900 he was appointed professor of mathematics in the Navy, a rank he retained until retirement in 1930. Eichelberger became director of the NAO in 1910. Eichelberger's career was compound, involving a number of twists and turns. His career stabilized only in 1900 when he became a professor of mathematics at the USNO.

Frank E. Ross moved from astronomy to the private sector and then back again. Born in San Francisco, Ross (1874–1960) was educated at Berkeley, earning a doctorate in mathematics in 1901. During his graduate career he served as a teaching assistant at Berkeley and a research assistant at the Lick Observatory. After brief stints at a military school and the University of Nevada, Ross joined the staff of the NAO in 1902. He served as a research assistant at the Carnegie Institution of Washington from 1903 to 1905 and then spent a decade as director of the International Latitude Station at Gaithersburg, Maryland. During these years, he managed to make respectable contributions to science. In 1915 Ross left astronomy for a position as physicist at Eastman Kodak where he remained until he was called to the Yerkes Observatory in 1924. Ross was promoted to full professor in 1928 and elected to the NAS in 1930.

Each of these sketches illustrates aspects of the compound career. Some careers moved forward and then backward, others attained major positions early, while some careers developed relatively late, reaching positions of eminence well past the age of fifty. The compound career could exhibit amazing variety between date of entry and the time an individual retired or left the field.

A Quantitative Approach to Career Patterns

The quantitative study of careers is based on a comparison of first and second positions held and then analysis of the second and the third positions. By position held I mean *institutional location,* not movement within the hierarchy of a single institution. In order to qualify for holding a second position an individual *must move to a new institution.* The first position is further defined in terms of date of entry into the profession. Frequently, *but not always*, it is the first full-time position after completion of highest earned degree (if any). Thus, graduate assistantships are not counted.

Comparisons between positions one and two and then two and three are organized by status (elite or rank-and-file astronomers), using the parameters of cohort and astronomy type (old astronomy or new astronomy). *Position type* is also examined. Positions are classified in terms of the responsibilities they entailed: teaching, research, administration, or a mixture. In addition, a major new parameter is introduced into the discussion: *institutional* potential (IP).

Institutional potential is a measure of an institution's ability to contribute to the research career of an astronomer by providing instruments, colleagues, assistants, a library and adequate funding, as well as time, so that significant research programs can be undertaken. The highest IP was assigned to private research institutions such as the Mount Wilson or Dudley observatories, in which there was state-of-the-art instrumentation, support staff and financial resources that permitted advanced research. Further, at research institutions, astronomers were not distracted by the demands of graduate instruction.

Private universities were ranked just below private research institutions. At Harvard, Chicago, and Princeton there were resources that made it possible to do innovative research. While finances were sometimes a problem, scientists in private universities were not dependent on state legislatures. Both undergraduate and graduate programs frequently made demands on the time of astronomers in private universities, but these institutions could afford to be selective in their admissions policies.

Public universities were assigned a position on the institutional potential scale below private universities. Here funding was more problematic than in private universities and, often, teaching responsibilities came to weigh heavily on individual scientists. At many state universities, the fate of the sciences depended on both administrative leadership and the political and social climate of the state. At the beginning of the twentieth century, the University of Michigan was able to plan for major expansion of its astronomy program, working closely with the state legislature and private donors to secure funding. At other state institutions, such as the University of Missouri, both the campus administration and the political and cultural leaders of the state continued to support models of higher education that emphasized the humanities at the expense of the sciences.

Federal research facilities such as the Naval Observatory and the Nautical Almanac Office are located next on the IP scale. Here research was, for the most part, mission-oriented. Astrometric data were intended for the use of the navy and merchant marine as well as surveyors and mapmakers. Even the field of celestial mechanics was organized to produce practical results. Further, the bureaucratic structure of federal science could inhibit rather than encourage innovation. This was especially true at the USNO, which was under the administrative control of a naval officer, whose understanding of astronomical research was often problematic. Federal science relied on financial support from Congress, in the form of annual appropriations. The appropriations process could be more political and unsettling than dealing with state legislatures. Patterns of out-migration suggest that for many astronomers, government research institutions were not desirable locations. Individuals often moved to more promising locations when the opportunity permitted.

Lowest on the IP scale are colleges (both public and private). Here teaching, rather than research, was the primary order of business. Limited resources were rarely diverted to create research facilities. Even where modest research facilities existed (for example at Dartmouth when Charles Young served as director of the observatory and professor of astronomy) time for research had to be squeezed from a full day of teaching, advising, and committee work. As fields like solar physics matured, the facilities at Dartmouth became inadequate and Young moved to Princeton, attracted in part by the promise of a large refractor. With few exceptions, astronomers at private or public colleges who were active in research produced data of a routine sort. They lacked instrumentation to take part in major astrometric or astrophysical investigations. As with government research facilities, patterns of out- migration suggest that for ambitious astronomers these were not necessarily desirable positions.

Theoretical justification for the metric of institutional potential rests on the exchange theory of recognition and advancement in science developed by Merton ([1960] 1973), Hagstrom (1965) and Storer (1966). This theory is generally accepted in the science-studies community (Ziman 1984: 72–73). As Warren Hagstrom suggested in his classic *The Scientific Community* (1965: 168), "Recognition is given for information, and the scientist who contributes much information to his colleagues is rewarded by them with prestige." This exchange involves, for the most part, papers published in the open literature. Institutions that ranked high on the IP scale provided greater opportunity for the production of scientific knowledge that could be exchanged for recognition. Recognition includes increased resources as well as career mobility. As individuals established distinguished research records they began to move through the reward system of the American astronomical community. The reward system is discussed in chapter 8.

The First Move. The collective biography was searched for individuals who moved from a first to a second position and then for those who went on to a third. For all three cohorts of elite scientists the mean was two moves (i.e., three positions) and for cohorts two and three of rank-and-file astronomers, the mean number of moves after the first appointment was one move (i.e., two positions). Thirty-one percent of elite astronomers held only one position during the course of their careers. The same was true for 72 percent of the rank and file. Among the rank and file, 29 percent moved at least once. For elite astronomers the percentage of those who made at least one move is 69. Here is further evidence illustrating significant differences between elite astronomers and the rank and file. The population that moved from a first to a second position includes 116 elite astronomers and 175 members of the rank and file.

Let us first consider the experience of elite astronomers. Was there a relationship between moving to a second position and the type of astronomy they engaged in? For the cohort that entered in or before 1859, there was no change in the type of astronomy (old, new, or mixed) as they moved to second positions. For cohorts two and three, the number working in astrophysical research institutions almost doubled as a consequence of moving. The data do not permit us to disentangle employers' demands from the interests of employees, but the pattern is clear.

As elite astronomers moved from a first to a second position, interesting trends emerge concerning institutional potential. For cohorts one and two, the first move entailed leaving private research institutions. Only in the third cohort did the pattern change. For members of cohorts two and three, private universities became increasingly desirable for second positions and, after 1900, federal research institutions became less attractive. The number of elite astronomers moving to colleges steadily declined after 1860.

Another perspective is provided through a discussion of position type. For elite astronomers in the first cohort, there was little significant change as they moved. For the 1860–99 cohort, however, there was a decline in second positions classified as either teaching or research, with the most striking loss being in teaching. The number of second positions that included both teaching *and* research increased considerably, presumably because of those who moved into academic settings. In the third cohort of elite scientists there was a significant move away from teaching as a result of taking a second position.

For the first cohort of elite astronomers, the tendency was to move to a similar type of institution. When second positions are compared with first, virtually the same number were working at observatories as in academic settings. The second cohort exhibits different patterns. Academic employment increased by approximately 20 percent with the greatest gain at the level of full professor. There was also substantial growth in the number who moved to a second position as an observatory director. Elite astronomers who entered after 1900 showed no change in the number working in observatory or academic settings. Within each category, the greatest increase occurred in the ranks of professors and astronomers. This suggests a relationship between moving to a second position and movement up the academic ladder.

The experiences of the rank and file indicate different patterns as a result of the first move. Across three cohorts, there was relatively little change in astronomy type. For this group, decisions concerning research (the old or the new astronomy) apparently were made early in the career. Unlike elite scientists, there is no indication that the rank and file were drawn to astrophysics as a result of taking a second position. In each cohort there was movement away from private research institutions. It was as if rank-and-file astronomers (and their initial employers) found that full-time research, often in highly competitive situations, was not the kind of work for which they were best suited. In cohorts two and three, the rank and file found positions in private universities, while the number who moved into public universities declined between 1860 and the end of the century. After 1900 this pattern was reversed. In the twentieth century there was also a decline in the number of rank-and-file astronomers who obtained second positions in government research institutions. There was an increase in those who moved to colleges in the second cohort, but their numbers fell after 1900.

In a comparison of the experiences of elite and rank-and-file astronomers as they moved to a second position, interesting differences emerge. Recall that the elite tended to remain in either observatory *or* academic settings as a result of the first move. By contrast, the rank and file tended to leave full time research and move to academic settings. For the rank and file, this proved to be a significant sorting process, involving changes in career direction. Employment in an observatory generally provided greater opportunities for research, and therefore recognition, than did many academic venues.

The Second Move. This population (sixty-five elite astronomers and seventy-three who were members of the rank and file) was about half the size of the group who made a first move. Conditioned by high rates of job mobility that were the norm for several decades after World War II, readers may find this trend remarkable. Before World War II, job mobility presented a very different pattern from that which obtained later.

Across three cohorts, elite astronomers were seldom tempted to change research fields as a result of moving to new positions. The research interests of these elite astronomers were fixed relatively early in the career and pursued single-mindedly, as they moved from one position to another. As elite astronomers made decisions about moving to a third position, the most attractive option was the private research institution. In the second cohort, it appears that private universities lost out to the new state universities, while in the third no elite astronomer accepted employment in a college. Overall, there was a tendency for employment in federal research institutions to slow down and, after 1900, decline as a result of moving to a third position.

Between 1860 and 1940, changes in position type (teaching, research, administration, or a combination) for elite astronomers were more frequent than for the rank and file and probably more significant for career development. The types of institutions in which these elite astronomers found themselves as a result of a second move mirror changes in position type. As members of cohort one moved into a third position, there was no change between academic and observatory locations. In the second cohort, however, there was a net loss in observatory employment, while academic employment grew by thirteen percent. This included an increase in the number of professors who also held the post of observatory director. For the third cohort this trend was reversed. After 1900, observatories gained (83 percent) as a result of the second move.

As rank-and-file members of cohorts one and two moved to a third position, there was virtually no change in the type of astronomy in which they were engaged. Only after 1900 did members of the rank and file find that a second move sometimes entailed shifting the focus of their research. In the first two cohorts, position type remained relatively constant for rank-and-file astronomers. After 1900 there was a slight increase in those who emphasized research to the exclusion of all other duties. From 1859 to the eve of World War II, members of the rank and file tended to remain in either academic or observatory positions as they moved into a third position. On balance, it appears the rank and file continued an established pattern in the move to a third position.

Comparisons between the elite and the rank and file who made second moves (that is, accepted third positions) are illuminating. Elite astronomers entered private research institutions in larger numbers than did the rank and file. For both groups, the total affiliated with public universities increased in the second cohort and fell in the third. After 1860, federal service became progressively less attractive for elite astronomers but more attractive for the rank and file. While rank-and-file astronomers who selected colleges for a third position declined across all three cohorts, they were still working in these institutions after 1900. No elite members of the third cohort held positions in colleges as a consequence of the second move.

This discussion suggests that elite astronomers demonstrated a keener sense of direction and purpose in moving from a first to a second or third position than did the rank and file. The elite tended to leave colleges and federal research institutions as a consequence of moving to a new position. The pattern of career differentiation between the two groups was most obvious after 1900, when the elite became deeply involved with astrophysical research at universities and private research institutions.

CAREER DURATION

Two topics are considered under the heading of duration. The first is career length and the second time in grade. Variation in career length serves to emphasize differences in the careers of the elite, rank-and-file astronomers, and women. Table 5.1 provides information on career length for each group. The left-hand column indicates the percentage of the population that had already left the field and the numbers in the columns to the right represent the professional age of the population.

The professional longevity of elite astronomers is the most striking fact revealed in table 5.1. At the end of approximately twenty years, 90 percent of the elite remained in the field. After about the same length of time, more than 70 percent of the rank and file were gone from the profession. For women, the differential was even greater. After twenty years less than 20 percent of the women remained. At no point do these data suggest the rank and file or women ever narrowed the gap between themselves and elite astronomers. Even at the mid-point, when half the original population had left astronomy, the differences remain immense. Thirty years separate the elite from the rank and file and thirty-three years for women. After twenty-nine years, only 10 percent of the women remain as compared to thirty-four years for the rank and file and fifty-six years for the elite. These differences are staggering. They help explain the power of the elite. While members of the other groups left astronomy, elite astronomers remained, persevered, and achieved.

Some data on time in grade were presented as part of the discussion of rudimentary and compound careers. The following discussion is based on an analysis of elite astronomers. The elite class was broken down into its basic components: those whose highest form of recognition was a star in *AMS* and those who were elected to the NAS. These groups were then divided by academic and observatory rank.

Table 5.1 Length of Career

Percent Left Astronomy	Professional Age (Years in Astronomy)		
	Elite	Rank & File	Women
10%	19	1	< 1
20%	23	2	< 1
30%	30	3	2
40%	35	5	3
50%	38	8	5
60%	40	12	9
70%	44	18	12
80%	49	26	18
90%	56	34	29

NOTE: Using the elite as an example, 10 percent had careers that lasted nineteen years, while 90 percent were still working in astronomy at professional age nineteen. For women, 30 percent had careers that lasted less than one year and 50 percent were gone from astronomy after careers that lasted five years.

Twenty-two astronomers destined for *AMS* stars and twenty-four who would be elected to the NAS began their careers with the rank of instructor. Those who went on to the Academy remained in grade for 3.5 years while the mean time in grade for the other group was 4.6 years. The same pattern holds true for the six *AMS* star awardees and three NAS scientists who started as assistant professors. The mean years in grade for the NAS group were 3.6 compared to 6.5 for the other. At the level of associate professor similar trends are evident, with those who had NAS potential remaining 4.6 years as opposed to 5.1 years for individuals whose honors peaked with *AMS* stars. These data indicate that NAS-bound astronomers moved through the academic ranks more rapidly.

Twelve individuals destined for *AMS* stars and eighteen who were bound for the Academy started their careers as assistants or computers in observatories. The NAS group remained in grade for 6.8 years while the others moved after 6.2 years. Here the pattern is reversed, but the difference is not great. Thirty-three *AMS* star candidates and four NAS astronomers began as assistant astronomers. The NAS astronomers remained in grade for 6.9 years, a full year less than the *AMS* group. No elite astronomers started their careers at the rank of associate astronomer. It is interesting to note that five of the *AMS* group and six NAS-bound individuals began as astronomers. Six scientists who would later attain NAS status began their careers as observatory directors. No *AMS* star candidates entered astronomy at that level.

One further indication of differentiation at the elite level is found in data on initial appointment as full professor. Four astronomers who would attain *AMS* stars and fourteen who were eventually elected to the Academy began their careers with the rank of full professor. This must have provided significant advantages on the road to *AMS* stars and/or Academy membership.

MARKET CONDITIONS

The Ph.D. Pool

The analysis of scientific careers requires a knowledge of market conditions. Both the supply and demand side of the equation should be considered. What follows is a qualitative overview of demand based on manuscript sources. It is not easy to develop a quantitative picture of the Ph.D. pool. The metric adopted here is Ph.D.'s who entered astronomy. This figure approximates, *but does not equal* Ph.D. production in astronomy. To be sure, this metric does not take into account the pool of potential astronomers who did not earn the Ph.D. That population is discussed later.

While it is probable that a degree of ritual hand-wringing was associated with the search for qualified young scientists, material discussed here suggests a much deeper level of concern. In the first decade of the twentieth century senior astronomers frequently remarked on the fact that "Good men to fill subordinate positions are very scarce."[1] Lewis Boss, director of the Dudley Observatory, reported to Simon Newcomb, former director of the NAO, who maintained a research staff even after retirement, "I do not know of any astronomer unemployed who would be suitable for the use you describe."[2] Boss indicated that in spite of a major grant from the Carnegie Institution of Washington that guaranteed the stability of the observatory, he had been unable to attract capable young assistants. Boss and Newcomb were not alone. The superintendent of the USNO, Admiral Colby M. Chester, commented that the Naval Observatory was "trying to secure men . . . for its own staff, with very poor results."[3] It did not appear that money was the issue. The admiral believed the USNO pay scale was competitive. From Illinois, Joel Stebbins informed William Hussey at Michigan concerning his desire for an assistant with a fresh Ph.D., for whom he would pay at least $1,000, but feared that given the "present prospects we may have to pay more to anybody at all." Stebbins reported

1. William Wallace Campbell to Simon Newcomb, 16 March 1903, p. 2. SALO, Newcomb Papers.

2. Lewis Boss to Simon Newcomb, 6 April 1903. LC, Newcomb Papers.

3. Admiral Colby W. Chester to William Wallace Campbell, 23 February 1904. SALO, Campbell Papers.

to director Campbell at the Lick Observatory that as far as he knew, only two individuals were awarded the Ph.D. in astronomy in 1907 and that competition was keen.[4] Joel Stebbins attributed difficulty in finding new recruits to boom times. "Most of our good students with mathematical ability seem to choose engineering."[5]

The supply did not improve during the second decade of the century. As Stebbins lamented, "Is it not striking how many openings there are for men to start in astronomy?"[6] Each year, observatory directors and department heads found themselves in competition for a very limited supply of candidates. William Wallace Campbell suspected that the problem could be traced to the policies of many college and university presidents, who were offering appointments only to candidates with the degree in hand. Thus graduate students were spending more time in school before entering the job market.[7] Some institutions even encountered difficulty recruiting graduate students. Princeton's Henry Norris Russell decried the fact that there were no good applicants for graduate fellowships.[8]

It is not clear what impact World War I had on the market. American involvement in the war was brief (nineteenth months) and demobilization rapid. Be that as it may, senior astronomers were singing the same old refrain in the early 1920s. Harlow Shapley, newly appointed director of the Harvard College Observatory (HCO) reported to George Ellery Hale that he was finding it hard to attract "clever and youthful astronomical collaborators here."[9] Caroline E. Furness, professor of astronomy at Vassar, lamented the fact that "There are not many graduate students in astronomy, including both men and women."[10] Edwin B. Frost, director of the Yerkes Observatory, speculated on the causes of the shortage. He believed that salaries in astronomy were not keeping pace with the cost of living. Campbell told Frost that astronomy was not the only science finding it difficult to recruit students. Mathematics and physics at

4. Joel Stebbins to William Hussey, 25 April 1907. UWMA, Stebbins Papers, and Stebbins to William Wallace Campbell, 20 June 1907. SALO, Stebbins Papers. At Northwestern University, the committee searching for a replacement for George W. Hough as director of the Dearborn Observatory were also finding market conditions discouraging. See Ormond Stone to Henry Crew, 15 February 1909, NWUA, Crew Papers.

5. Stebbins to Campbell, 3 June 1909. SALO, Stebbins Papers.

6. Stebbins to Hussey, 14 February 1910. UWMA, Stebbins Papers.

7. William Wallace Campbell to Philip Fox, 18 May 1914. SALO, Fox Papers.

8. Henry Norris Russell to William Wallace Campbell, 11 April 1916. SALO. Russell Papers.

9. Harlow Shapley to George Ellery Hale, 3 October 1921. Copy in Adams Papers, MTWA.

10. Undated fragment, Furness Papers, VCA. Probably 1920 or 1921. Harlan True Stetson, director of the Astronomical Laboratory at Harvard University, reported the same situation. Stetson to William Wallace Campbell, 23 February 1921, SALO, Stetson Papers.

Berkeley were also having problems. Chemistry continued to attract students, Campbell said, because they believed well-paying jobs were available.[11] In 1923, with demand brisk and no apparent growth on the supply side, Harlow Shapley indicated that some institutions were so desperate that they were offering large salaries, but with little apparent success.[12]

The first signs of change appeared in 1924. While lamenting how "extremely difficult" it was "to get young men to enter astronomy at the present time," the associate director of the Lick Observatory, Robert G. Aitken, indicated there were several very talented individuals doing graduate work at Berkeley. "We hope to be able to hold on to some of them."[13] The situation turned around in a relatively short time. By 1929 the influx of students into the Berkeley graduate program was taxing available resources. There was a surplus of applicants for teaching and research fellowships and some qualified individuals had to be denied support.[14] Ten years later it was apparent that the American astronomical community had moved from a supply-side crisis to a crisis of demand. In one short decade, Ph.D. production caught up with and then outstripped demand.

Data on Ph.D.'s who entered astronomy confirm the trends inferred from literary sources. During the decade of the 1880s, Ph.D.'s taking first jobs in astronomy grew by 200 percent. Gains of approximately the same order of magnitude were registered during the 1890s. For the first decade of the new century, this indirect measure of Ph.D. production declined by more than half its previous value and for the next twenty years (1910–30) remained virtually flat. It would appear, from this metric, that Ph.D. production stalled-out after 1900 and remained flat until the end of the 1920s.

Material discussed in chapter 4 helps to explain these patterns. The decline of the pre-1900 Ph.D. programs (Yale, Columbia, Virginia, and Hopkins) was balanced by new graduate departments at the Universities of California and Chicago. It would seem, however, that the combined output of these new departments was not sufficient to meet demand for the first thirty years of the new century. During the 1930s, Ph.D.'s entered astronomy in numbers that were about 70 percent ahead of the figures

11. Edwin B. Frost to William Wallace Campbell, 19 April 1922 and Campbell to Frost, 25 April 1922. SALO, Frost Papers.

12. Harlow Shapley to Robert G. Aitken, 6 July 1923. SALO, Shapley Papers.

13. Robert G. Aitken to John A. Miller, 18 September 1924. SALO, Miller Papers. Aitken reported the same situation to Edwin B.Frost, 29 January 1924. SALO, Frost Papers. See also Aitken to Henry Norris Russell, 9 March 1926, and Aitken to Edward A. Fath, 20 March 1926. SALO, Russell Papers and Fath Papers.

14. Robert G. Aitken to Caroline E. Furness, 21 January 1929. SALO, Furness Papers.

for the period from 1910 to the end of the 1920s. Graduate programs at Harvard, Michigan, and Princeton were primarily responsible for this increase. After 1900, increasing demand coupled with problems of supply, stimulated the importation of young astronomers from Europe.

The Non-Ph.D. Pool

Given available data, there is only one way to develop a measure of the non-Ph.D. pool. For 54 percent of the elite and 14 percent of the rank and file and women combined, data exist on the occupational location of these individuals *before* they assumed a first position in the American astronomical community. Given the limited data, the figures should be seen as a rough approximation of the size and ranking of the several sources from which astronomers were recruited. While the combined data represent just under 20 percent of the population of the American astronomical community, information on the elite accounts for more than half the cases and, for elite astronomers, may be considered virtually complete. We can say that, for the elite, just over half began careers outside astronomy and were recruited into the science.

For all three cohorts of elite scientists, those recruited from secondary school teaching formed the largest single group. Almost 43 percent of those recruited into astronomy in the first cohort came from secondary school teaching. The second largest group (32 percent) were drawn from various military-related sciences including topographic mapping, hydrography, and the like. Professors of mathematics ranked third (14.3 percent) as a source from which elite-bound astronomers were recruited in the first cohort.

Secondary school teaching continued to top the list (23.5 percent) of locations from which elite astronomers were recruited in the period 1860–99. College or university professors of mathematics (17.6 percent) and physics (15.7 percent) ranked second and third. Engineering (13.7 percent), legal or business careers (11.8 percent), and military-related sciences (9.8 percent) round out the leading sources. Even in the post-1900 cohort, secondary school teaching dominated (45.5 percent) followed by engineering and mathematics (18 percent each).

Between 1859 and the coming of World War II, the percentage of elite astronomers who began their professional lives outside astronomy declined. For the first cohort the figure stood at 88 percent, fell to 65 percent of the second cohort, and was only 18 percent of those who entered after 1900. The pattern suggests the growing importance of credentialling for those who would some day be ranked as members of the elite.

For first cohort of rank and file and women for whom there is information, 35 percent were recruited from outside. Military related sciences led the way, accounting for a third of the number. Engineering, mathematics, secondary school teaching, and legal or business careers each account for about 17 percent of this group. Available data indicate that approximately 15 percent of the women and rank and file in the second cohort were recruited from other occupations. A full 40 percent came from secondary school teaching while 34 percent were called from college or university posts where they taught mathematics. Physics ranked a distant third (8 percent) followed by engineering and legal or business careers with 6 percent each. Available records suggest that just over 10 percent of the post-1900 cohort of rank-and-file males and women entered from outside astronomy. Individuals recruited from mathematics departments (45.8 percent) account for the largest share followed by secondary school teaching (25.3 percent) and physics professors (19.3 percent). For the elite, rank and file, and women, time spent outside the astronomical community before recruitment ranged from less than one year to more than twenty.

Assuming these data reflect the general contours of the population from which non-Ph.D. astronomers were drawn, they are suggestive. Secondary teaching provided a major and continuing source of elite astronomers as well as members of the rank and file and women, even into the twentieth century. This cannot be explained by the linear progressive model of scientific careers with its assumption that Ph.D. credentialling was standard practice even before the end of the nineteenth century. The role of military-related sciences as a seed bed for future astronomers underscores the centrality of the federal government as a leading patron of nineteenth-century science. The importance of mathematics as a significant component of the pool after 1860 is another interesting finding. Assuming that most of these individuals went into the traditional fields of astrometry and celestial mechanics rather than astrophysics, it suggests that formal training in astronomy did not necessarily play a critical role in shaping the careers of those who worked in the old astronomy. Further, it is surprising that physics played such a relatively minor part. It might be expected, given the need for cross-fertilization in astrophysics, that trained physicists would have been in greater relative demand. Finally, the distribution by status is interesting. Both mathematics and physics played a greater role as sources for the rank and file and for women than for the elite.

This discussion of non-Ph.D. sources from which astronomers were recruited provides further evidence of the permeable nature of careers in the American astronomical community in the years before World War II. There was substantial recruiting from a variety of scientific and nonscientific occupations Clearly, the linear progressive model with its rigid demands for education and prescribed career patterns simply does not reflect the realities of careers in American astronomy.

Employment Patterns

Forty-nine institutions were grouped into four categories and then analyzed in order to establish changing employment patterns. The four categories are factory observatories (10), major universities (10), second-order universities (16), and colleges (13).[15] Using a simple calculus of gains and losses in staff size, each category was examined by decade and assigned to one of three classes: growth, decline, or static. The resulting patterns suggest the overall contours of employment during the decade. To be sure, these forty-nine institutions do not exhaust the list of those providing employment for astronomers. The list does, however, include the most important institutions and can be considered exhaustive for the categories of factory observatories and major universities.

Because the decade 1859–1868 serves as the benchmark, the discussion begins with the period 1869–78. Two categories (major universities and colleges) registered growth during this decade while second-order universities remained stationary and factory observatories declined in staff size. During the decade 1879–88 important changes were under way. The factory observatories entered a thirty-year period (1879–1909) of sustained growth, while the major universities began two decades (1879–98), during which the size of their astronomy departments and observatory staffs remained virtually level. Second-order universities, however, grew during the ten years beginning in 1879 and so did colleges. Both colleges and second-order universities entered a static period following 1889 but grew again in the decade after 1899. The early years of the new century were also a time of growth for the major universities after a twenty-year holding period.

The years 1910 to 1919 were not good for American astronomy. Employment in factory observatories and second-order universities declined and remained level in the major universities and colleges. The years of selective prosperity after World War I witnessed growth for the universities and colleges, but the great research factories found themselves faced with limited resources that did not permit growth. During the 1930s the factory observatories experienced a decade of decline and colleges could do no more than remain static. However, at both the major and second-

15. The ten factory observatories include Allegheny, Dudley, HCO, Lick, Lowell, Mount Wilson, NAO, Smithsonian Astrophysical, USNO and Yerkes. Under the heading of major universities are listed California, Chicago, Columbia, Northwestern, Harvard, Michigan, Princeton, Texas, Virginia and Yale. The sixteen second-order universities include Arizona, Brown, California Institute of Technology, Case, Cincinnati, Connecticut Wesleyan, Cornell, Illinois, Indiana, Johns Hopkins, Minnesota, Missouri, Ohio State, Ohio Wesleyan, Pennsylvania, and Wisconsin. The final category includes thirteen colleges: Amherst, Carleton, Central High School in Philadelphia, Dartmouth, Georgetown, Hamilton, Swarthmore, and Williams as well as Mount Holyoke, Radcliffe, Smith, Vassar, and Wellesley.

order universities, the Depression decade saw astronomy grow. These employment trends rest on changing funding decisions and thus provide a rough measure of policy objectives as four types of institutions made choices about the allocation of resources for research and teaching programs in astronomy.[16]

CAREER MANAGEMENT

Moving to a Second Position from a University

From data drawn from the collective biography, individuals were identified whose first position was at an institution representing each of the three groups of Ph.D.-producing schools discussed in chapter 4. Yale was selected to represent the declining institutions, Michigan those universities that were moving to the forefront in Ph.D. production, and the University of California at Berkeley as an institution that had long been the leader in granting astronomy doctorates. Since the Lick Observatory is so closely connected with the graduate program at Berkeley, it also is examined. The discussion provides insights into career management as individuals made decisions concerning second positions.

This population was divided into two groups: those who spent their professional lives at the institution where they received their first appointment and those who moved on to a second job. The data were analyzed in ways that permit us to see the calculus of gains and losses involved in the move to position two.

Between 1859 and 1940 eighteen individuals started their scientific careers at Yale (fourteen men and four women). Although the statistics are differentiated by gender, a full discussion of women will be reserved for chapter 9. Four men and one woman left Yale for a second position. This gives the institution a mobility index of .28. The mobility index (MI) is simply the number who moved divided by the total. Three women and ten men spent their entire careers at Yale. Four astronomers destined for elite status found first positions at Yale and three never moved. Of the ten rank-and-file astronomers who started at Yale, 30 percent moved on. Of the mobile men whose careers began at Yale, one moved to an institution with greater institutional potential (IP), while three lost IP as a result of moving. Two, however, traded IP for rank and two made moves that involved no change in rank. For this group, the mean number of moves

16. Geiger (1986) provides valuable discussions of the growth and funding of scientific research in the twentieth century. Kohler (1991a) examines foundation support for research in the natural sciences from 1900 to the end of World War II. After World War I, the CIW found it difficult to keep funding for the Mount Wilson and Dudley Observatories at an adequate level. For many years the Mount Wilson budget was flat and the CIW developed plans to phase out support for the Dudley as soon as the great catalogue was completed.

after the first job was two, for a total of three positions over the course of a career.

Between 1859 and 1940 a total of twenty-one individuals began their careers at Michigan (twenty men and one woman). Seven moved on, giving a MI of .33. All of the astronomers who achieved elite status before 1940 left Michigan. In this sense, the department can be seen as one whose fortunes rested on its rank-and-file members. Before World War II, Michigan served as a way station for those on the way to elite status. Of those who were outward-bound from Michigan, two improved their situations by moving to locations with a higher IP; but four traded down, losing IP but gaining rank, and one made a lateral move. Five of the seven gained in rank as a result of the move and two made parallel moves. For this population of mobile astronomers the mean number of moves after the first job was 1.6, a figure in accord with data presented earlier.

Seven men and two women had their first jobs in the University of California at Berkeley astronomy department. Two men and one woman moved on giving a MI of .33. California managed to retain all three beginners who would later achieve elite status. Of the mobile males, one gained and one lost IP as a result of moving to a second job. In one case the trade-off involved improved rank while the other made a parallel move. The mean number of moves after the first position was two for this group.

The Lick Observatory is closely related to the graduate program at Berkeley, providing research facilities and dissertation supervision for students working in observational astrophysics and astronomy. To be sure, not all the Lick staff carried an equal share of responsibility. Thirty individuals (twenty men and ten women) started their scientific careers as members of the Lick staff. Twenty (seven women and thirteen men) moved on. This yields a MI of .67, the largest of any institution considered in this section. The outward-bound included two future members of the elite and eleven rank-and-file astronomers. Lick was able to retain 60 percent of those beginners who would one day be numbered in the elite and a quarter of those who occupied rank-and-file status. The mobile population averaged two moves beyond the first position. Trade-offs between IP and rank are clear in the case of those who left the Lick. Nine of the thirteen lost IP but improved their rank. Four made parallel moves to institutions of similar IP and at a similar rank.

These data indicate that those who moved to a second position frequently opted for improved rank at an institution less well equipped for research. Career management involved weighing alternatives. Before World War II, for example, ambitious young astronomers destined for elite status moved from Michigan to better their chances to do research. In comparison, California was able to hold all the beginners who would one day achieve elite status. This is not sufficient to explain the preeminence of Berkeley as the leading producer of astronomy doctorates after 1900, but it does account for one aspect of its strength.

Moving to a Second Position from a Research Institution

Four research institutions have been selected for discussion. These include the Mount Wilson Observatory, the U.S. Naval Observatory (USNO), the Nautical Almanac Office (NAO), and the Harvard College Observatory (HCO). Before the mid-1920s, the HCO was a research institution and never lost that function, even after some of its staff became active in graduate and undergraduate education. Thus, the HCO has been included.

The NAO was the site of first jobs for forty-four individuals (forty-two men and two women) between 1859 and 1940. The mobile population (fourteen men and one woman) produced an MI of .34, the highest for any institution considered in this section. The NAO lost 83 percent of those beginners who later would achieve elite status. For the rank and file, the turnover rate was 25 percent. The mean number of positions after beginning at the NAO was 1.7.

Of the mobile population, five improved their opportunities for research and one sacrificed IP. For eight the move to a second position was lateral. Four improved, but four lost, rank. Six made a parallel move in rank. Five of this group were willing to sacrifice either rank or IP in order to move while eight made parallel moves to institutions of similar IP and six retained the same rank. The pattern of parallel movement is explained, in part, by the tendency of NAO personnel to move to the USNO. This change did not generally improve their rank and made no difference in IP. It would seem, however, that many believed the observatory provided more opportunity for career development than the Almanac Office.

One-hundred and ten individuals began their careers at the USNO (106 men and four women), and eighteen men moved on to second jobs. For the USNO the MI is .16. A significant number of elite-bound beginners left the USNO (66.7 percent) in search of better opportunities, but only 11 percent of the rank and file followed a similar course. The mobile population averaged two moves after initial appointment at the USNO.

While there are some missing data, twelve of those who moved sacrificed IP, but nine attained a higher rank. Only one of the group was able to move to an institution with a higher IP. Four individuals made parallel moves in terms of IP and five moved to other institutions with the same rank. Three sacrificed rank in order to move. Since a loss in IP can only mean that these individuals moved into colleges, we can infer something about their career management decisions. It would appear that for twelve of the eighteen, college teaching, with its limited opportunities for research, was preferable to remaining in a federal research institution.

Harvard was the site of the first job for 185 individuals (116 men and 69 women). Twenty-three (18 men and 5 women) moved on to a second position yielding an MI of .12. The mobile population included 39 percent of elite-bound beginners and 11 percent with rank-and-file status. The Harvard operation rested on rank-and-file scientists, plus a larger number of elite astronomers than at any other institution considered in this section. The average number of moves after initial appointment for this group was 1.8.

Of the mobile men, twelve gave up IP, but nine gained in rank as a result of leaving Cambridge. Four made parallel moves to institutions of similar IP and seven to second jobs at the same rank. The picture is not complete because of missing data. However, it does appear that the pattern in moving from Harvard was to trade up in rank at the expense of IP.

Sixty-nine individuals (17 men and 52 women) had their first job at Mount Wilson. Fourteen (evenly divided by gender) moved on. The MI for Mount Wilson is .20. Of the group of beginners destined for elite status only one (out of five) moved, while half the rank and file secured second positions. The average number of moves after the first position for this group was 1.5. Mount Wilson, standing at the top of the IP scale, was an impossible act to follow, so it is not surprising that all the mobile men lost IP. Five of the seven, however, improved their rank as a result of the move while two lost. No one made a lateral move.

Data discussed in this section indicate several different patterns for those who sought a second position in moving from a research institution. There seemed to be a sense of urgency in the move from the NAO or USNO. Frequently, individuals were willing to accept either parallel moves or a loss of IP. Moving from Harvard suggests a different career calculus. IP was sacrificed for improved rank. At Mount Wilson the pattern was again different. Here it was the rank and file who moved on while elite-bound beginners remained. Those who moved could not, by definition, achieve a higher IP, but they frequently improved their rank. Depending on the type of institution to which they were initially appointed, a variety of career management options were open to individuals.

Moving to a Second Position from a College

Fifty-four astronomers (twenty-two members of the elite and thirty-two who belonged to the rank and file) moved from a first job in a college to a second position. Since the IP assigned to colleges is the lowest ranking on the Institutional Potential scale, the move to a second position from an initial appointment in a college must reflect relatively unambiguous decisions concerning the management of a career. Here differences between elite and rank and file are striking.

For the first cohort of elite astronomers, who entered the profession in or before 1859, the experience of moving to a second position from a first appointment in a college was complex. To be sure, in the age of preprofessional science, opportunities were limited. Two gained in rank and two lost as a result of moving to a second position. Four individuals made lateral moves at the same rank. Six moved to other colleges, while only two managed to improve their chances to do significant research by going to institutions with greater IP.

The pattern shifted dramatically for the second cohort of elite scientists, who entered the American astronomical community between 1860 and the end of the nineteenth century. Two lost rank as a consequence of the move; nine, however, improved their rank. The most significant aspect of the experience of this group of mobile astronomers involves IP. In all but one case, these individuals moved upward, gaining resources for research.

Four elite astronomers began their careers in colleges in the period between 1900 and the eve of World War II. Three were willing to sacrifice rank in moving to a second institution, and one gained. In all four cases the move was to an institution that provided greater opportunities for research. Clearly, for the second and third cohorts of astronomers who would achieve elite status, the imperative was to secure a second position that provided resources for research. Many were willing to trade rank for greater IP or to make a lateral move in order to gain resources necessary to assure a research career. For these three cohorts of elite astronomers, the average number of moves was slightly greater than the median discussed earlier. This suggests that some elite astronomers had difficulty attaining a position that provided adequate resources and continued the search to a third position and beyond.

The experience of the rank and file differed from the elite. In cohort one, only two rank-and-file astronomers moved from a first position in a college. Both made career decisions that took them to the same type (IP) of institution and at the same rank. Thirteen individuals in cohort two moved to a second position from a college. Four improved their rank, two lost as a result of moving, and seven made parallel moves. Six of the thirteen, however, did improve their chances for research, while for seven of the group, the move was to an institution with the same IP. These data can be interpreted as career choices between teaching and research. The group appears to be evenly split between the two as they selected second positions.

The number of rank and file who sought to move from an initial appointment in a college increased for the third cohort. Of this group six improved their rank, two lost, and nine occupied similar rank as a result of taking a second position. More strikingly, eight improved their ability to engage in research while nine moved to a second position at a similar institution. The nine who moved to another college were, in effect, selecting teaching over research while the eight who achieved a higher IP indicated greater interest in research. The number of positions held by this group of rank-and-file astronomers was somewhat larger than the average discussed earlier. This suggests that, like the elite, these rank-and-file scientists took longer to find satisfactory positions.

Comparisons between elite and rank-and-file astronomers who moved from colleges to a second position are instructive. Fewer elite (N=22) than rank-and-file (N=32) scientists began their careers in college settings and more elite astronomers moved to a second position offering greater potential for research. The number of elite scientists whose first jobs were in colleges declined sharply after 1900, while for the rank and file the figure grew. Clearly, an initial appointment at a college was considered a disadvantage by astronomers who would one day either be starred in *AMS* or elected to the NAS. Most moved to second positions that provided greater research opportunities. On the other hand, rank-and-file astronomers appeared about evenly divided as they selected second positions in similar collegiate settings or moved to institutions that offered greater opportunities for research.

Leaving Astronomy

One hundred and thirty-six individuals (11.3 percent of the population) left astronomy for other occupations. This includes 35 elite scientists and 101 members of the rank and file and women. For the elite, the greatest exodus (21 individuals) occurred during the second cohort, and for women and the rank and file the largest number left (69) between 1900 and 1940. Economic motives appear important in explaining why individuals left astronomy, but they do not tell the whole story. If some sought greater financial rewards in business, there were others who believed their talents would be more fully utilized in physics, geophysics, mathematics, or meteorology. A few sought positions of power and status as deans and presidents of colleges and universities.

Of the cohort that entered in or before 1859, eight elite scientists left astronomy. Three went into academic administration while two moved into physics or geophysics. The rest were divided between business and military or government-funded science. The second cohort witnessed the largest out-migration of elite astronomers. Eight left for careers in business and seven for government service in sciences other than astronomy. Two each entered physics or geophysics and academic administration. Politics and secondary school teaching attracted the other two. Of the six members of the elite who left astronomy after 1900, two opted for business careers and the rest were equally divided between physics or geophysics, government-related science, planetarium work, and secondary school teaching. On balance, the tendency was for elite scientists to move into other sciences or academic administration.

Among the larger population (rank and file plus women) only two individuals left astronomy from the first cohort. One entered business and the other became a military scientist. The number rose to thirty for the 1860–1899 cohort. Ten moved to careers in business, while five entered government service in various sciences not related to astronomy and five migrated to physics or geophysics. Four of those who left during this period took up secondary school teaching and three sought teaching or research posts with the military. Two left astronomy to be married and one to take up a career as an academic administrator.

Between 1900 and the eve of World War II, sixty-nine members of this group left astronomy. As with the previous cohort, the largest loss (twenty-five individuals) was to business with sixteen shifting to careers in physics or geophysics. Eleven obtained posts as administrators in colleges or universities and eight left the field because they married. Four moved to government-related science careers and three took up secondary school teaching. One entered politics and one became involved in planetarium work. For the second and third cohorts, business ranked as the most attractive alternative to a career in science, with about a third of those who left entering the world of trade and commerce. If we add those who moved to academic administration, the figure swells to more than 50 percent.

When these groups are compared, it appears the elite had a higher percentage of out-migration (21 percent) than did women or the rank and file (9.7 percent). Apparently, those destined for membership in the Academy and/or stars in *AMS* were more willing to experiment with a variety of careers. Further, it appears that the elite frequently moved to other science-related careers, rather than leaving science altogether, as was the case with the rank and file or women. Some of these out-migrants thought better of their decision and returned to Urania's domain.

Returning to Astronomy

Of the 136 astronomers who left for other occupations, forty-three (31.6 percent) reconsidered and returned. This included seventeen elite astronomers and twenty-six members of the rank and file and women. Those who came back to astronomy included almost half the elite out-migrants, but only a quarter of the rank and file and women.

Of elite astronomers from the first cohort who left and then returned, two resigned positions as military scientists, while the other two left business and academic administration in favor of astronomy. For the most part, the profiles of their reentry positions matched the jobs held before leaving astronomy. One important difference is that no one in this or later cohorts returned to positions in private research institutions. For the second cohort of elite scientists, eleven out-migrants returned. Their activities outside astronomy suggest no clear patterns. The majority returned to research positions in academic settings. This contrasts with the situations from which they left astronomy, primarily in government research institutions and colleges. Apparently, the returning group that would one day achieve elite status managed to improve their careers as a result of having been out of astronomy. Only two elite astronomers returned after 1900.

What is most striking about the experiences of the second (17) and third (9) cohorts of women and rank and file who came back is that, unlike elite astronomers, they did not improve their careers by returning. For the second cohort, about the same number returned to appointments in colleges and government research institutions as had resigned similar positions. None of these returnees were offered appointments in private research institutions, though two left such positions. The third cohort is even more striking. None returned to a position in a private research institution and there was a sharp increase in those who came back to positions in colleges. In cohorts two and three, the profiles of jobs held before returning to astronomy are roughly similar. In both cohorts, academic positions proved more attractive to the returnees than did employment in observatories. It may be that many of these individuals were burned out on research and sought reentry positions in academic settings that stressed either teaching or a mixture of teaching and research. Returning astronomers came from physics or geophysics (three in each cohort), while two individuals were secondary school teachers in the second cohort. The rest were scattered across a wide range of occupations.

A look at the experiences of two distinguished astronomers who left and then returned help us appreciate the complexity of the process. The career of Frank E. Ross, who left astronomy for a position with Eastman Kodak, was discussed earlier. His reentry was difficult, compounded, perhaps, by his personality and style of decision making. Charles St. John (1857–1935) provides a second well-documented example of an individual returning to full-time research from a career that involved academic administration.

Ross began the search for a position in astronomy in 1921. William Wallace Campbell, director of the Lick Observatory, was apparently trying to secure funding for a photographic zenith tube that Ross was to design and operate as a member of the Lick staff. Ross, however, demanded a number of concessions, including the right to divide his time between Lick and Kodak for at least two years. Campbell, perhaps with misgivings, requested approval from the university regents. Ross then thought better of the plan and declined to leave Kodak. In January 1923 director Campbell made Ross a firm offer: the position of associate astronomer at $3,300 beginning 1 July. Ross hesitated and finally wired Campbell that he could not possibly accept the offer, citing the health of his wife and her sisters as well as one of his children. It is not clear whether Ross was attempting to improve the Lick offer or using it to negotiate with other institutions. At all events, he tried to keep Campbell interested by assuring the director that he would leave Kodak and come to California in 1924. Campbell, however, was loosing patience. The negotiations came to an end and later, when Ross again sought employment at the Lick, he was not taken seriously.[17] Eventually, Ross secured a position at Yerkes.

Charles St. John was born in Michigan and educated at a state normal school. He earned a doctorate in physics at Harvard in 1896. In 1898 he was appointed associate professor of physics and astronomy at Oberlin College, became full professor the next year, and in 1906 dean of the college. In the winter of 1908, George Ellery Hale set his sights on St. John to fill the position of solar physicist at Mount Wilson. St. John was open to Hale's enticements. "Yours just at hand and it sets my pulses going with the fullness of the spring," he scrawled in response. But St. John apparently had reservations. Would the Carnegie Foundation guarantee his pension even if he left teaching? And, more important, would he be able to make a greater contribution to the world as an administrator or as a researcher? Hale assured St. John there would be no difficulty with the pension and then elaborated on the second issue. "The more I think of the question the more I firmly believe that you would make no mistake in casting your lot with us. . . . Full reflection has made me feel certain that contributions to pure science should have a more lasting value than any results that can be accomplished in such a position, for example, as you now hold. We have a marvelous opportunity here."[18]

St. John confessed that he "looked forward with great anticipation to the years of work with you and [Walter S.] Adams, and already I begin to feel relief from the time consuming matters that demand so much attention from an administrator. I begin again to feel the satisfaction that comes from intellectual advances." However, St. John had not anticipated the impact his resignation would have on the Oberlin community. The president told St. John that the event was "a great and far-reaching disappointment" and feared "the faculty and officers of the college will think he has failed in holding a man even with the honor of the deanship." Clearly, the moral atmosphere of Oberlin had infected St. John, for he worried about deserting his Christian colleagues. As the date of departure for Pasadena approached, St. John complained to Hale that he had to

17. The negotiations can be traced as follows: William W. Campbell to Frank E. Ross, 11 October 1921; Campbell to Ross, 7 November 1921; Campbell to Ross, 12 June 1922; Ross to Campbell, 23 May 1922; Ross to Campbell, 26 May and 1 June 1922, and Robert G. Aitken to Campbell, 24 February 1926. See also Campbell to Ross, 15 June 1923; Ross to Campbell, 23 January 1923; Telegram, Ross to Campbell, 2 May 1923, and Ross to Campbell, 9 May 1923. All citations are to the Ross Papers, SALO.

18. Charles E. St. John to George Ellery Hale, 20 June 1908; Hale to St. John, 10 February 1908 and St. John to Hale, 19 March 1908, p. 1. All in the MTWA, Adams Papers.

spend a great deal of time explaining the career opportunities at Mount Wilson, so that friends will "yield gracefully to my going."[19] In the end, St. John left Oberlin and devoted the rest of his life to preparing an authoritative set of wavelength tables for the solar spectrum.

The interpretation of biographical data presented in this chapter rests on the conception of the scientific career as a *social process.* Before 1940, careers in American astronomy exhibited considerable variety; some were rudimentary, others were compound. Movement from a first to a second or even third position suggests a remarkable range of behavior. There were differences in the experiences of elite and rank-and-file astronomers within and between cohorts. Pluralism appears to mark the process of career development in the American astronomical community. On the basis of this discussion it should be clear that the linear progressive model of careers does not apply to American astronomy before World War II.

Any deep understanding of what constitutes a scientific career must relate career patterns to developments in the community. For example, institutional change that affected the specialist research community, such as the emergence of research universities and private research facilities like the Dudley and Mount Wilson observatories, presented attractive career options. As a consequence, federal research institutions became less desirable employment sites for elite astronomers in the second and third cohorts. Change in the research community also was reflected in the choice between the old and new astronomy. Across three cohorts, the pattern of change indicates that an increasing number of individuals opted for the new astronomy. Some made the choice in graduate school; others decided only when they moved from a first to second position. While astrometry and celestial mechanics remained viable career options, astrophysics became increasingly attractive. The appearance of major astrophysical research institutions, toward the end of the nineteenth century, reinforced this trend. The relationships between career and community forms a complex, dynamical system.

19. St. John to Hale, 19 March 1908, pp. 1–2, and St. John to Hale, 10 April 1908, p. 2. All in the MTWA, Adams Papers.

6 ✦

CAREER MANAGEMENT IN SCIENCE

This chapter explores aspects of career management. The activities of mentors, patrons, and sponsors are examined through case studies. Patrons tended to operate at the level of the generic community. Their activities encompassed both specialist research communities (the old and new astronomy). On the other hand, mentors and sponsors were almost always located in the research community. Aspects of career mobility, such as achieving the position of observatory director, are examined in detail; so is the taxonomy of the failing career. Career management strategies are surveyed across three cohorts of astronomers. The chapter closes with a discussion of factors that influenced the most fundamental of career decisions, as individuals opted for either the old or the new astronomy.

INFLUENCES ON THE CAREER: MENTORS, PATRONS, AND SPONSORS

As a result of the rise of the new feminism in the 1960s, the concept of mentor has been redefined. Today, the functions of patron and sponsor are subsumed under the heading of mentor (Levinson et al. 1978; Collins 1983; Riley and Wrench 1985). In historical analysis, however, it is useful to distinguish between these activities; before World War II they functioned in different ways in the career development process. The mentor can be defined as *a trusted advisor and counselor.* The mentoring relationship generally develops during the years of graduate training and extend into the early career of the mentee. There are exceptions. Harlow Shapley acquired Frederick Seares as a mentor while an undergraduate at Missouri. For women, it was frequently the case that mentoring relationships were established while they were undergraduates at one of the women's colleges, and their female mentors remained more important than male dissertation supervisors encountered later. Further, it was possible to have more than one mentor. For example, Joel Stebbins, depended on both George C. Comstock of the University of Wisconsin, with whom he did graduate work before moving to Berkeley, and William Wallace Campbell of the Lick Observatory, who directed his dissertation. Mentors are replaced by patrons and sponsors as a career unfolds.

Patron and sponsor mark the ends of a continuum. Power is the key to defining the role of patron. Patrons were powerful and highly visible individuals in the American astronomical community whose advice, judgment, and assistance were highly prized. *Patronage involves a relationship based on the ability of the patron to help a client achieve specific goals, such as an academic appointment, a position in an observatory, or resources to support research programs.* The intervention of a patron was often important in advancing the client's career.

Sponsors have much less power than patrons. They are most often colleagues, called upon to write letters of recommendation. *Sponsors take responsibility for another, insofar as they vouch for the abilities, knowledge, or good judgment of an associate.* In this sense, the word of a sponsor is a pledge, offering surety that a given individual is worthy of consideration. In a scientific community there are many more sponsors than patrons. During the course of a career, scientists seek many sponsors. They will also act as sponsors for others. Patrons tend to operate at the level of the generic community, dispensing favors without reference to the research tradition represented by the client. Sponsors, on the other hand, are located in the disciplinary community. The complex social network involving mentors, patrons, and sponsors reminds us that individuals did not achieve success simply on the basis of hard work or native intelligence. As members of a community, astronomers were politically and socially dependent on one another.

Both mentoring and patronage relationships entail complex webs of social obligations. Mentees and those who benefit from the actions of patrons owe certain responsibilities and duties. In a sense, these are similar to acts of homage and fealty in the world of medieval feudalism. An examination of manuscript collections indicates the range of obligations.[1] At the most obvious level, individuals indebted to patrons and mentors employed flattery. They invited the mentor or patron to their institution to lecture, commented enthusiastically on their publications, and enquired after their health and that of their families. They also entertained patrons

1. Documentation of these points would prove excessive. The responsibilities of clients can be traced in a number of manuscript collections. Of special value are the Schlesinger papers at Yale, the Stebbins papers at Madison, and the Holden and Campbell collections in the Shane Archives of the Lick Observatory at the University of California at Santa Cruz.

and mentors during visits. Further, many who depended on patrons or mentors often assigned their published works to students. Where the adoption of textbooks for large survey classes was an issue, this could involve significant monetary rewards. In some cases, clients were in a position to supply mentors and patrons with candidates for admission to graduate programs; and, less often, patrons and mentors are presented with important research accomplished with their aid and encouragement. Historians sometimes forget the reciprocal nature of the mentor/patron relationship. While not as complex as that which obtained among politicians, scientists, and churchmen in early modern Europe, it none the less required those who benefited from the actions of patrons and mentors to respond according to the unwritten social rules that defined the relationship.

Mentors were called upon for all kinds of advice and counsel, not all related to astronomy. Joel Stebbins, for example, confessed to George C. Comstock of the University of Wisconsin that he had no idea how to dress for the formal events associated with an international congress of astronomers to be held in connection with the St. Louis World's Fair. Speaking with both genuine concern and perhaps a little wry humor (considering the salary of an instructor) Stebbins wrote, "They surely can't expect the young fellows to put on much style."[2] We have long since lost the world in which correct evening dress is an issue of great moment, but to a young man who grew up in Lincoln, Nebraska, at the end of the Victorian era, it was a question of considerable weight. Aside from members of his own family, Comstock was probably the only person to whom Stebbins could turn for such advice.

Four years later we find Stebbins requesting advice from his dissertation supervisor, William Wallace Campbell, on sources of funding to support a highly innovative research program in photoelectric photometry. Working in cooperation with members of the physics department at the University of Illinois, Stebbins was on the verge of a major breakthrough. There was, however, a hidden agenda; Stebbins was also interested in improving his position at Illinois. Then, as now, university administrators took notice when faculty members won recognition in the form of external funding. As Stebbins put the matter, external support would "help in educating our authorities to understand that our work is worth while."[3]

George Ellery Hale, one of the most powerful figures among second generation astrophysicists, entered the field with only an undergraduate degree from the Massachusetts Institute of Technology. At no point did Hale ever take a formal course in astronomy, astrophysics, or instrumentation for astronomical research. He learned by doing and with the help

2. Joel Stebbins to George C. Comstock, 22 November 1903. UWMA, Comstock Papers.

3. Joel Stebbins to William Wallace Campbell, 16 August 1907, p. 2. SALO, Stebbins Papers.

of two mentors (Wright 1966: 48–72). Here we have an unusual case: mentors who were not formally associated with the mentee in an educational context.

Among the Hale papers in the archives of the Yerkes Observatory are two rich collections. Both date from 1886, the year Hale entered MIT. The larger includes approximately one hundred letters from John A. Brashear (1840–1920), Pittsburgh telescope-maker and patron of the Allegheny Observatory. "Uncle John," as Brashear was affectionately known in later life, ranked, next to the Clarks of Cambridgeport, Massachusetts, as America's most respected telescope-maker. The eighteen-year-old Hale entered into an extensive correspondence with Brashear concerning the design and construction of telescopes and spectroscopes for solar research. Brashear provided important technical resources for the development of Hale's career.[4] Hale's other mentor was Charles A. Young, one of the founders of solar physics in America. The second collection at Yerkes is of the letters exchanged between Hale and Young in which they explore many topics from instrumentation to current research problems.[5] In addition to an extensive correspondence, Hale occasionally visited both of his mentors.

Hale's selection of mentors bespeaks his maturity and judgment. For example, had Hale approached Samuel P. Langley at the Smithsonian, he would have found America's premier solar physicist distracted by administrative concerns, as well as by a consuming interest in the physics of flight. Langley would have been a poor choice. By the same token, had he asked help from the Clarks, Hale would have found an optical firm that tended to specialize in traditional instrumentation. The fabrication of instruments for spectroscopic investigations was never the Clarks' forte. By luck or foresight, the MIT freshman chose mentors who would be of inestimable benefit to his career.

The archives seldom document the intellectual foundations of a mentoring relationship. These foundations most often were laid down early in the graduate career. Fortunately, one example has been recorded. Late in life, Harlow Shapley reflected on his experience at Princeton, working with Henry Norris Russell. In a burst of rare and luminous insight Shapley (1966: 39) recalled: "It's strange on thinking it over [,] how happy I was when he'd [Russell] come in and I'd say, 'I've laid out another corpse for you to look at.'" The corpse was an analysis of a binary star system based on photometric data. Shapley remembered that Russell would "just go over to the plot . . . and walk up and down . . . he was so excited about it." Shapley remarked, perhaps with uncharacteristic nostalgia, "Those

4. Directors Papers, Box 1, Folder 1. UCYOA.

5. Directors Papers, Box l, Folder 5. UCYOA.

were the happy days for him—and for me." These, and similar experiences during Shapley's Princeton years, laid foundations so deep that the two men (separated by less than a decade in age) formed a friendship that lasted a lifetime.

Patronage could take many forms. Sometimes mentors asked leading patrons to help their students. William Hammond Wright, director of the Lick Observatory in succession to Campbell and Aitken, sought the aid of Henry Norris Russell. He wanted Russell to use his power and prestige on behalf of several recent products of the California program. "We are sending a bunch of youngsters to the Philadelphia [American Philosophical Society] and Washington [NAS] meetings in April [1939] and . . . would appreciate it very much if you could manage to make a few remarks after the presentation of their papers."[6]

Showcasing talented young scientists at the April meetings of the American Philosophical Society (APS) and the National Academy was nothing new. Senior scientists had been arranging these events for half a century or more, but the matter had to be handled skillfully. It was not enough to see that talented young people were given places on the program. The way they were received by the scientific community depended on how the discussion of their papers was orchestrated. If left to chance, innovative research might elicit hostility or simply be ignored. Hence, Wright requested Russell's aid in opening the discussion on a positive note. "I am of course not suggesting that your criticisms be favorable rather than otherwise . . . but these boys [Gerald Kron, Horace Babcock, Arthur Wyse, and Ira Bowen] are quite inexperienced, and anything you may be able to do to high-light the critical points of their papers will be gratefully appreciated." Wright promised to send the papers to Russell and to try and arrange for the four to meet with him before their presentations. Further, director Wright requested Russell to intercede for Horace Babcock, who was seeking a grant from the Penrose Fund of the APS. Wright also urged Russell to find time to meet with Daniel Popper, another recent Berkeley Ph.D. "We regard Dan as one of the best men we have turned out here. I should like very much for him to know you." Between the wars, Russell was a powerful patron whose approbation and assistance could materially advance the careers of young scientists.

In each generation there were a few leading patrons, astronomers who were respected throughout the community. Benjamin A. Gould occupied such a position in the late nineteenth century. Gould was approached by Thomas Jefferson Jackson See, a young Missourian with a Berlin Ph.D., hanging on for dear life to a position in the University of Chicago astronomy department. See clashed with George Ellery Hale and the university administration. He was about to take a leave of absence to work at the

6. William H. Wright to Henry Norris Russell, 13 March 1939, p, 1. SALO, Russell Papers.

Lowell Observatory, but must have realized that it was improbable that he would ever return to Chicago. It is important to notice that See approached Gould. Patrons almost never volunteered their services. The former director of the Argentine National Observatory felt the best way he could help See was by getting him on the program at the April meeting of the NAS. Gould contacted his associate, Seth C. Chandler and together they approached the president of the Academy, who presented the request to the Council. The matter was agreed to and See was slated to make a presentation of his research on 21 April 1896. Gould cautioned that the presentation must be descriptive and empirical, "anything speculative will be omitted."[7]

After Gould's death, See would claim a special relationship with the senior astronomer and offer, as proof, extracts from letters destroyed in a fire that burned all See's possessions. An examination of Gould's Letter books, in the Dudley Observatory Archives, indicates that Gould wrote no such letters and suggests the so-called extracts were forgeries.[8] Anyone who uses published or manuscript materials by the ubiquitous Dr. See must employ caution and all the critical tools of historical analysis.

In the 1880s, Edward S. Holden attained the status of patron within the American astronomical community. Holden left the directorship of the Washburn Observatory at the University of Wisconsin in 1885 and assumed the presidency of the University of California. When the Lick Observatory was handed over to the University, Holden moved from Berkeley to Mount Hamilton as the first director. No stranger to the role of client, Holden was deeply indebted to Simon Newcomb, who earlier acted as his patron.

As Holden's reputation grew, so did the number of clients who sought his aid. Herbert C. Wilson relied on Holden to secure him a post at Madison, but Holden's move to Berkeley caused the plan to misfire. Wilson importuned Holden to help him, either with an appointment at the Lick or by using his influence with the new director at Madison. William C. Winlock also sought Holden's aid in seeking the directorship of the University of Minnesota Observatory.[9] Winslow Upton (1853–1914), teaching at Brown University, asked Holden's help in moving from an institution with limited research facilities and a heavy commitment to teach-

7. Benjamin A. Gould to Thomas Jefferson Jackson See, 6 March 1896 and Gould to See, 25 March 1896. DOA, Gould letter book I.

8. Compare the material See attributes to Gould in Box 7 of the See Papers at the Library of Congress with the letter from Gould to See dated 27 November 1896. DOA Gould letter book II. Lankford (1980) discusses the life of See.

9. Herbert C. Wilson to Edward S. Holden, 5 November 1885, 11 November 1885, and 8 February 1886. SALO, Herbert C. Wilson Papers. See also William C. Winlock to Edward S. Holden, 14 October 1885, 6 November 1885, and 17 December 1885. SALO, William C. Winlock Papers.

ing. Francis P. Leavenworth (b. 1858) requested Holden's assistance. He wanted an observatory position, but one with better pay than Virginia provided. Leavenworth remained a client and was indebted to Holden for a position at the University of Minnesota. The young double star expert desired, however, to escape the heavy teaching burdens of a growing state university, where he lectured on astronomy as well as algebra and trigonometry. When Holden's position as Lick director became embattled, Leavenworth offered support for his patron and asked for a position on Mount Hamilton. He assured Holden, "You would find in me no disposition toward insubordination." Holden did not offer his client a position, but did press for Leavenworth's promotion to full professor.[10]

Edward S. Holden was a petty man with an overblown ego that masked deep insecurities. He became a powerful patron in late-nineteenth-century American astronomy, yet being his client could be dangerous. Holden might not play by the rules if his personal interests were at stake. His dealings with the young Armin O. Leuschner illustrate the dark side of the client/patron relationship. Leuschner served as tutor to Holden's son Ned, for which he received room and board. He also worked with the Lick transit circle. When a vacancy in the Berkeley mathematics department was announced, Holden did not bother to pass on the information to Leuschner, nor did he inform the young astronomer of an instructorship at Michigan. Perhaps even more shocking was his refusal to publish a paper by Leuschner based on his work with the transit circle. Holden, like T. J. J. See, had a tendency to arrange the facts to suit his own needs and in this case informed other astronomers that Leuschner had taken no part in research programs on Mount Hamilton. Leuschner eventually received an appointment at Berkeley, but five years later the young astronomer sought an appointment at the Lick Observatory. This time Lick astronomer John Martin Schaeberle (b.1853), interceded with Holden and received a promise that Leuschner would be given serious consideration. Later Schaeberle read in the newspapers that the appointment had gone to William Hussey.[11] Clients, then, had to be careful. Although most patrons were honest, there were exceptions like Holden.

Sponsorship is nowhere more clearly illustrated than in campaigns mounted to secure professorships at the U. S. Naval Observatory. These senior appointments, technically known as professors of mathematics in the United States Navy, while few in number, carried excellent salaries and benefits. Professorships were hotly conteste. The appointments were presi-

10. Winslow Upton to Edward S. Holden, 22 October 1885. SALO, Upton Papers; Francis P. Leavenworth to Holden, 29 January 1886; Leavenworth to Holden, 17 June 1895, and Leavenworth to Holden, 13 April 1897. SALO, Leavenworth Papers.

11. John M. Schaeberle to Armin O. Leuschner, 11 June 1895, 28 June 1895 and 13 September 1895. BLUCB, Leuschner Papers.

dential gifts, and could be obtained only by political maneuvering at the highest levels of the government (White 1958). Applicants had to compete for the favor of congressmen and senators as well as the secretary of the navy, who would, in turn, present their claims to the president. Indeed, at the USNO, all appointments beyond that of computer were political in nature.

In April 1897, Aaron N. Skinner (1845–1918) wrote excitedly to his friend Ormond Stone, director of the University of Virginia's McCormick Observatory. Two professorships of mathematics were vacant and while all of the assistant astronomers at the USNO had applied, Skinner was the only one to have the endorsement of the senior staff. His candidacy was also supported by the superintendent. Skinner told Stone he was asking friends "to make a plain statement of my services in order that the Sec'y of the Navy may have evidence that especially in the last 3 or four years I have been of essential service in giving the Observatory a name [for] good work" and reported that Sherburne W. Burnham of the Yerkes Observatory was bringing pressure on members of congress.[12] Skinner's goal was to obtain the patronage of politicians, but the means to that end involved the sponsorship of scientific colleagues. Apparently members of congress based decisions on the number and status of sponsors an applicant could enlist.

Stone began writing letters at once. Skinner responded quickly, thanking him and noting that the campaign now had the support of Hale and Edward Emerson Barnard of Yerkes. He also enclosed a list of the Illinois congressional delegation (Skinner's home state) with their Washington addresses, and urged Stone to contact as many as he could. Ten days later, Skinner indicated that he had been interviewed by the secretary of the navy, who made it clear that he disapproved of the patronage system and wanted the law concerning the appointment of professors of mathematics amended. The interview left Skinner in a quandary. He felt the secretary was favorably disposed to his candidacy, but did not want to see an appointment made until the law was changed. Skinner, however, feared that President McKinley would make an appointment in the near future. At this point, all Skinner could do was lengthen his list of sponsors. In May, observatory directors Leavenworth (Minnesota) and Doolittle (Pennsylvania) added their weight to the campaign.[13]

Matters dragged on through the summer of 1897. Skinner reported to Stone, "No it is not too late. The more pressure the better."[14] If this cam-

12. Aaron N. Skinner to Ormond Stone, 6 April 1897, pp. 1–2. UVA, Stone Papers. A copy of Skinner's formal application was attached to this letter. See also Skinner to the Secretary of Navy, 8 March 1897. UVA, Stone Papers.

13. Skinner to Stone, 9 April 1897, 19 April 1897 and 20 May 1897. UVA, Stone Papers.

14. Skinner to Stone, 9 September 1897. UVA, Stone Papers.

paign failed, Skinner would be well positioned to compete for the next vacant professorship. The matter was resolved by year's end; Skinner achieved the rank of professor of mathematics in the U. S. Navy, with a presidential commission. Early in the twentieth century, these professorships were abolished (Peterson 1989), but during the second half of the nineteenth century they were sought after in campaigns that demanded much time and effort from sponsors.

NEGOTIATION AND THE MANAGEMENT OF CAREERS

At certain key points in the course of the scientific career, individuals find themselves in situations where they must negotiate. These situations most often involve moving from one institution to another. Negotiation is a sociopolitical process aimed at reaching an agreement satisfactory to the parties involved. The process involves compromise. Parties engaged in negotiations test the limits of one another's power as they seek to maximize advantage while making few concessions. Frequently, negotiations result in the offer of a position. The agreement that has been negotiated forms the contractual basis for the new position and defines the social, intellectual, and economic responsibilities of the contracting parties. A close reading of the manuscript sources suggests that there were definite rules governing the negotiating process.

As part of his plans for revitalizing Yale astronomy, Frank Schlesinger planned a southern hemisphere station, to extend research on proper motions and parallaxes to stars invisible from North American observatories. The success of the plan rested on attracting the right individual to direct the station. During the course of negotiations with University of Virginia astronomer Harold Alden, Schlesinger articulated his view of the process. The Yale director admitted "It would naturally be much pleasanter for me to say that I accept the points you make without reservation, just as it would have been more pleasant for you to have done so in reply to my letter, *but experience in such matters has shown that we cannot be too explicit and that in no other way is it possible to avoid later some degree of disappointment on one side or the other, or both*."[15] A full and mutually agreed upon description of the responsibilities of employee and employer guaranteed future harmony and productivity.

Before World War II it was customary for a potential employer to request permission from the current employer before opening negotiations with an employee. In part this custom reflected the high degree of civility that once prevailed in academia, but it also had something to do with market conditions in the early decades of the twentieth century. At a time

15. Frank Schlesinger to Harold Alden, 11 September 1924, p. 3. Emphasis added. YUA, Astronomy Department Papers.

when able young astronomers were in short supply, the community did not look kindly on raiding. When informed that the loss of a staff member would hamper the work of an institution, the potential employer usually withdrew.

Two case studies illustrate the complexities of the negotiating process. The first involves Schlesinger's attempt to hire Harold Alden to direct the Yale southern station and the second focuses on the carer of astrophysical theorist Donald Menzel, as he moved from Ohio State University to the Lick Observatory and then to Harvard.

In addition to being one of the most powerful patrons in the interwar period, Frank Schlesinger was a remarkable entrepreneur. From the time he assumed the directorship of the Allegheny Observatory in 1905, Schlesinger organized and directed large-scale projects in astrometry. These ventures involved the wholesale acquisition of data by photography and the reduction and analysis of data by a staff of assistants using new technologies. Schlesinger was committed to ambitious production schedules that involved careful, long-range planning. The Yale director specified the yearly output expected of Alden and his staff. He laid down these requirements in the context of a seven-year observing program to which he demanded Alden's assent. This was the minimum length of time Alden had to commit to directing the station. Schlesinger set ten years as the upper limit of Alden's stay. Cost-of-living adjustments were a second area Schlesinger proposed to negotiate.[16] Alden gave careful consideration to Schlesinger's points. His response involved three items of concern. Alden required assurance that he would have the right to publish the fruits of his South African labors under his own name. In addition, he requested clarification concerning the methods to be used to measure cost-of-living differentials between the United States and South Africa. Finally, Alden wanted to know under what circumstances (i.e., medical or family emergencies) he could return to the United States before the end of the contractual period.[17]

Schlesinger sought to clarify these points. He also attempted to dispel the tense atmosphere surrounding the negotiating process. "I know of no one in this country to whom I would sooner entrust the conduct of our station than to you, and it will be a matter of great personal satisfaction and relief to me when our negotiations have come to a favorable termination." It took two more rounds before Alden and Schlesinger reached full accord.[18] An unanticipated consequence of Alden's move involved his isolation from the mainstream of the American astronomical community.

16. Schlesinger to Alden, 3 September 1928. YUA, Astronomy Department Papers.

17. Alden to Schlesinger, 3 September 1928. YUA, Astronomy Department Papers.

18. Schlesinger to Alden, 11 September 1926, p. 3; Alden to Schlesinger, 12 September 1924; Schlesinger to Alden, 19 September 1925. YUA, Astronomy Department Papers.

Later, he apparently believed that by accepting the South African position he had removed himself from consideration in the contest for honors and awards within the community.

The early career of Donald Menzel illustrates several facets of the negotiating process. Menzel, trained by Russell at Princeton (Ph.D., 1924), was one of the first of a new breed: an astrophysical theorist grounded in quantum mechanics who kept abreast of current developments in physical theory. Menzel served as an instructor at Iowa for a year and in 1925 moved to Ohio State as assistant professor. Neither institution offered opportunities for research and both positions entailed heavy teaching loads.

After checking with Menzel's current and previous employers, Robert G. Aitken of the Lick Observatory approached Menzel. The first move was purely hypothetical. "Would you be willing to consider at all a position at the Lick Observatory, with duties mainly in the spectrographic department?" Acting director Aitken made it clear that Menzel would be part of a team, devoting much of his time to established programs. Aitken requested that Menzel, if interested, furnish him with information on his experience as an observer. The senior astronomer also wanted to know about Menzel's "personal situation, that is whether you are married . . . and other items that might be of interest to me in considering you for the place."[19] Lick astronomers and their families formed a tight-knit community on a remote mountain top and Lick directors were careful to consider personalities in making staffing decisions. In this case, Aitken had grounds for concern. The chairman of the Department of Mathematics and Astronomy at the University of Iowa (Menzel's first employer) recommended him, but with reservations. Menzel was described as possessing an "active mind" and being "exceptionally industrious" but also as "a little more aggressive than they [colleagues at Iowa] like." The Iowa chairman assumed Menzel was Jewish.[20] At a time of growing anti-Semitism, such comments sometimes appeared in the private communications of American astronomers.

Menzel responded with a carefully crafted letter. It is of great significance that in the first sentence he redefined the position as one that involved "spectrographic *and astrophysical work*."[21] From this point, both parties to the negotiations accepted the addition of astrophysics as part of the job description. Menzel went on to detail his experience with equipment in a physics research laboratory, as well as his work at Princeton with the 23-inch refractor. He provided Aitken with personal information, including his wedding plans, and went on to ask for a description of

19. Robert G. Aitken to Donald Menzel, 24 April 1926. SALO, Menzel Papers.

20. H. L. Rietz to Robert G. Aitken, 23 March 1926. SALO, Menzel Papers.

21. Menzel to Aitken, 29 April 1926. Emphasis added. SALO, Menzel Papers.

life in the Mount Hamilton community. Menzel also asked about salary and the rate of advancement in both rank and pay. When Aitken replied, he specified the rank, duties, and responsibilities associated with the position, and indicated the degree of freedom Menzel would have to pursue his own investigations. Only at this point did he ask Menzel for a *vita*.[22]

Menzel remained at Lick for six years. During this time he rapidly gained visibility. In 1932, Harlow Shapley, who had known Menzel since the early 1920s, began negotiations that would lead to his appointment to the recently formed Harvard astronomy department. From the outset, Shapley bent the rules to suit his purposes. He wrote Aitken and Menzel at the same time, and presented his action to the Lick director not in the form of a request to open negotiations but rather as an accomplished fact.[23] The offer came as a complete surprise to Menzel. He indicated his desire to accept, but he felt obliged to make a show of seeking advice from Aitken. While Aitken graciously accepted Shapley's bending of the rules, he indicated that no counteroffer would be made. Shapley assured the director that Menzel would be permitted to complete any projects he brought to Harvard, but the affair ended on a sour note. Exasperated, Aitken made it clear that he was offended by the way Shapley handled the negotiations.[24]

The negotiation process can be looked at from another perspective. When scientists received an outside offer, they had to decide whether to leave or to negotiate. If they negotiated, goals included improved support for research as well as increased salary and rank. The experience of Joel Stebbins is especially instructive. Stebbins was dissatisfied with his rate of advancement at the University of Illinois. Salary and working conditions were acceptable and the observatory was given solid support, but the young scientist was passed over for promotion in spite of his contributions to both teaching and research. As he told his Lick mentor, William Wallace Campbell, "I am rather sorry that I did not have the offer of a position elsewhere, for it so happened that the men who spoke of leaving were the fortunate ones this time." In 1911 he reported to Campbell that the administration at Illinois could not make up its mind concerning the future of astronomy. Should it be separated from mathematics and made a department? If so, would Stebbins become a professor in the new department? The president was not encouraging, so Stebbins went in search of

22. Aitken to Menzel, 6 May 1926, and Aitken to Menzel, 20 May 1926. SALO. Menzel Papers.

23. Harlow Shapley to Robert G. Aitken, 20 June 1932. SALO, Shapley Papers.

24. Menzel to Aitken, 24 June 1932. SALO, Menzel Papers; Aitken to Shapley, 27 June, 9 July 1932, and 21 July 1932. See also Shapley to Aitken, 14 July 1932. SALO, Shapley Papers.

an outside offer. Campbell moved behind the scenes, and in the spring of 1912 Stebbins was invited to the University of Virginia for an interview. The mere invitation had the desired effect on the administration at Urbana. Funds were provided for a 30-inch reflector and Stebbins was promoted to professor with a salary of $3,000. Stebbins was relieved, since he did not want to leave Illinois, especially for an institution like Virginia, which was not strong in the physical sciences.[25]

Harlow Shapley was not always successful in his attempts to lure scientists to Harvard. In 1928 he sought permission from Mount Wilson director Walter S. Adams, to enter into negotiations with Seth B. Nicholson, to whom he wanted to offer the Willson professorship. The request touched off a flurry of activity. In the end, Adams and the President of the Carnegie Institution, the parent organization of Mount Wilson, decided Nicholson was too valuable and his salary was increased by about a third. Adams and the Carnegie Institution of Washington (CIW) took this action only when they were sure that funds were available to adjust other salaries at Mount Wilson. Before World War II, many universities and private research institutions were guided by the rule that outside offers should not skew salary scales.[26]

Often negotiations lasted a long time. In the case of Nicholson the matter dragged on from May until August, while Stebbins' campaign at Illinois took several years. Perhaps the shortest negotiation on record is that between Henry Norris Russell and the president of Princeton. On a Saturday in the spring of 1927 Russell "was asked to consult with some people who offered me a rather good position." Russell reported the offer to the president on Monday, and it was immediately countered with a research professorship and "a liberal increase in salary." The finance committee of the trustees was scheduled to meet on Tuesday and by Wednesday had approved the counteroffer.[27]

Charles Donald Shane, who would direct the Lick Observatory during its post–World War II era of growth, started his career at Berkeley and never left except for service at Los Alamos during the war years. Shane used outside offers in a very different way than Stebbins or Russell. He turned them all down, thus demonstrating his loyalty to the University of California. Rather than using an outside offer as a means of improving his position, Shane built a reputation for loyalty and service, which would,

25. Joel Stebbins to William Wallace Campbell, 13 July 1906, 12 January 1911, p. 3, and 10 June 1912, p. 1. SALO, Stebbins Papers.

26. Harlow Shapley to Walter S. Adams, 28 May 1928, Walter S. Adams to John C. Merriam, 27 July 1928, Merriam to Adams, 6 August 1928 and 7 August 1928. MTWA, Adams Papers.

27. Henry Norris Russell to Vesto M. Slipher, 8 May 1927. LOA, V. M. Slipher Papers.

in time, provide the political capital needed to become a power in the institution.[28]

FAILED CAREERS

The discussion begins with what is perhaps the most subtle form of failure: the realization, generally late in life, that a scientist's work has been rendered nugatory as a result of technical or conceptual innovation. The career of Danish-born Christian H. F. Peters illustrates this form of failure. Peters was trained at Berlin, where he earned the doctorate under Johann Encke (1791–1865), and he later worked with J. K. F. Gauss at Göttingen. Involvement in radical politics led the young scientist to seek refuge in the United States, where he was associated with the Coast Survey and the Dudley Observatory before accepting a professorship at Hamilton College in 1858. At Hamilton, Peters specialized in discovering asteroids and in visually mapping the ecliptic. The application of photography to astronomy, in the last quarter of the nineteenth century, rendered visual mapping of the sky obsolete. As Lewis Boss reported, "I spoke to Dr. Peters about his star observations about 6 months before his death. He seemed to feel discouraged about them, and assumed that photography had made it doubtful whether his observations would ever be worth publication."[29] Only a few sheets of the great map ever appeared. Although a member of the NAS, at the end of his life Peters felt he had made few lasting contributions to science.

For Wisconsin-born Sidney D. Townley (b. 1867), failure took a very different form. Educated at Madison and Ann Arbor (Ph.D., 1897), Townley taught at Berkeley as an instructor and worked as an assistant at the Lick Observatory before moving to the International Latitude Station at Ukiah, in California. In 1907 he joined the Stanford faculty where he spent the remainder of his career. Townley, a product of the Michigan department at its nadir under Asaph Hall, Jr., was not well trained in either astrometry or celestial mechanics, and had little understanding of astrophysics. The life sciences were given first priority at Stanford, and the administration would not entertain Townley's request for even "modest astronomical equipment."[30] With no instrumentation, Townley's career was devoted primarily to teaching.

Sometimes failure can be traced to an angular personality. Holger

28. Charles D. Shane, Diary, vol. 5, p. 18. SALO, Charles D. Shane Papers. See also George Ellery Hale to William Hammond Wright, 26 July 1920. SALO, Hale Papers.

29. Lewis Boss to Edward S. Holden, 12 January 1893. SALO, Lewis Boss Papers.

30. Sidney D. Townley to William Wallace Campbell, 23 March 1911. See also Townley to Campbell, 4 June 1898. SALO, Townley Papers.

Thiele (1879–1946) was educated at Copenhagen and joined the Lick staff in 1917. An able observer, he did not get along with director Campbell. Thiele was secretive about research projects, refusing to discuss them with colleagues. Apparently, he made exaggerated claims, interpreting his observations in ways Campbell and other members of the Lick staff considered "wild." At length, Campbell recommended Thiele's appointment be terminated. It was simply not enough to be a good observer. Lick astronomers also had to have "that judicial and philosophic quality of mind which goes with the successful investigator."[31] Thiele served as a computer in the Berkeley Astronomy Department from 1923 to 1930, and then left astronomy altogether.

Some careers fail because they lack clear definition. Sebastian Albrecht (b. 1876) was educated at Madison and Berkeley (Ph.D., 1906). He served four years on the Lick staff and then went to the Argentine National Observatory. Two years later he was at Ann Arbor and in 1913 joined the Dudley Observatory were he remained for the rest of his career. Albrecht was a transitional figure. Trained in the old astronomy, he endeavored to achieve competence in the analysis of high-dispersion spectrograms. Unfortunately, Albrecht applied astrometric thinking to astrophysics. He sought to determine the spectral class to which a star belonged with an accuracy of two decimal places. Once he joined the Dudley staff, most of Albrecht's energies were devoted to work on various star catalogues. As time permitted, he continued to apply idiosyncratic techniques to the analysis and classification of stellar spectra, but as Lick director Campbell cautioned, Albrecht was alone in his approach. No other astrophysicist had taken up the methods Albrecht developed.[32]

In the late 1920s, the CIW moved to phase out support for the Dudley Observatory. The reasons for this decision have never been subjected to historical analysis, but it appears the CIW viewed Director Benjamin Boss as an inept administrator and second-rate scientist. Financial considerations must also have played a part. At all events, Sebastian Albrecht's future was in jeopardy. Over fifty years of age, without a clear professional identity and only a modest research record, Albrecht begged CIW president John C. Merriam for a post as research associate, in order to continue his spectroscopic investigations.[33] Merriam turned the request over to

31. William Wallace Campbell to Holger Thiele, 7 February 1921 and 21 July 1921; Campbell to the President of the University of California, 13 August 1921, pp. 1–2. SALO, Thiele Papers.

32. William Wallace Campbell to Sebastian Albrecht, 25 November 1919; Albrecht to William H. Wright, 1 June 1922. SALO, Albrecht Papers, and Joel Stebbins to William Wallace Campbell, 10 November 1920. SALO, Stebbins Papers.

33. Albrecht to John C. Merriam, 29 October 1928; Merriam to Walter S. Adams, 3 November 1928. MTWA, Adams Papers.

Walter S. Adams at Mount Wilson. Adams expressed sympathy for Albrecht, but objected to appointing him a CIW research associate. Albrecht's work was not of the first rank and he had "antagonized a good many people" during the course of his career.[34] The phase-out lasted longer than anticipated and Albrecht was able to continue at the Dudley Observatory until at least World War II.

James D. Maddrill (b. 1880) illustrates a unique form of failure. I know of no other case in which the doctorate (1907) was conferred *before* the dissertation had been completed. Just how Maddrill convinced William Wallace Campbell and the graduate dean at Berkeley to enter into such an arrangement is unclear. In any case, Maddrill assumed the post of director of the Ukiah International Latitude Observatory with a Ph.D. but still had to complete the dissertation in astronomical spectroscopy. Campbell continually put pressure on Maddrill, who responded with excuses that included physical and mental illness as well as the distraction of a failed love affair. Maddrill apparently made little progress and in 1912 left Ukiah for Berkeley where he hoped to complete his thesis.[35] It is not clear whether this was ever done. At all events, Maddrill's astronomical career came to an end and he disappeared from view.

Some careers failed because of inadequate institutional support. Henry C. Lord (1886–1925) earned a star in *AMS* but was never able to command the institutional resources to fulfill his potential as a researcher. With an undergraduate degree from Wisconsin, Lord accepted appointment at Ohio State University (1891) and was soon made director of the Emerson McMillin Observatory (1895) and five years later professor of astronomy. The instrumentation at Columbus permitted some research, but the size of the telescope (12.5-inches) and poor observing conditions, placed serious limitations on astrophysical investigations. Lord reported to Campbell that he was planning to move into solar physics, if he could convince the observatory donor to supply funds for a spectroscope. McMillin declined, citing financial reversals suffered as a result of the San Francisco earthquake.[36]

Sometimes careers fail because the ambitions of scientists out strip their intellectual abilities. These individuals are not taken seriously by their peers. The career of Dinsmore Alter (b. 1888) illustrates this form of failure. After teaching several years at the University of Alabama, Alter earned a California Ph.D. in 1916. He joined the faculty at Kansas in 1917

34. Adams to Merriam, 13 November 1928, pp. 1–2. MTWA, Adams Papers.

35. James D. Maddrill to William Wallace Campbell, 18 May 1908; Campbell to Maddrill, 22 May 1908 and 15 October 1908; Maddrill to Campbell, 6 May 1912. SALO, Maddrill Papers.

36. Henry C. Lord to William Wallace Campbell, 26 May 1904 and 12 June 1904. SALO, Lord Papers.

and left astronomy in 1936. At Kansas, Alter pursued statistical investigations in climatology and sought funds to build a major research observatory. His research met with skepticism from colleagues and rejection notices from editors. Donors ignored his pleas for money. Discouraged, Alter left for a career in planetarium administration.[37]

ON BECOMING AN OBSERVATORY DIRECTOR

The epitome of success in an astronomical career is an observatory directorship. A directorship combines the power to guarantee resources necessary to carry out one's own research with the authority to define the research of the observatory staff. Further, directorships are positions of great prestige and visibility in the astronomical community and confer status on the occupant, thus increasing opportunities for the accumulation of honors and awards. Before World War II, observatory directors tended to remain in office a long time. For example, from its founding in 1839 to the appointment of Harlow Shapley in 1921, the Harvard College Observatory had only four directors. Shapley was the fifth. After World War II, the time scale used to measure the tenure of observatory directors became logarithmic.

Being a candidate for an observatory directorship was a complicated business. Before World War I, contenders had to maintain the pose of a disinterested noncandidate, in keeping with nineteenth-century traditions of gentility, while in fact, they were hard at work behind the scenes organizing support and preparing position papers indicating the direction in which they would take the institution. The campaign of William Wallace Campbell for the Lick directorship illustrates some of the strategies employed by noncandidates.

James E. Keeler, second director of the Lick, died unexpectedly in the summer of 1900. This was a tragic loss to astronomy and a serious blow to the institution. Keeler made great strides toward improving the morale of the Lick staff, badly damaged during the later years of the Holden regime. He also brought the 36-inch Crossley reflector on line, something Holden and his staff had failed to do. Keeler took the lead in developing an observing program for the Crossley and actively pursued night work. Here too, he stood in sharp contrast to the previous director, whose clumsy attempts at using the telescopes on Mount Hamilton proved failures.

The day after Keeler's funeral, Campbell was summoned to a con-

37. Dinsmore Alter to William Wallace Campbell, 2 August 1920, 28 September 1920, 16 October 1920, 9 February 1927, and undated prospectus by Alter, "Proposed Program for an Observatory and for Astronomical Research at the University of Kansas." SALO, Alter Papers. The Astronomy Department Papers in the Yale University Archives also document Alter's career. Apparently, Alter sought out Schlesinger as a patron.

ference with Benjamin Ide Wheeler, president of the University of California. Campbell urged Wheeler to write Simon Newcomb for advice. Earlier, Newcomb had advised the Lick Trust and was no stranger to the institution. Campbell lost no time in writing to Newcomb himself. His carefully crafted letter provided the senior scientist with all the information needed, should Newcomb choose to nominate Campbell for the directorship. President Wheeler directed Campbell to consult with the Lick staff to determine their choice, but, as he reported to Newcomb, "Before I had time to do so, four of the five men on the staff came to me and volunteered the information that they considered me to be the man for the place." From that time on, Campbell told Newcomb, "I have not been able to get any other recommendation from them."[38]

The young spectroscopist assured Newcomb that he was not a candidate, at least "in the ordinary sense. I shall not apply for, nor work for, the appointment, either directly or indirectly." Campbell was content to leave the decision to President Wheeler. If he "thinks I am the man for the place, I shall be glad to serve." Campbell went on to stress his loyalty to the observatory, the university and the state. He then assured Newcomb that he understood the instruments and the staff and "*what they can do.*"[39]

In 1900, divisions between the old and new astronomy were still a very real part of the American astronomical community and the Lick directorship would be a plum for one of the factions. Indeed, the only member of the Lick staff who was cool toward Campbell was Richard H. Tucker, the sole practitioner of the old astronomy on Mount Hamilton. In dealing with this issue Campbell played the statesman, pointing out to Newcomb that instrumentation on Mount Hamilton was best suited for astrophysical work, and that the new director should have experience with astrophysical research. But Campbell went on to express larger views. The next director should be a versatile scientist, "in sympathy with both the old and the new astronomy."[40] In fact, this characterization fit Campbell perfectly. Campbell also reminded Newcomb that the director should be a skilled administrator. The community on Mount Hamilton numbered about fifty and the director was responsible for all aspects of life on the mountain. It was not enough to be an able research scientist. The Lick director would have to be a capable manager of human resources as well.[41] Campbell was the successful candidate and guided the fortunes of the Lick for more than two decades.

38. William Wallace Campbell to Simon Newcomb, 8 October 1900, p. 2. LC, Newcomb Papers.

39. Campbell to Newcomb, 8 October 1900, p. 3. Emphasis in original. LC, Newcomb Papers.

40. Campbell to Newcomb, 8 October 1900, p. 3. LC, Newcomb Papers.

41. Campbell to Newcomb, 8 October 1900,, p. 4. LC, Newcomb Papers.

Reputation and visibility were important elements in competing for directorships; so were powerful patrons and, often, the recommendation of the retiring director. After eighteen years at the University of Illinois, Joel Stebbins moved to Madison, as director of the Washburn Observatory and professor of astronomy at the University of Wisconsin. Two years before his retirement, George C. Comstock began discussing the possibility of Stebbins as his successor. Stebbins had studied with Comstock, and the Wisconsin director continued to serve as a mentor, even after the young man moved to Berkeley to study for the Ph.D. In the negotiations, salary became an important issue. As Stebbins indicated, "The authorities at Wisconsin ought to be willing to make the directorship of the observatory as attractive as the corresponding positions at Michigan or at Illinois, and with present tendencies I do not see how this could mean less than the equivalent of $5,500 per year in the near future." Matters relating to sabbaticals and pensions also had to be clarified. Stebbins then went on to outline the course of action he would follow if appointed. "It is not my ambition to direct a large observatory with activities in various fields . . . [rather] I should expect to concentrate on a few things, certainly on my present line of work for some time to come." Then Stebbins adroitly turned this to his advantage. "You would probably find few other candidates of the desired calibre whose work would not depend upon a good sized telescope."[42] Perhaps the administration would be willing to consider a larger salary for a director who was not going to demand expensive new instrumentation. During the course of the negotiations, Stebbins was able to secure funds to transfer his photometric work to Wisconsin and pay the salary of an assistant, C. M. Huffer (b. 1894).[43] He moved to Madison in 1922.

Arthur B. Wyse (d. 1942) provides a different kind of example. He was offered the directorship of an important observatory at an exceptionally early age. Wise was one of the group of bright, but inexperienced young Ph.D.'s that Lick director Wright sent east in 1939 to present papers at the American Philosophical Society and the National Academy. He took the Ph.D. in 1934 and was immediately awarded the Martin Kellog Fellowship for postdoctoral research at Lick. By 1935 he achieved the rank of assistant astronomer. Had he not been killed on a mission for the navy, Wyse might have succeeded to the Lick directorship after the war.

In the spring of 1941 the dean of the College at the University of Pittsburgh wrote Lick director William Hammond Wright enclosing a list of candidates for the directorship of the Allegheny Observatory and asking for comments. Besides Wyse's name, the list included eight others. Af-

42. Joel Stebbins to George C. Comstock, 22 January 1922, p. 2. UWMA, Comstock Papers.

43. Stebbins to Comstock, 16 March 1922. UWMA, Comstock Papers.

ter the war, many of these individuals would rise to positions of prominence, often as observatory directors. The list included Lyman Spitzer (b. 1914), at Princeton, Case astronomer J. J. Nassau, and Fred L. Whipple of Harvard. Wright methodically commented on all the names but rated Wyse at the top. On a trip East, Wright stopped and discussed the future of the Allegheny Observatory with the authorities at the University of Pittsburgh.[44]

In time, Wyse was offered the directorship, but declined. His stated reasons, curiously enough, involved nonscientific issues. Wyse objected to university policies, especially the tenure regulations and faculty privileges. The provost replied that a faculty senate was being created and that other changes were underway that would give the faculty more power. Apparently, he also increased the offer, including a higher salary, a rent-free house, and moving expenses. Wyse still was not impressed.[45] Director Wright kept University of California president Robert G. Sproul informed as the negotiations proceeded. Available records do not indicate that the administration at Berkeley made any promises concerning the Lick directorship in order to persuade Wyse to reject the Allegheny offer.[46] At any event, the prospect was eliminated by his tragic death a year later.

Otto Struve was made the director of the Yerkes Observatory in part to keep him from accepting a professorship at Harvard. Struve joined the Yerkes staff in 1921 and eleven years later, at age thirty-five, became professor and director. So valued was Struve that the administration, in spite of institutional policies to the contrary, made remarkable efforts to retain him. In 1926 Yerkes Director Frost exercised the privilege of an employer by telling Robert G. Aitken at Lick that Struve was not on the market and that his loss would cripple the Yerkes spectrographic program.[47] Frost tried to arrange observing time for Struve at the Lick, Mount Wilson, and the Dominion Astrophysical Observatory in Canada. The efforts made by the Yerkes director apparently did not satisfy Struve. In 1929 he wrote Walter S. Adams, asking about the possibility of a position at Mount Wilson or with the new observatory that would operate the 200-inch. Struve was politic, denying any dissatisfaction with Yerkes. Rather it was a question of research opportunities. He needed bigger telescopes and darker skies. In 1932 the dean of physical sciences at Chicago, Henry Gale, re-

44. Straton C. Crawford to William Hammond Wright, 31 March 1941. The other names on the list were Fletcher G. Watson, Jr., Harold L. Alden, Dinsmore Alter, Robert H. Baker, and Keivin Burns. Wright to Crawford, 3 April 1941, p. 2. SALO, Wyse Papers.

45. Arthur B. Wyse to Rufus H. Fitzgerald, 26 May 1941; Fitzgerald to Wyse, 23 June 1941. SALO, Wyse Papers.

46. William Hammond Wright to Robert G. Sproul, 23 May 1941. The letter is marked "Confidential." SALO, Wyse Papers.

47. Edwin B. Frost to Robert G. Aitken, 24 March 1926. SALO, Frost Papers.

ported to Walter S. Adams that Harvard was trying to lure Struve away with a salary of $8,000. Chicago was determined to keep him, even if that meant building a new observatory.[48] In the end, Chicago entered into a cooperative agreement with the University of Texas that resulted in the McDonald Observatory. Struve became director of both institutions.[49]

Not all candidates for observatory directorships achieved their goal. William A. Rogers (1832–98) was a member of the first cohort. Trained at Brown University (A.B., 1857), he devoted his life to astrometry and the study of standards of measurement. He taught at Alfred University until 1870 when he accepted an appointment at the Harvard College Observatory (HCO) to take charge of astrometric research. Rogers left Harvard for Colby College in 1886. This move was the result of a failed campaign for the directorship of the Washburn Observatory at Madison.

In 1885, Edward S. Holden nominated Rogers as his successor at Madison. The news brought a ritual response from Rogers. He would not consider the position unless he received a unanimous call from the regents. He went on to promise Holden that if elected he would finish Holden's research program and see that it was published, before beginning his own. Rogers then entered into a frenzy of activity. He asked for letters of support from President Eliot of Harvard and the presidents of Brown, Johns Hopkins, and Michigan. Senior astronomers such as Pickering at Harvard, Gould, just returned from Cordoba, and Young at Princeton also were enlisted.[50]

However, all was not smooth sailing. Alice Lamb, an assistant at Washburn who would later marry Milton Updegraff (1861–1938), reported to Holden that while the faculty supported Rogers the regents had other ideas. Lamb feared George Comstock would be appointed and that this might lower the prestige of astronomy at Madison.[51] To diffuse tension, perhaps, the regents postponed the election. Rogers provided John Bascom, president of the University of Wisconsin, with an eleven-page letter detailing the policies he would implement if elected. Rogers was not successful. The regents compromised, with Asaph Hall, Sr., of the Naval Observatory taking the post of scientific advisor and Comstock appointed associate director.[52]

48. Henry Gale to Walter S. Adams, 22 March 1932. MTWA, Adams Papers.

49. Gale to Adams, 19 April 1932. MTWA, Adams Papers.

50. William A. Rogers to Edward S. Holden, 16 October 1885; Rogers to Holden, 12 November 1885. SALO, Rogers Papers.

51. Alice Maxwell Lamb to Edward S. Holden, 4 January 1886. SALO, Washburn Observatory Papers.

52. Rogers to Holden, 9 February 1886; Rogers to John Bascom, 12 May 1886. SALO, Rogers Papers.

Rogers moved rapidly to save face. He quickly secured a position at Colby College. He was deeply hurt by the outcome of the Madison campaign. Apparently it had become a contest of some visibility in the American astronomical community. Rogers found Colby supportive of his research on standards of measurement and felt appreciated in his new situation.[53]

CAREER MANAGEMENT: CHOICE AND CHANCE

The experiences of seven individuals illustrate the diversity and complexity of career management in American astronomy between 1859 and World War II. The variety of careers and the ways in which they departed from the linear progressive model has been a central theme of chapter 6. Here the focus is on the management of individual careers, emphasizing tactical choices, as well as the ways in which chance restricted or advanced careers.

George Washington Hough illustrates the permeable nature of careers in American astronomy. Like most members of the pre-1859 cohort, Hough had a variety of active scientific interests. He was at home in meteorology, astrometry, planetary astronomy, and double star work. A man of wide experience, Hough understood both the economic marketplace, and the competitive world of science (Lankford 1979). His career management strategies were based on diversification rather than specialization, including the ability to pursue a career outside the astronomical community.

Hough graduated from Union College in 1856 and entered astronomy as an assistant at HCO. By 1860, the young Hough, an assistant at the Cincinnati Observatory, was recording his first double star discoveries.[54] Moving to the strife-torn Dudley Observatory, Hough was soon running the institution in the absence of its director. In 1862, he assumed the directorship. Hough sought to cooperate with other astronomers, observing asteroids for James C. Watson and Neptune for Simon Newcomb. When the Dudley closed in 1874, Hough entered business, building and marketing a variety of self-registering instruments for the study of meteorology. In 1879 Hough became director of the Dearborn Observatory and later professor of astronomy and meteorology at Northwestern University. Hough was versatile enough to remain in control of his career, whether in or out of the astronomical community.

In 1930, Robert G. Aitken succeeded William Wallace Campbell as fourth director of the Lick Observatory. Aitken owed his success as much

53. Rogers to Holden, 14 September 1886 and 6 September 1887. SALO, Rogers Papers.

54. Hough, Diary, 15 March 1860. NWUA, Hough Papers.

to chance as to his own efforts at career management. He was in the right place at the right time and was able to impress director Holden with his enthusiasm, loyalty, and potential. Educated at Williams College (A.B., 1887), Aitken became an astronomer by chance. Although professor of classics at the University of the Pacific, he taught a class in descriptive astronomy and was in charge of the university's modest observatory. By the early 1890s, Aitken was in correspondence with Holden and soon mastered the literature in celestial mechanics. He came to the Lick on a one-year appointment in 1895 and remained for the rest of his professional life. His work in double star astronomy led to his election to the NAS (1918). Without formal training, Aitken demonstrated great ability as an observer and, under the tutelage of Edward Emerson Barnard, developed into the premier double-star astronomer of his generation.[55]

The careers of some astronomers were stillborn. Charles A. Borst (b. 1851) never had the opportunity to take charge of his career, and his famous lawsuit against C. H. F. Peters was a form of revenge for blocking Borst's career. Borst had been a student of Peters at Hamilton College and, after graduation, became an assistant in the Litchfield Observatory, working on Peters' great star map. Their relationship proved difficult. Borst was ambitious and Peters demanding, indeed dictatorial, in the fashion of a European institute director.

Early in his association with Peters, the director urged Borst to undertake a special project. At first, Peters suggested an introductory text, but later the plan was changed. The young assistant was to comb the European and American journals and collect star positions, which he would reduce to a common epoch and then publish as a synoptic catalogue. After three years at Litchfield, Borst came to see the importance of such a project to his career. "Everything I had done up to that time was absorbed entirely in the Doctor's work. I had no credit, [or] publicity, at all; and the salary was so very small[.] I debated for some little time and came to the conclusion that I must either try and do something for myself . . . which would be recognized . . . [and lead to] a more lucrative position; or give up science and go into something else."[56]

As the catalogue neared completion, Peters demanded to see the manuscript. Indeed, the imperious director made it clear that Borst could not publish under his name because he was only an assistant. This was the last straw. Borst "tried to have a fair talk" with Peters, reminding him that it

55. Aitken's entry into astronomy is documented in his early letters in the SALO. See, for example, the following: Robert G. Aitken to Edward S. Holden, 12 September 1893, 22 May 1894, 28 May 1894, 17 September 1894, 5 January 1895, and 13 May 1895. Aitken sums up his early career in a "Statement Relating to the Lick Observatory" (1910). SALO, Aitken Papers. For biographical information see Willem Van Den Bos (1958).

56. Transcript of the Trial, 1889, pp. 47–49; 51–52. HCA, Peters Papers.

was at his suggestion that he had undertaken the project. But Peters would not grant permission, insisting the manuscript catalogue was his property.[57] In the end, the matter became a celebrated lawsuit to determine ownership of the catalogue, but it was too late to salvage Borst's scientific career. He left astronomy for the rough and tumble world of business.[58]

Paul W. Merrill (1887–1961), long-time staff member at Mount Wilson, wrote (1923: 548) that the basis for a "worth-while research career in astronomy . . . [rested on] an insatiable curiosity concerning the secrets of the stars, and unbounded enthusiasm for personal investigation [research]." Merrill might well have been describing Mount Wilson's founder, George Ellery Hale. Hale was a man in a hurry, driven by the desire to make brilliant discoveries in the field of solar physics, to create new forms of instrumentation, and, later, to reorganize American science. Two years after graduating from MIT, he asked Holden to support his campaign for a post at the University of Chicago. The young man was quite honest. "At present I am not well qualified for the position, but possibly I might be of some service after such a course as I hope to begin next fall at the Johns Hopkins. Nothing would suit me so well as to be connected with a department of spectroscopic astronomy, if they would only decide to establish one" at Chicago.[59]

Hale never found time to attend Hopkins, and his attempt at graduate study in Europe was brief and unsuccessful. Yet this did not block his career. Many years later, Hale confessed "I never had any instruction in advanced astronomy, . . . have no mathematical ability, and any graduate student in astronomy or physics could pass an examination beyond my range." According to Hale, his strength lay in "an intense enthusiasm for research . . . coupled with optimism enough to carry me through much discouragement."[60] Hale, it seemed, had the right stuff to drive a distinguished research career, even if he lacked formal credentials. This drive, coupled with his use of mentors and patrons, helped Hale develop his career at a rapid rate.

For a decade after taking the doctorate, Samuel A. Mitchell devoted a great deal of energy to career management. After graduation from Hopkins, he spent time at Yerkes and acquired Edward Emerson Barnard as a patron. In 1898 Mitchell moved to Columbia University as a tutor, and was promoted to adjunct professor of astronomy ten years later. A review of Mitchell's job-seeking activities between 1898 and 1908 is instructive. In 1898 the president of Knox College wrote concerning a professorship

57. Transcript of the Trial, 1889, pp. 53–54. HCA, Peters Papers.

58. Transcript of the Trial, 1889, p. 74. HCA, Peters Papers.

59. George Ellery Hale to Edward S. Holden, 9 February 1891, p. 2. SALO, Hale Papers.

60. George Ellery Hale to Robert G. Aitken, 6 December 1915, p. 1 SALO, Hale Papers.

of mathematics. At the same time, Mitchell was being considered for a position at his undergraduate school in Canada. As if to hedge his bets, Mitchell was also in contact with the principle of Central High School in Washington, D.C. In 1900, his patron, Barnard, urged Mitchell to apply for the position at Missouri eventually filled by Seares. The next year found Mitchell negotiating with Queens College in Canada concerning the position that would be vacant when his undergraduate professor retired. In the fall of 1901, William Eichelberger wrote from the USNO urging Mitchell to apply for the post of assistant astronomer.[61]

The Columbia faculty were aware of Mitchell's activities. Professor John K. Reese was more than willing to recommend Mitchell for a position at Connecticut Wesleyan but insisted that he wanted to keep Mitchell at Columbia. Barnard reported that he had recommended Mitchell to the president of Brown University, and in 1903 Eichelberger was still urging him to consider the USNO. Hobart College indicated interest in the young Columbia tutor. In 1905 William Wallace Campbell virtually offered Mitchell a position on Mount Hamilton. In 1906 West Virginia University asked whether Mitchell would be interested in a professorship of mathematics and astronomy, and the next year saw an inquiry from the International Latitude Observatory in Maryland. In 1908, Mitchell's former Hopkins physics professor, John S. Ames, sponsored him for the position at Princeton, which eventually went to Henry Norris Russell. Campbell also lent his weight by writing the president of Princeton.[62]

Mitchell had been well treated by Columbia. It is difficult to secure comparative data, but it appears that his salary was somewhat above those of his peers with comparable experience. For the academic year 1900–1901 Mitchell was paid $1,100. Two years later this was increased to $1,300 and by another $100 for academic 1903–4. For the next two years his salary was $1,500, rising to $1,700 in 1907–8. President Butler informed Mitchell in the spring of 1908 that he would be promoted to an adjunct professorship with tenure effective 1 July and that his salary would be $2,000 per year.[63]

61. President of Knox College to Samuel A. Mitchell, 9 August 1898; Secretary of Queen's College to Mitchell, 4 October 1898; Principal of Central High School to Mitchell, 26 September 1898; Edward Emerson Barnard to Mitchell, 5 May 1900; N. F. Dupris to Mitchell, 3 August 1901; William S. Eichelberger to Mitchell, 20 September 1901. UVA, Mitchell Papers.

62. John K. Rees to Samuel A. Mitchell, 31 May 1902; Barnard to Mitchell, 15 September 1902; Eichelberger to Mitchell, 17 March 1903; Norman E. Gilbert to Mitchell, 13 July 1903; William Wallace Campbell to Mitchell, 12 May 1905; W. L. Fleming to Mitchell, 3 December 1906; Superintendent, U.S. Coast Survey, to Mitchell, 2 December 1907; John S. Ames to Mitchell, 3 March 1908; Campbell to Mitchell, 19 March 1908. UVA, Mitchell Papers.

63. Salary data from the contracts that renewed Mitchell's appointment at Columbia, box 1, Mitchell Papers. Nicholas M. Butler to Mitchell, 2 March 1908. UVA, Mitchell Papers.

Diverse management strategies in the third cohort are represented by two individuals: John A. Pitman and Harlow Shapley. John H. Pitman (b. 1890) graduated from Swarthmore College in 1910 and remained as an assistant in the Sproul Observatory, while studying for an M.A. He spent 1911–13 at the University of California working on a doctorate, but returned to a position at Swarthmore without completing the dissertation. After leaving Berkeley, he made little progress toward the degree. In 1919 Campbell closed the file on Pitman. Unless Pitman made a strong case for another extension, supporting it with a recommendation from Swarthmore Observatory director John A. Miller, he would be dropped as a doctoral candidate.[64] Pitman apparently did not feel the Ph.D. was necessary for the kind of career he envisioned, and it appears that the decision did not limit his chances for advancement at Swarthmore. He was promoted to assistant professor in 1918 and associate in 1928. There is no indication, however, that he was made full professor before World War II. Pitman was a man with limited, tightly focused ambition. He wanted a position at his alma mater and he got it.

Harlow Shapley quickly found a place on the fast track. He did not earn an undergraduate degree until age twenty-five, but the M.A. followed the next year and by twenty-eight Shapley had a Princeton Ph.D. and a place on the staff of the Mount Wilson Observatory. During the years at Mount Wilson he produced a stream of papers. These investigations provided a new and effective method for measuring cosmic distances. At thirty-six, Shapley became the fifth director of the Harvard College Observatory.

With Frederick H. Seares as Shapley's mentor when he was an undergraduate and Henry Norris Russell as the one who guided him in his doctoral program, the young Missourian was better prepared than most Ph.D.'s. From Seares he acquired a taste for research and from Russell a knack for picking important problems. Russell also had the sensitivity and skill to help Shapley (1966: 36, 23, 43) develop a sense of intellectual independence. It was Seares who arranged for him to join the staff at Mount Wilson, where Shapley was able to use the 60- and 100-inch reflectors. When Russell decided against the Harvard directorship, he recommended Shapley. With mentors like Seares and Russell, plus his own burning ambition, Shapley found career management relatively easy. To be sure, the young Missourian was endowed not only with brains and quick wit but an ego that was larger than life. Shapley occasionally admitted to a "blunder," but he never could imagine (at least after his training at Princeton) failure. No where is his ego more evident than in his recounting of his decision to become a candidate for the Harvard directorship.

64. John A. Pitman to William Wallace Campbell, 21 March 1913, 13 November 1915, and 1 November 1917; Campbell to Pitman, 2 September 1919. SALO, Pitman Papers.

Each day Shapley (1966: 70–71) walked home for lunch from the observatory offices on Santa Barbara Street. On the day he heard of Edward C. Pickering's death, Shapley walked in deep thought. "Should I, or should I not? Should I curb my ambition [?]" The problem, of course, was how to maintain a research career if he became the head of a major observatory like HCO. "Finally I said to myself, 'All right, I'll take a shot at it.'" In all probability, Shapley was misremembering the actual circumstances. Russell's decision to remain at Princeton was a key element in the chain of events that led him to Harvard. But the point remains: Shapley was driven by a remarkable ego.[65]

OLD ASTRONOMY OR NEW? A FUNDAMENTAL CAREER MANAGEMENT DECISION

One last topic remains to be considered. How did individuals make fundamental career decisions as they chose between the two research communities? Before 1859 astronomy consisted of celestial mechanics and astrometry, with a few secondary areas including descriptive studies of lunar and planetary phenomena, comets, or nebulae and clusters. The choice for the first cohort lay between celestial mechanics and astrometry, or observational work that was considered marginal. With the rise of astrophysics the situation changed; individuals had a choice.

Manuscript sources suggest two factors were important in the decision-making process: the institution of highest degree and the first position. From the resources of the collective biography, quantitative data were assembled that permit an examination of these factors. How important was the institution from which astronomers earned their highest degrees in determining the type of astronomy in which they worked? What role did the first position play in choosing between the old and new astronomy?

Undergraduate and graduate education provides intending scientists the opportunity to sample and decide just where their talents and tastes lie. Paul Herget (1908–81), one of the last representatives of the old astronomy to win election to the National Academy, recalled (Herget 1977: 20) his experiences at the University of Cincinnati (Ph.D., 1935). He had little interest in physics but did attempt a course in quantum mechanics. It was not a rewarding experience. "I really wasn't prepared or motivated . . . and furthermore it struck me that they were pretty far out. So I never took an interest in that." However, this experience did permit Herget to sharpen his taste in science and shape his career accordingly. As he

65. In fact, Shapley indicated interest in the Harvard directorship at least a year earlier. See Smith (1982: 78).

recalled, "I guess also I had a kind of conscientious feeling that I'd better do what I'm able to do and not fritter my time away on things . . . [in which] I had no assurance that I'd get anywhere." By trial and error, many young scientists must have learned to distinguish between areas in which they would probably never get anywhere, and those in which they could maximize their creative potential. Another example is provided by the German-born astrophysicist Rupert Wildt (1905–76). Wildt indicated that early in his career he had a good deal of experience with both types of astronomy. "These experiences have taught me a great deal of respect for the exacting demands of modern astrometric work, but at the same time have convinced me that my heart is in astrophysics."[66]

Cohort was also a factor in the decision-making process. Perceptions of which types of astronomy were more desirable changed between 1859 and 1940. Early in the twentieth century, Richard H. Tucker of the Lick Observatory found that "most of the prospective candidates [for staff positions] want to do astrophysical work" rather than take part in an arduous program of observational astrometry using the transit circle. In the 1920s, Frank Schlesinger lamented that "the superior attractiveness of astrophysical work has nearly monopolized the services of our most promising young men."[67]

In 1894 a young and enthusiastic William Wallace Campbell described the attraction of observational astrophysics. "I regret having to give some time to the *old* work, & want to get out of it entirely. In spectroscopy one can cut loose from traditions and roam as free as he likes." Waxing poetic, the future Lick director concluded, "There is so much waiting to be done, and the nights are few & short."[68] As suggested in chapter 3, the personal attraction of astrophysical work was one of the new astronomy's most compelling advantages.

Table 6.1 compares the type of astronomy represented by the institution at which the highest degree was earned with that of the first full-time position. Two-thirds of those trained in institutions where the old astronomy (OA) predominated took first jobs in institutions that can be characterized as specializing in OA. A slightly larger percentage of those trained in the new astronomy (NA) found first jobs in institutions in which NA was emphasized. Less than a quarter of the group trained in OA changed to NA as part of moving to a first job while only 18 percent trained in NA

66. Rupert Wildt to Samuel A. Mitchell, 9 June 1942, p. 2. UVA, Samuel A. Mitchell Papers.

67. Richard H. Tucker to Charles W. Frederick, no date (ca.1909), p. 1. SALO, Tucker Papers. Frank Schlesinger to Robert G. Aitken, 7 July 1926, p. 1. YUA, Astronomy Department Papers.

68. William Wallace Campbell to James E. Keeler, 4 April 1894, p. 2. UPAIS, Allegheny Observatory Papers.

Table 6.1 Astronomy Type at Institution of Highest Degree v. Astronomy Type at First Position

Astronomy Type at Institution of Highest Degree	Astronomy Type at First Position		
	Old Astronomy	New Astronomy	Mixed Astronomy Type
Old Astronomy	66%	27%	7%
New Astronomy	18%	68%	14%
Mixed Astronomy Type	15%	50%	35%

changed. Half the individuals educated in institutions that represented a mix of both NA and OA tended to move into astrophysics while 15 percent found positions in institutions that emphasized the old astronomy. Chicago and Berkeley are perhaps the most obvious examples of mixed graduate programs.

It would seem, however, that in determining the type of astronomy to which individuals devoted their careers, the first position is of greater importance than the institution which granted the highest degree. The experience of Frederick H. Seares illustrates the importance of the first job. As Seares's European studies drew to a close he realized the need to come "face to face with an important decision—as to whether I shall stick to the purely mathematical and theoretical side [the old astronomy] or whether I shall give my attention more to observational work [astrophysics]." While perhaps better prepared for work in celestial mechanics, Seares noted that a decision favoring the more traditional field would lead to a position teaching astronomy and mathematics in a university, thus excluding the possibility of an observatory position. Seares also pondered the role chance played in shaping careers. "Our decisions in such matters are always so influenced by practical considerations as to what there is available in the way of positions that I am on the outlook [*sic*] for observatory positions as well."[69] In the end, Seares went to the University of Missouri where he embarked on a career in observational photometry. When Seares moved to a second position, it was at a private research institution (Mount Wilson), devoted to astrophysical investigation.

Quantitative data suggest that the first job had a decisive impact on overall career patterns. On balance, the type of astronomy (AT) that defined the first position is more important than the AT of the institution awarding the highest earned degree. Table 6.2 compares the AT of the first job with that of the career evaluation. About 83 percent of those whose

69. Frederick H. Seares to William Wallace Campbell, 30 December 1900, p. 3. SALO, Seares Papers.

Table 6.2 Astronomy Type of First Position v. Astronomy Type Career Evaluation

	Astronomy Type Career Evaluation		
Astronomy Type at First Position	Old Astronomy	New Astronomy	Mixed Astronomy Type
Old Astronomy	83%	15%	2%
New Astronomy	15%	79%	6%
Mixed Astronomy Type	57%	26%	17%

first job was at an institution specializing in OA went on to careers in either celestial mechanics or astrometry. Notice the differentials between tables 6.1 and 6.2. Fewer individuals whose first job was in an OA institution later switched to NA (15 percent), than moved from a graduate or undergraduate program in OA into NA (27 percent; see table 6.1). The same general tendencies hold true for those whose first job was in a NA institution. Almost 80 percent of this group remained active in astrophysics. The relationship breaks down with the group whose first job was at an institution involving both the old and new astronomy. In this case, the largest group (57 percent) followed careers in OA.

If we consider an educational institution as providing a context in which a *choice* between the old and the new astronomy was possible, and the first job a context in which the type of astronomy pursued was determined by the *needs* of the employer, then we can develop an appreciation of the balance between the two factors. A comparison of tables 6.1 and 6.2 suggests that demands of employers in first positions had a greater impact. If this was the case, *chance* (the availability of positions) plays a greater role than *choice* in this most fundamental aspect of career management.

Chapters 5 and 6 should be understood as a unit; both are concerned with the careers of astronomers across the professional lifetimes of three cohorts of scientists. Chapter 5 examined the collective (quantitative) experiences of three cohorts of American astronomers as they managed their careers. Chapter 6 shifted the focus to individual biography. For many astronomers, the actions of mentors, patrons, and sponsors had much to do with career success or failure. It is clear that, while talent counted, so did the support of powerful allies. It would be pleasing to imagine the broad domain of science as a meritocracy. However, this is not the case. This issue will be explored further in chapters 7 and 8.

7 ✦

POWER AND CONFLICT IN A SCIENTIFIC COMMUNITY

When historians discuss power in modern science, they tend to look at the politics of government patronage (Smith 1989) or focus on struggles between competing theories (LeGrand 1988). Social constructivists (Woolgar 1982) relate power to the construction of scientific knowledge in the laboratory. This chapter examines the nature, forms, and uses of power in the American astronomical community. There are significant differences between the approach developed here and more traditional views of power in modern science.

The archives contain materials illuminating the activities of a self-conscious power elite: observatory directors. The activities of this power elite were directed toward two primary goals. The first involved interobservatory competition that often centered on issues of scientific competence. The second encompassed the complex topic of institution-building. As might be expected, conflict frequently accompanied the exercise of power.

Definitions of power are deeply colored by theory and ideology. It is important to make clear the conception of power on which the following discussion rests. Power involves resources that can be mobilized in order to achieve clearly defined goals (Clegg 1989: 2). Power should not be seen as a thing or possession. As British sociologist Stewart R. Clegg (1989: 207) insists, power is most fruitfully understood as a network of relationships: a complex, dynamic web of individuals, groups, and resources. The scientific community provides both resources that can be mobilized to achieve various goals *and*, in a larger sense, is itself a product of the exercise of power (Clegg 1989: 64). Further, the cultural and social context in which a scientific community is located helps define the forms and uses of power. American culture provided the models for new forms of organization, as well as new roles for observatory directors. In turn, these organizations and roles became closely linked to new definitions of the nature and scope of power in American astronomy.

In American astronomy, the exercise of power took many forms. In the second half of the nineteenth century, the astronomical community witnessed a prolonged struggle between the old and new astronomy. Eventually, this contest led to the ascendancy of astrophysics. Beginning in the 1890s, the development of graduate programs resulted in the addition of new resources to the community and provided the institutional basis for defining educational credentialling in astronomy. The discussion of careers in chapters 5 and 6 suggests a third arena for the exercise of power. A complex labor market emerged in which the tastes of both employees and employers had to be matched. In this context, mentors, patrons, and sponsors exercised power over careers. In addition, competition between observatories escalated. This competition occurred on at least two levels. The first involved the race to produce novel results and add to the store of scientific knowledge. The second entailed the search for funds to support large-scale research institutions.

ORIGINS AND FORMS OF STRUCTURAL CHANGE

To understand the pressures that led to restructuring observatories and a redefinition of the role of director, it is necessary to recall changes in scale (the growth of various classes of observatories from the 1860s to 1940) discussed in chapter 5. These changes were exemplified in the activities of Simon Newcomb at the Naval Observatory (USNO) and Nautical Almanac Office (NAO), and Edward C. Pickering at Harvard. Newcomb (1882: vii-xiv) assembled a corps of assistants and collaborators in order to push forward ambitious projects in celestial mechanics.[1] Pickering, who became director of the Harvard College Observatory (HCO) in 1877, inaugurated major research in photometry, and by the 1890s, was using photography for both photometric and spectrographic studies that included virtually the whole sky visible from the latitude of Cambridge. A Peruvian station allowed Pickering to expand his coverage to the southern heavens, and improved instrumentation extended the research to fainter objects (Jones and Boyd 1971: chaps. 4–10; Lankford 1981a: 23–9; Plotkin: 1993).

The introduction of photography marked a turning point in astronomical research. The rate of data acquisition increased many orders of magnitude. In areas like photometry and spectroscopy, photography pro-

1. For a parallel discussion of the Royal Greenwich Observatory as a knowledge factory, see Schaffer (1988). Smith (1991) also provides important perspectives on industrial science at Greenwich. There is no comparable discussion for the Naval Observatory.

vided innovative methods for the production of scientific knowledge and led to novel products such as spectroscopic binaries and new data bases, including the great photographic *durchmusterungen* of stellar magnitudes and spectra.

These internal developments placed great strain on traditional forms of observatory organization and the role of observatory director. No longer could institutions be directed in the autocratic manner of C. H. F. Peters at Hamilton College. Neither would it do to operate observatories as if they were little more than collections of independent craftsmen, as was the case at the Naval Observatory until the 1860s. As Simon Newcomb recalled, when he came to the USNO, there was no coordination of research programs. Each senior astronomer collected data on stars he "thought best to observe" without consulting other observers, and with no direction from the observatory superintendent. As the observatory grew in size, this laissez-faire approach to research led to chaos. Under Newcomb's prodding (Campbell 1916: 4), the superintendent developed uniform observation programs that tied the work of the senior staff together.

As the means of production in American astronomy were transformed, there was increasing pressure to implement new organizational practices that would maximize the efficient use of instruments and observers. This process, known as rationalization, is characteristic of the shift from an economy of independent craftsmen and stable local markets, to an economy dominated by machine production and expanding markets. At the institutional level, the response to changes in the mode of production, and the structure and nature of markets takes the form of bureaucratic, hierarchical organizations in which individuals are assigned specific duties and responsibilities (Weber 1947).

In addition, developments in the larger social and cultural environment proved significant for the astronomical community. After the Civil War, the production and marketing of new products through innovative technologies and strategies led to the reorganization of American business (Chandler 1962). This, in turn, provided the model for institutional reorganization in the astronomical community. Both science and industry sought to rationalize production in order to reduce costs and achieve greater efficiency. The result was large-scale structures. In astronomy, this led to the factory observatory with its expanded labor force of semiskilled and unskilled workers (most often women), generally supervised by male astronomers. A well-defined division of labor obtained in factory observatories. In these hierarchical structures there were distinctions between those who collected data, measured plates, and reduced the information obtained, and those who discussed the material and published the results. At the top stood the observatory director, a role that increasingly resembled a chief executive officer in the corporate community. Often there were associate and assistant directors as well as division directors with line administrative responsibility.

An unanticipated consequence of the expansion of American astronomy, and the creation of the great factory observatories, was a decline in the research activities of smaller institutions. In the business community, new production technologies often reduced barriers to entry and permitted new companies to enter the market (Jenkins 1977: 4). In astronomy, the reverse was true. New astronomical research technologies demanded capitalization on a very large scale.

By the 1890s the trend was clear. A few large, heavily capitalized institutions working with state-of-the-art instrumentation, controlled astronomical research in the United States. The result was a form of oligopolistic competition between giants. Institutions that could not command the capital to invest in large telescopes, or pay a staff of assistants, simply could not compete. At best, smaller institutions produced routine data; they could seldom produce novel products that demanded large telescopes and numerous assistants.

The appropriation of resources from the larger social and cultural environment for the restructuring of institutions is a process familiar to students of the sociology of organizations. It is known as *institutional isomorphism.* As new technologies transform production, markets expand and competition increases; successful economic institutions tend to incorporate elements of rationality and coordination into increasingly complex structures. The goal is to attain the highest possible level of efficiency, in terms of both human and technical resources (Sawyer 1958; Chandler and Galambos 1970). The organization itself becomes a central concern of managers, as does individual responsibility and accountability. Institutions that restructure in response to changing conditions tend to be more competitive (Hawley 1950; Meyer and Rowan 1977; DiMaggio and Powell 1983). Institutions which survive this process come to resemble each other.

Institutional transformation has other consequences as well. Organizations are brought in line with patterns and practices in the larger society, and come to share a common vocabulary with that society. This provides increased legitimacy. For observatory directors seeking support from the business sector, isomorphism provided important advantages. Directors not only understood the language of rationalization, efficiency, and large-scale production, they also could demonstrate to potential donors that their institutions were efficient and up-to-date. The same applied to observatories that received public funds; those that reorganized in line with the practices of the larger society were able to justify increased support from state legislatures.

In fact, institutional reorganization frequently meant increased competitive advantage over institutions that had not brought their organizations into line with prevailing practice (Meyer and Rowan 1977: 350–53). For observatories, the change often meant continuous rather than episodic support. With their institutions in stronger competitive positions, directors were able in enhance their power.

The directors of the great factory observatories were children of the second half of the nineteenth century, and grew up in a society that held efficiency, organization, and the entrepreneur in high esteem (Miller 1952; Diamond 1955).[2] The role of entrepreneur provided the American astronomical community with a new model and new rules that helped define the uses of power.

Business historian Arthur H. Cole spent much of his career studying entrepreneurship. For Cole, the entrepreneur represented the active, aggressive businessman. Cole (1959: 7) defined the entrepreneur as one who was engaged in "purposeful activity undertaken to initiate, maintain or aggrandize a profit oriented business." The Harvard historian was quite explicit in emphasizing hard work, competition, and innovation as basic to the character and personality of the entrepreneur (Cole 1959: 12–14). The successful entrepreneur also possessed administrative and managerial skills.

The personalities of successful scientists often show marked entrepreneurial characteristics.In a study of eminent physical scientists (including astrophysicists), psychologist Anne Roe (1951: 160) reported that "single-minded concentration on the field" was a common theme running through all the interviews she conducted. Roe concluded that "intense and continuous work is probably one of the most important factors in their present eminence." She found the drive of physical scientists remarkable. Roe (1951: 182; 1953a; 1953b) traced this appetite for hard, dedicated work, to internal psychological pressures.

In addition to an appetite for work, the entrepreneurial personality included a strong aggressive component. Roe (1951: 208) found that a tendency toward aggressive behavior was characteristic of the personalities of leading physical scientists. Indeed, when comparing scientists in different domains, Roe observed (1951: 232) that physicists were much more aggressive than biologists. Roe concluded that, in the group of physicists she studied, there was "a general acceptance of aggression as a permissible element in behavior." In a highly competitive environment aggression can be a valuable survival trait.

2. The third director of the Yerkes Observatory, Otto Struve (1897–1963), who was born in imperial Russia, is the exception. He did, however, learn about the organization and administration of large-scale science from his father, Ludwig, director of the Kharkov Observatory. On Otto Struve see Krisciunas (1992).

Sociologist Randall Collins provides some illuminating ideas concerning the structural basis for aggression and competition in science. It may be that in America before 1940 the characteristics discussed by Collins were most fully developed in the physical sciences. Scientists, Collins argues, begin their careers with positions to defend. These positions are a function of the ideas and techniques they were exposed to in graduate school (or in the early phases of their careers) and are closely tied to the views of mentors and patrons. Successful individuals are bound to defend certain positions and attack others as a consequence of the baggage they carry into the field. Further, Collins suggests that as careers unfold, productive scientists must continually make choices that involve them in conflict. These choices include research goals and the selection of allies, sponsors, and patrons, as well as decisions about the internal politics of their fields and the politics of science as played out on the national and international stage (Collins 1975: 496).

Collins (1975: 482) suggests that scientists whose careers are moving toward eminence must assume aggressive political roles. Most often these roles are associated with specific institutional locations. Scientists find it necessary to defend the legitimacy of their own institutions and attack the legitimacy of competitors. Nowhere are these imperatives more evident than in defining the role of observatory director or directing competition between observatories.

Of course, the motives of scientist-entrepreneurs differ significantly from those of businessmen. Cole (1959: 16–17) argues that in business, the entrepreneur is driven by a quest for security (wealth), prestige, power, and social service. This cluster of motives can be used to define *homo economicus*. *Homo scientificus* devotes most of its energy to maximizing knowledge and garnering increments of prestige and power that follow from important contributions to knowledge. This, in turn, leads to the acquisition of resources that individuals can mobilize for a variety of ends.

Entrepreneurs in both the corporate and scientific communities devote a great deal of energy to the creation of novel products. In the business world, success means opening up new markets or capturing a larger share of an established market. In science, product novelty means making contributions to knowledge beyond the routine or expected. If accepted, novel products in science lead to increased visibility for the entrepreneur and, in time, the acquisition of new increments of power and prestige.

In both science and business, competition in the creation and marketing of novel products is keen. Businessmen have an advantage; they can call upon experts to advertise their products. Scientists can not be so open, but have to seek aid from allies, in order to keep their products before the scientific public. From time to time scientists use the print media. Percival Lowell, scion of the Boston economic and social elite, was master of the art of publicity. His success angered other observatory directors and earned Lowell a reputation for unprofessional behavior (Hoyt 1976: 295). As Lowell wrote to a staff member, "I look forward with great anticipation to the photographs you will be able to make during the next opposition [of Mars]. . . . We will get out something [reports to the print media] to make others sit up."[3]

On balance, there are remarkable similarities between the drives to produce novel products in the worlds of business and science. Product innovation led to profits in business and new knowledge in science. Both enhanced the power of entrepreneurs who made them.

THE OBSERVATORY DIRECTOR AS CHIEF EXECUTIVE OFFICER

Simon Newcomb (Dryer 1923: 149) suggested that most astronomers looked to the great Victorian Astronomer Royal, Sir George Biddle Airy, as the man who "introduced production on a large scale into astronomy." However, by the 1890s, the English astronomer Herbert H. Turner (1861–1930) was writing enthusiastically of developments in the United States. Turner characterized the Harvard College Observatory under Pickering's direction as a "perfect organization" to collect data on an "immense scale." Ever one for the mathematical metaphor, Turner (1891: 191) suggested that at Harvard astronomical knowledge was growing in three dimensions. "The field for investigation is increasing with the square, and the mass of information to be collected with the cube of the radial extension."

Astronomers could get some idea of the scale of operations at Harvard when they learned that 9,000 photographs were made by its telescopes during the year 1890. Two decades later the Harvard director held a meeting of the Royal Astronomical Society spellbound as he described Annie Jump Cannon's staff working on the spectral classification of more than two hundred thousand stars. Pickering (1913: 286) referred to this as "a beautiful problem in scientific management." In time, other observatories would challenge Harvard. As *The Observatory* (Anon. 1913: 264) noted, Yerkes astronomers boasted the 40-inch refractor was "the busiest great telescope in the world," observing stellar objects by night and the sun by day, thus gathering scientific knowledge around the clock.

Compared to observatory directors in previous eras, the CEO of a corporate science factory had many more duties and responsibilities.[4] Chief among these was raising capital that would permit the institution to remain

3. Percival Lowell to Carl Otto Lampland, 13 October 1904. LOA, Lampland Papers.

4. I do not mean to deny that factory observatories developed in Europe as well as the U.S. But they were by no means identical. The difference between large-scale astronomical research institutions in Europe and America is a matter of scale. Mount Wilson or the Lick were simply larger and more complex organizations than Paris or Greenwich. See chapter 11 for a discussion of the differences between European and American astronomy.

competitive. Funds were needed for instrumentation and related equipment, including buildings, payrolls, and the publication of results. As the plant grew in size, routine maintenance became a significant budget item. At remote mountain sites like the Lick, housing for staff and their dependents, plus a school and a post office, added to the cost of doing science.

Directors raised funds in a variety of ways. Lick astronomer Edward Emerson Barnard discovered the fifth moon of Jupiter in 1892 using the great 36-inch Clark refractor. Both Lick director Holden and CEOs at other factory observatories were quick to realize that the economic potential of this event was at least as great as its scientific value. As Pickering wrote from Harvard, "I hope it [the discovery] will result in procuring for your Observatory valuable sympathy and energetic support in California, where public attention must certainly be strongly aroused by so remarkable an event."[5] Public sympathy could be converted into political support for the Lick budget.

Pickering himself employed a variety of fund-raising techniques (Jones and Boyd 1971: 219–31). One of the most interesting involved the use of a public relations expert, George V. S. Michaelis, president of the Boston Publicity Bureau. Pickering employed the bureau to maximize favorable newspaper coverage of work at the Harvard College Observatory. Apparently media coverage brought at least $20,000 in gifts to the Harvard coffers in 1902.[6] As Sir David Gill, director of the Observatory at the Cape of Good Hope and a major figure in the international astronomical community at the end of the nineteenth century, remarked (Forbes 1916: 374), "Pickering is the fellow to pick up money—and he uses it well when he gets it."

Pickering's successor, Harlow Shapley, expanded on the methods of his predecessor. Elected to the National Academy of Sciences in 1924, Shapley immediately began mobilizing resources available through the Academy. In the year of his election, Shapley approached the directors of the Smith Fund for $1,000 to study meteor trails captured on plates exposed at Harvard. Two years later the grant was renewed. While minuscule by late twentieth-century standards, these grants provided funds to push forward an important area of investigation in solar system astronomy.[7]

5. Edward C. Pickering to Edward S. Holden, 12 September 1892. SALO, E. C. Pickering Papers.

6. George V. S. Michaelis to Edward C. Pickering, 13 March 1902. HUA, Observatory Director's Correspondence, File 2. The folder labeled "Publicity Bureau" contains numerous letters between Pickering and Michaelis. For a discussion of funding nineteenth-century astronomy, see Miller (1970).

7. The scientific significance of this research is discussed by Ronald Doel (1996). The history of NAS trust funds as a source of support for research in nineteenth-century American science has yet to be analyzed. For an introduction to the topic, see Lankford (1987b).

Shapley's ability to mobilize Academy resources went hand in hand with his rise to power in the NAS. In 1926 he succeeded his mentor, Henry Norris Russell, as chair of the Draper Fund. The next year found Shapley requesting money from the Gould Fund. In 1928 Shapley secured money from the Academy to underwrite publication of his monograph on star clusters. Elected to the governing board of the NAS in 1929, Shapley later turned down the chairmanship of the Smith Fund, pleading potential conflict of interest. In the 1930s he continued to garner resources from the Academy.[8]

Observatory directors had to excel in a second area. They needed to be skilled engineers, able to envision new instrumentation and communicate their ideas either to optical craftsmen like the Clarks or Brashear, or work with in-house technicians. Virtually all of the great factory observatory directors lavished time and attention on the development of new instrumentation. Those who failed at the task found their positions seriously undermined. Holden's failure with the corrector lens designed to adapt the 36-inch Lick visual refractor for photographic investigations, coupled with his inability to bring the Crossley 36-inch reflector into operation, contributed to an already serious situation on Mount Hamilton and eventually led to his downfall (Osterbrock 1984b).

As Henry Norris Russell (Brouwer 1945: 124) pointed out, the successful observatory director had to be able "to organize a program of research so that results of the highest precision can be obtained efficiently and at a minimum cost of labor and money." To achieve these ends, the director had to possess skills in both "engineering and economic management."

The successful entrepreneur in charge of a great factory observatory also had to have executive ability. The demands of fund-raising and designing new instrumentation, coupled with developing and supervising research programs, required the ability to organize time and resources in the most efficient way. It also placed a premium on inter-personal skills, including communication. In addition, the successful CEO of a factory observatory had to delegate authority to subordinates.

The files of the second director of the Mount Wilson Observatory, Walter S. Adams, reveal the range of administrative duties that demanded

8. These activities are documented in the Director's Files of the Harvard College Observatory. I am indebted to HCO director Irwin Shapiro for permission to examine post-1920 materials. The letters in questions are Harlow Shapley to Whitman Cross, 7 December 1924; Assistant Secretary of the NAS to Shapley, 20 January 1926; Shapley to the Treasurer of the NAS, 25 January 1927; Local Organizing Committee for the fall 1927 meeting of NAS to Shapley, 2 June 1927; Shapley to Raymond Pearl, 5 May 1928 and 2 November 1928; Home Secretary of the NAS to Shapley, 24 April 1929, and Assistant Secretary of the NAS to Shapley, 29 April 1931. HUA, Harlow Shapley, Director's File.

his attention. Adams was responsible for the broad outlines of research programs in solar physics, stellar spectroscopy, and galactic astronomy and consulted frequently with staff members concerning their work. Through an assistant director he controlled staff publications. He also was responsible for setting salaries, making scientific appointments, and filling vacancies in the computing section, as well as in the technical support staff that fabricated new equipment and kept the telescopes in working order. Adams was in frequent contact with the president of the Carnegie Institution of Washington, the parent and primary funding agency of Mount Wilson.

In addition, Adams had to deal with several publics. Frequently, senior scientists desired to use the unrivaled facilities at Mount Wilson for special investigations. To those whose projects were considered worthy, Adams assigned time on one of the telescopes. Other astronomers simply wanted a desk in the library at the Santa Barbara Street headquarters in order to spend a sabbatical catching up on the literature and discussing research with the staff. He also received requests for the loan of plates to astronomers at other institutions. Director Adams always had to walk a tight rope, balancing the demands of the Mount Wilson staff and the needs of the astronomical community, whose leading members he could ill-afford to alienate.

The press and general public often contacted the Mount Wilson director with questions covering a wide range of issues. Keenly aware of the value of favorable press coverage, as well as the negative response of the Carnegie Institution to publicity that put the observatory in a bad light, Adams sought to educate science writers to the complexities of astrophysical research. He also dealt with the lay public whose letters ranged from those of the simply curious to ones from flat-earth cranks.

A final quality demanded of the successful observatory director in the world of corporate science involved political skills. Indeed, without such skills, it would have been difficult to succeed as a fund-raiser, technical innovator, or chief executive officer. In fact, the distribution of political skills among the directors of the great factory observatories was far from uniform. Pickering at Harvard or Newcomb at the NAO were consummate political animals, polished, patient, able to understand other points of view, yet ruthlessly single-minded in pursuit of cherished goals. Holden, the first director of the Lick, was a complex individual whose political skills were not equal to the demands of the job or the psychological pressures generated by his own overweaning ego. His flight from Mount Hamilton into exile as librarian at West Point provides dramatic testimony to the fate that awaited those who did not possess the necessary political ability.

George Ellery Hale and William Wallace Campbell make an interesting contrast. Their political skills differed, but each was an effective CEO. Hale was an executive of great vision and energy, whose strengths were fund-raising and selecting able lieutenants. William Wallace Campbell, third director of the Lick, was a man fiercely dedicated to his institution with a leadership style that emphasized hierarchy, organization and the importance of the chain of command. Gruff, given to outbursts of temper, but always willing to drive himself as hard as he drove the staff, Campbell inspired either intense loyalty or deep hostility. He showed great political skill dealing with the administration of the University of California, its Board of Regents, and the political leaders of the state, as well as with the staff on Mount Hamilton.

PRODUCT LINES

A survey of the great corporate astronomical research institutions in the early twentieth century will deepen our understanding of the new shape of American astronomy. The Harvard College Observatory under Pickering was, like the Ford Motor Company, dedicated to mass production. At Harvard the primary product lines included photometric and spectroscopic *durchmusterungen.*

George Ellery Hale resembled the iron and steel magnate Andrew Carnegie in his obsession with new and more effective technologies for the production of novel results. At Yerkes and then Mount Wilson, Hale concentrated on solar physics and stellar spectroscopy. Yerkes developed secondary lines including photographic mapping of the Milky Way, the study of radial velocities, double stars, and stellar parallaxes. The 60- and 100-inch reflectors at Mount Wilson permitted photometric and spectroscopic investigations of nebulae and star clusters, an important secondary line of production.

Lick concentrated on two primary lines: stellar radial velocities and double stars. Stellar and nebular spectroscopy made up a secondary line for the California institution. Its long-time director, William Wallace Campbell, had a style in many ways similar to the corporate executives of the American railroad industry, emphasizing hierarchy and organization.

At Flagstaff, Percival Lowell concentrated on solar system astronomy. A secondary line involved nebular spectroscopy. Lowell represented the flamboyant captain of industry of the precorporate period. He used every means at his disposal to keep his novel product line before the professional and general public.

At Allegheny and then Yale, Frank Schlesinger concentrated on the study of stellar parallaxes and the use of photography to acquire astrometric data of high precision. In both instances, he took rundown institutions and modernized them. Under Schlesinger, Allegheny and Yale moved into the mainstream of American astronomy. In this sense, he resembled the corporate executive with a talent for turning around a failing enterprise, making it productive and profitable.

Virginia's McCormick Observatory, during the directorships of Ormond Stone and Samuel A. Mitchell, remained an undercapitalized plant, used by its directors for personal advantage (reputation, visibility, and prestige). Virginia needed a Schlesinger but did not find one until after World War II. Like the Ford Motor Company after 1920, it was an ill-managed corporation whose product lines were routine and lackluster. Stone concentrated on visual observations of nebulae and Mitchell used the great refractor for photographic astrometry, concentrating on parallax work.

The Smithsonian Astrophysical Observatory specialized in solar research. Indeed, under its second director, Charles G. Abbot, the SAO became so narrowly focused on studies of the solar constant that it was hardly more productive than Virginia, even though better staffed and financed. Abbot was a CEO who did not keep abreast of consumer demand. His product declined in appeal after World War I.

AGGRESSIVE ENTREPRENEURSHIP

The entrepreneurial tendencies of the leaders of the American astronomical community are nowhere more clearly illustrated than in columns of figures printed in the 1902 *Yearbook* of the Carnegie Institution of Washington (1904: li). The CIW was, after the Rockefeller Foundation, the second of the great twentieth-century philanthropic foundations created in the United States.

From its inception in January 1902 through 31 October 1903 the CIW received applications totalling $2,200,398. While the number of applicants was not specified, they represented at least twenty-seven fields in the physical, biological, and earth sciences as well as social and behavioral sciences, and the humanities. Applications from scientific disciplines constituted 65 percent of the total funds requested ($1,427,863). Within this group, astronomy led the way with proposals amounting to $567,750. This was 40 percent of the total requested by *all* the sciences. Second were the biological sciences with requests totalling $468,000 (33 percent) followed by the earth sciences ($223,700) whose applications were 16 percent of the total. The American physics community submitted requests for $37,350. This was behind chemistry ($90,500) but ahead of engineering ($27,040). Mathematics brought up the rear with applications amounting to $13,523.

If *all requests* (science and nonscience) are considered together, astronomy still led the way (26 percent), followed by the biological sciences

(21 percent). The social and behavioral sciences were in third place with requests adding up to $343,083 or 16 percent of all funds requested.

These figures are remarkable. They suggest that astronomy was, indeed, the most entrepreneurial of the sciences in the Age of Little Science, measured not by size of the discipline (chemistry and biology were much larger), but in terms of aggressive behavior toward funding sources, self-confidence, energy, and involvement in large-scale projects demanding the corporate organization of resources and a factory approach to the collection and analysis of data. Astronomy's leading researchers were clearly animated by the entrepreneurial spirit. They were hard-driving scientists committed to large-scale research projects that demanded continual infusions of new capital. Their rush on the CIW during the first twenty-one months of its existence provides dramatic documentation of the astronomical entrepreneur in action

OBSERVATORY DIRECTORS: THE USES OF POWER

Directors of the great astronomical knowledge factories were sovereign, all-powerful individuals from whom there was no appeal except through reason. The power of a director was seldom openly discussed. A dramatic exception is found in connection with the turbulent events that marked the first decade of the Lick Observatory.

In the early 1890s, a young Johns Hopkins physics Ph.D., Henry Crew, was a member of the Lick Observatory staff, assigned to assist in developing a program in spectroscopy using a concave Rowland diffraction grating. The project was not successful and Crew ran afoul of director Holden, as well as members of the staff (Osterbrock et al. 1988: 81–82, 91–92). In December 1891, Crew wrote a long and apparently bitter letter to one of his undergraduate professors, Princeton astronomer Charles A. Young. In the letter he detailed the situation on Mount Hamilton and painted Holden in a most unflattering light. Young was slow to respond, assuming that anyone so deeply at odds with the management of a research establishment would resign and seek a more congenial position. When, in February 1892, Young learned Crew was still on Mount Hamilton, he penned an exceptionally strong reply. The senior scientist chastised Crew. If he could not convince Holden of his point of view though rational discussion, then "your only course as a subordinate was to accept his judgement. . . . The Director must decide, not the assistant, and it is the assistant's duty to carry out his instructions, cheerfully [and] faithfully."[9]

Observatory directors were always under special pressures, virtually unknown to other astronomers. They had to devote time and energy to

9. Charles A. Young to Henry Crew, 28 February 1892, pp. 1–2. SALO, Young Papers.

administration *and* maintain active research programs. The young Harlow Shapley acknowledged these pressures when he responded to a congratulatory note from Lowell director Vesto M. Slipher. "I have sworn not to let the administrative end of this place [HCO] spoil my joy for research."[10]

Solutions to this problem varied greatly. At Mount Wilson, Adams kept up an active program, including monthly observing runs on either the 60- or 100-inch telescope. His research was clearly distinct from that of other staff members. At Harvard, Shapley had a great deal to do with shaping the direction of research for many staff members. His interests included studies of nebulae and galaxies, star clusters and variable stars. He continued the tradition of photographic *durchmusterungen* established by Pickering and completed the Draper catalogue of stellar spectra. Shapley also encouraged research in astrophysical theory.

Perhaps the most extreme situation can be was at Allegheny and then Yale under Schlesinger. At those institutions, the director's research program involved *all* staff members. At Allegheny the program included the study of eclipsing binary stars and stellar parallaxes. At Yale, Schlesinger dropped binaries but continued work on parallaxes and added photographic astrometry. The Dudley Observatory presented virtually the same picture. Astrometry held center stage and there was little room and less encouragement for independent projects by members of the staff.

Directors operated in different ways. A comparison of the styles of Harlow Shapley, who became director of HCO at the age of thirty-six (1921), and Otto Struve, who assumed the reins at Yerkes at age thirty-five (1932), is instructive. The Russian-born Struve was heir to a great astronomical dynasty. Shapley, the Missouri farm boy, embodied the classic American tradition of the self-made man. Both men were exceptionally hard-working individuals. Harvard-trained astrophysicist Jesse L. Greenstein (1977–78: 40), whose first appointment was at Yerkes, recalled that Struve "bawled us all out for being lazy . . . [but] I couldn't feel any resentment to somebody who worked twice as hard and twice as many hours as I did, and Struve literally did that." Struve was, in style, an autocrat. Yet, Greenstein remembered, "he was a great leader." Struve developed excellent relations with Robert M. Hutchins, the president of the University of Chicago. He also brought visiting astronomers from around the world in order to keep the Yerkes staff abreast of the latest developments, especially in astrophysical theory. Berkeley Ph.D. Horace Babcock, who also served under the Russian-born director, recalled (1975–77: 18) that "Struve and his personality rather dominated Yerkes. He was a very strong, extremely energetic director. And he did so many things everybody wondered how he could possible achieve it all."

10. Harlow Shapley to Vesto M. Slipher, 13 January 1922. LOA, V. M. Slipher Papers.

No one doubted Shapley's energy; and he was an autocrat but cut from very different cloth than Struve. At the foundation of Shapley's personality was a deep and complex vein of insecurity. In some strange way this world-class scientist could never escape the culture of Missouri. Shapley was, by turns, a tyrant and a father figure. He presented himself very differently to men, women and graduate students (Payne-Gaposchkin 1984: chap. 11). Leo Goldberg (1978: 12) recalled of his graduate days at HCO, Shapley "seemed to run the place with the 'divide and rule' principle as far as the staff were concerned." Some were his favorites, others were not. Graduate students loved him. He was always available for an encouraging word and reserved Saturday afternoons for volley ball or ping pong games and there were frequent socials at the director's residence. However, Jesse Greenstein (1977–78: 32) recalled that there were "such dreadful things about the stress within the Observatory [HCO under Shapley] that I modeled all my management" of the astronomy program at the California Institute of Technology "on the opposite—love your colleagues, think you're all great, you're all in the best place in the world."[11]

The Shapley and Struve styles both produced important science, but the price tags attached to each were quite different. Perhaps the most compelling illustration of these differences comes from comparing the wartime experiences of HCO and Yerkes, and the condition of the institutions in the years immediately following World War II. At Harvard, many astronomers left the observatory for war work. Those who remained had great difficulty in continuing their research. Struve, on the other hand, made heroic efforts to keep the nucleus of his research staff at Williams Bay. Soon after the end of hostilities (DeVorkin 1982b: 623), "Yerkes was fully functioning once more, boasted a staff of more than 30 persons, and was better equipped than ever to pursue astronomical research." Harvard did not fare as well. A number of staff members never returned, and those who did were deeply divided over the observatory's growing dependence on military contracts (Kidwell 1992). Further, Shapley became a lame-duck director, whose left-of-center political activities lost the observatory support from the university administration.

Horace Babcock was the son of a distinguished Mount Wilson astronomer who would himself one day administer that institution. Com-

11. While it is difficult to document, I have often wondered about the impact of the culture of the University of Missouri, Shapley's undergraduate institution, on his style. Then as now it was a harsh, competitive environment, involving bitter interpersonal struggles for minimal resources. Missouri was never a place to learn collegial or fraternal patterns of behavior. Lankford (1980) deals with some of these issues in the context of the Missouri undergraduate education of Thomas Jefferson Jackson See. On T. J. See at Missouri, see Peterson (1990).

paring leading directors at the end of the 1930s, he rated Adams highest. The Mount Wilson director was characterized as less dictatorial than his counterparts at Yerkes or Lick. Adams sought to encourage staff members through conferences. Babcock (1975–77: 6) recalled that "a staff member's field of scientific investigation was mapped out in discussions with the Director" and "there were very many consultations." Adams preferred an approach that emphasized persuasion.

Heber D. Curtis, who took over the directorship at Allegheny when Schlesinger moved to Yale in 1920, provides an example of a director who used his power to change institutional goals in order to provide time for staff research. As he reported to William Wallace Campbell, Allegheny under Frank Schlesinger had become "a parallax machine without much chance for individual work." Curtis set about changing things. He planned to reduce parallax research to about "six-tenths of its present scope, so that everyone can have some time for himself."[12] Curtis also planned to modernize the Keeler reflector in order to move into solar spectroscopy. His plans were frustrated by lack of adequate funding, and in 1930 Curtis moved to Ann Arbor.

Observatory directors used their power to enhance the reputation and image of their institutions. Good public relations were considered essential for funding. A bad press, most directors feared, would lead to a decline in support. No one was more aggressive in managing the press than Campbell of the Lick. Many of his administrative decisions were taken with the print media in mind.

Mars, with its surface features and the possibility of intelligent life, attracted a great deal of media attention in the early years of the twentieth century. Indeed, if asked about astronomy, it is probable that the newspaper-reading public would have responded with comments about the red planet. Observations of Mars became a great bone of contention between leading observatories. Given these circumstances, it is not surprising that Campbell turned a deaf ear to a request from William Henry Pickering, brother of the Harvard director and a leading student of the moon and planets. Pickering proposed simultaneous observations of Mars by six major observatories around the world.[13] Campbell feared the press would turn the project into an inter-observatory contest, perhaps to the disadvantage of the Lick.

Campbell sought to protect his institution in other ways. In June 1920, he vetoed a scheme to repeat the debate between Mount Wilson astronomer Harlow Shapley and Heber D. Curtis of the Lick, on the scale

12. Heber D. Curtis to William Wallace Campbell, 22 July 1920, p. 2. SALO, Curtis Papers.

13. William Wallace Campbell to William Henry Pickering, 25 August 1913, p. 2. SALO, W. H. Pickering Papers.

of the universe. "It was fine and appropriate [at the April 1920 National Academy meeting] in Washington, but here in California I am rather on my guard against encouraging the public, especially the newspaper[s] . . . to draw the conclusion that there is anything but the greatest harmony prevailing between the Mount Wilson and Lick Observatories." Campbell believed that a repetition of the debate at Berkeley "would be pretty sure to get considerable notoriety in the newspapers, and the chances would be overwhelming that many of our constituents would get the wrong impression."[14]

Observatory directors did not have to answer to stockholders, but in most other ways their activities resembled those of CEOs in the business sector; they administered large-scale organizations whose future depended on their skills and judgment in the exercise of power.

INTEROBSERVATORY CONFLICT

Cooperation was the stated goal of observatory directors, and they made frequent reference to it in their formal correspondence. These references may have served a subtle psychological purpose. They can be read as reminders to themselves and others of the need to hold conflict in check. The stated values of the community stressed cooperation, but the economic, sociological, and psychological factors discussed earlier alert us to the potential for conflict. The situation was especially critical when directors believed the reputation and public image of their institutions were at risk. Challenges to institutional sovereignty and legitimacy could not be tolerated.

When he assumed the directorship of HCO, Harlow Shapley received letters of congratulation from CEOs at the other great corporate knowledge factories. Responding to William Wallace Campbell, Shapley referred to letters from "The leading astronomers of America" who had been "very generous in their welcome of a young man to Professor Pickering's place." Then, as if indicating that he knew the rules of the game, Shapley acknowledged "the spirit of good will and cooperation that prevails throughout our science."[15]

Cooperation and good will were necessary in order to facilitate the exchange of data between individuals at competing institutions. For many astronomers, the availability of data collected at other observatories was necessary for making significant contributions to knowledge. When a Lick

14. William Wallace Campbell to Armin O. Leuschner, 2 June 1920. BLUCB, Department of Astronomy Papers.

15. Harlow Shapley to William Wallace Campbell, 28 November 1921. SALO, Shapley Papers.

staff member returned from an examination of the Mount Wilson plate collection, he reported to director Robert G. Aitken that there was a wealth of data that would facilitate his study of Comet Halley. Aitken immediately wrote director Adams about the use of the Halley plates. His approach, however, was circumspect. "If you have the slightest hesitation" about loaning the materials "please do not send the plates." While Aitken affirmed his commitment to "mutual cooperation" he insisted that "I do not believe in disturbing the records of another man's work or making use of them if the other man is planning to use them in any way."[16] When a Harvard staff member needed data on radial velocities of faint variable stars, Shapley assured director Adams that he should feel no "special obligation" to send the information "if it seems to be to the disadvantage" of Mount Wilson staff members.[17]

By the 1930s deep distrust existed between Shapley's HCO and the great California observatories—the Lick and Mount Wilson. In part, tension can be traced back to Shapley's years at Mount Wilson and in part it reflected competition between Harvard and the West Coast institutions. Indeed, according to Greenstein (1977–78: 59; DeVorkin 1980: 604), the East-West division among the major research institutions even pitted West Coast astronomers against Yerkes and, from the late 1930s, its sister institution, the McDonald Observatory in Texas.

Sometimes directors felt competition between observatories was a byproduct of media publicity. As Edward S. Holden wrote George Ellery Hale, "There is a disposition in some newspaper accounts to set your large telescope [the Yerkes 40-inch refractor] as a somewhat hostile rival to ours [the Lick 36-inch refractor]." Partly as a question and, perhaps, a warning, Holden indicated that "I take it for granted that you are entirely blameless in this." The senior scientist concluded, "Rivals we shall always be, of course, but friendly ones I hope."[18]

Large-scale, routine observing programs appeared attractive targets for interobservatory cooperation. Through careful planning, the quantity of data could be increased and the quality improved. Further, the cost to individual institutions would be reduced. Cooperation rested on a division of labor, as well as mutually agreed upon guidelines for organizing and carrying out programs. Solar work, the study of stellar parallaxes, and special occasions like the 1909 return of Comet Halley provided ideal opportunities for cooperation between observatories.

There were, however, fundamental difficulties associated with devel-

16. Robert G. Aitken to Walter S. Adams, 23 July 1928. MTWA, Adams Papers.

17. Harlow Shapley to Walter S. Adams, 10 October 1928, p. 1. MTWA, Adams Papers.

18. Edward S. Holden to George Ellery Hale, 28 October 1892, p. 2. UCYOA, Director's Papers.

oping cooperative programs. As George Ellery Hale delicately put the matter, it was a question of insuring "perfect liberty to all who participate" so that hard-driving, talented and lucky individuals could "secure every advantage" that may result from "personal initiative."[19] In short, the problem was how to cooperate and yet allow highly competitive astronomers room for individual discovery. Clearly, no director would take part in a program that reduced the role of his observatory or its staff members to mere data gathering, without the chance of finding something novel and interesting to add to the store of knowledge. As a result, Hale (Wright 1966: 260–61) concluded that a plan for cooperation had to be built on very general lines to ensure individual initiative and institutional sovereignty. In the end, Hale's plans led to the formation of the International Union for Cooperation in Solar Research (1905).[20]

THE POLITICS OF LARGE TELESCOPES

As Holden indicated to Hale, large telescopes could provide grounds for conflict between leading observatories. Sometimes it was the print media that forced conflict on unwilling directors. On other occasions, directors themselves offered to do battle.

In 1888, soon after the Lick Observatory opened, Director Holden published a paper on the ring nebula in *Lyra*. His aim was to compare what he saw using the 36-inch telescope with earlier observations. Although the discussion was circumspect, Holden reported (1888: 385, 387–88) that on one occasion he and the assistant gave up mapping stars in and around the ring nebulae because the task "seemed to be endless." Of course, they recorded more objects than had been seen using the reflectors of Lassell, Lord Rosse, or the younger Herschel. Holden concluded that "the great 36-inch refractor is in fact and at present what Mr. Lick's deed required that it should be, 'the most powerful in the world.'"

In the entrepreneurial world of astronomical research, Holden's claim would not go uncontested. Percival Lowell, whose Flagstaff observa-

19. George Ellery Hale to Henry Crew, 4 February 1904, p. 2. NWUA, Crew Papers.

20. Just before World War I, the Astronomical and Astrophysical Society of America created a committee on stellar parallax research. Its charge was to coordinate the activities of the nine major parallax-producing institutions. The committee, whose work was frustrated by competing observational methods and results that differed substantially between institutions, was never able to achieve a high level of cooperation. The letters of participants reveal considerable frustration. Powerful observatory directors apparently blocked the efforts of the committee. As a consequence, stellar parallax research remained a field marked by waste and duplication. In spite of the attempts of the AASA, it proved impossible to rationalize methods of production. See for example, "Report of the Committee on Stellar Parallaxes," ca. 1915. UVA, Mitchell Papers. The Mitchell Papers provide excellent resources for further research on this topic.

tory was equipped with a 24-inch Clark refractor, soon challenged all comers with his observations of fine detail on Mars, Venus, and Mercury (Hoyt 1976: chaps. 10–12). When other observatories failed to replicate Lowell's findings, he characterized their instruments as optically defective or situated where the atmosphere was too turbulent and polluted to permit effective use. In either case, their products were inferior to those produced on Mars Hill.

Holden (1897: 92) did not remain silent for long. "I have no hesitation in saying that such markings as are shown by Mr. Lowell did not exist on *Venus* before 1890." Further, the Lick director argued, "It is my opinion that they do not now exist on the planet, but that they are illusions of some sort." Thus Lowell and his observatory became embroiled in conflict that sometimes moved beyond technical issues to the ability and character of Lowell himself.[21]

One of the most curious aspects of the conflict was a competition to see which instrument would reveal the faintest stars. The location selected was the region following the star delta *Ophiuchi*, first observed in the 1880s as a test for the 26-inch refractor at the U. S. Naval Observatory. According to Lowell (1905: 57), observers at Washington recorded 63 stars. Lick Observatory astronomer Richard H. Tucker reported a total of 161 stellar objects in the same field. Lowell announced that under less than ideal conditions he counted 172 stars, a 7 percent increase over the Lick. Lowell did not pretend that these observations were exhaustive. He strongly hinted that under better conditions many more stars could be observed.[22]

Soon the controversy shifted from the "space penetrating power" of large telescopes to more complex issues (Hoyt 1976: chap. 9; DeVorkin 1977). Lowell sought to prove that the atmosphere of Mars was similar to that of Earth. This was a necessary part of his campaign to demonstrate that the red planet was inhabited (Crowe 1986: ch.10). As a result, the Lowell Observatory moved from visual observations of surface markings to planetary spectroscopy. Technical difficulties associated with both the acquisition and interpretation of spectroscopic data added to the complexity of the debate. Soon the Lick and Lowell observatories would clash over spectroscopic studies of the Martian atmosphere.

As conflict grew, Lowell sought allies. He turned not to other observatory directors but to individuals who were marginal to the astronomical

21. The best biography of Lowell remains Hoyt (1976). Professor David Strauss of Kalamazoo College is at work on a biography of Lowell.

22. The plate that was published with this paper seems to represent the Lowell observations with star images (circles) that are darker and a little larger than the those that depict the USNO and Lick observations. This may be the fault of the engraver in England or it may be that Lowell supervised the preparation of the plate and sent it to England to be printed.

community. Perhaps conditioned by the dynastic politics of Boston's social and economic elite, Lowell sought highly visible individuals with connections to the business community and the federal government. In early 1908, Lowell described a luncheon given him in Washington by the inventor Alexander Graham Bell. Among the guests were the present and immediate past superintendents of the Naval Observatory, the director of the Weather Bureau, and the elderly Simon Newcomb. Lowell, who came armed with copies of spectrograms made on Mars Hill, sought to convince the luncheon guests of his findings. It is amusing to imagine Lowell, a splendid raconteur, expounding on the interpretation of spectra as the others studied the prints. No one in the room had any practical experience with or physical understanding of astronomical spectroscopy.

Writing with evident gusto, Lowell reported to Vesto M. Slipher that "The Weather Bureau . . . is now our warm friend" and that its director had assured Lowell "that we had now got the Naval Observatory." When the superintendent asked for a copy of the spectrograms, Lowell replied "semi-jokingly 'yes, if you will be very good.' He instantly said that he would and that he would do anything I wanted. So you see," Lowell concluded, "the L. Obsy. is forging ahead."[23]

Lowell's choice of allies made the struggle extremely complex. He confronted leading members of the power elite in the American astronomical community by using techniques and political strategies that appeared to them unprofessional. As Hale remarked to Campbell, "I fully share your opinion that Lowell is doing a great deal of injury to astronomy." Hale particularly disliked Lowell's "absolutely unscientific method of dealing with his material and of stating his case. This is bound to do much harm with the public."[24] Hale was referring to Lowell's use of mass-circulation magazines rather than scientific journals.

At the end of 1907 Lowell reiterated his claims concerning the power and superiority of his telescope and the Flagstaff site. In the December issue of *The Century Magazine* he extended the comparison to include the Yerkes telescope. The Yerkes director, Edwin B. Frost, had no desire to become involved in a public debate and warned Lick Director Campbell "Lowell is a rather agile party to deal with, and I think he will enjoy entering into any kind of a scrap that you may (in his opinion) open up."[25]

In Campbell, Lowell met his match. Early in his career, Campbell established himself as a tough fighter. In 1894 the young Lick astronomer

23. Percival Lowell to Vesto M. Slipher, 8 January 1908, p. 1. LOA, V. M. Slipher Papers.

24. George Ellery Hale to William Wallace Campbell, 20 May 1908, p. 1. SALO, Hale Papers.

25. William Wallace Campbell to Edwin B. Frost, 1 September 1908 and Frost to Campbell, 16 September 1908, p. 2. SALO, Frost Papers.

became embroiled in a controversy over the composition of the Martian atmosphere with the venerable father of astrophysics in Britain, Sir William Huggins. Campbell (DeVorkin 1977: 41), when urged to moderate his combativeness by Hale, responded that "it is not right for any man, however young, who happens to collide with the 'authorities,' to lie down and let them pass over him." Far from lying down, Campbell was given to frontal assaults.

Campbell's response to Lowell appeared in a mass-circulation journal, *The Outlook*. By turning to the popular press, Campbell sought to meet Lowell on his own ground. From Boston, Lowell telegraphed his assistant on Mars Hill for information to refute Campbell. He got that and more besides. Slipher, perhaps venting pent-up anger at Campbell that went back almost a decade, wrote "We have stood enough of such [criticism] in recent years . . . [and] the time has come when it seems to me we might now make a stand and have it out."[26]

Slipher's bark was worse than his bite. He proposed that Lowell challenge Campbell to yet another contest. Using similar spectrographs attached to the Lick 36-inch and the Lowell 24-inch refractors, Slipher suggested spectroscopic observations of the same star, using the same plates, developed in a similar manner. "If he [Campbell] declines this proposal he practically admits he has no case."[27] Lowell rejected Slipher's idea. "I think nothing would be gained by it . . . [Campbell] would give such good excuses for not trying it that you would lose rather than gain in the result."[28]

Campbell also struck back in *Popular Astronomy*. For the first half of the twentieth century, *PA* served many of the same functions that *Sky & Telescope* does today. It was read by amateur and professional alike. For Lowell, who found the professional journals closed to him, *PA* became an important outlet. Campbell sought to refute Lowell's claims that the Flagstaff instrument could see more faint stars than the 36-inch on Mount Hamilton. Observing only the first third of the delta *Ophiuchi* field, Lick astronomers counted 35 percent more stars than were claimed in Lowell's published observations. At that point, Campbell wrote (1908: 562), "I ordered the work stopped; there are endless ways in which the telescope and the observers' time can be put to better use." It is probable that the isolation of the Lowell Observatory from the mainstream of astronomy during the interwar years can be traced, at least in part, to these public squabbles over the prowess of large telescopes.

26. Vesto M. Slipher to Percival Lowell, 13 October 1908, postscript dated Wednesday A.M October 14, p. 4. LOA, V. M. Slipher Papers.

27. Ibid.

28. Lowell to Slipher, 19 October 1908, p. 1. LOA, V. M. Slipher Papers.

THE POWER ELITE ASSEMBLED

Peter Van de Kamp, who moved from Virginia to become director of the Swarthmore College Observatory at the end of the 1930s, recalled that between the wars, American astronomy was controlled by three men, all observatory directors. Shapley at Harvard, Russell of Princeton, and Schlesinger from Yale were, in Van de Kamp's words (1977–79: 47), "the generals." The Dutch-born Van de Kamp continued, "They ran astronomy, in a way. Their opinion was very important; when it came to appointments and that sort of thing, they were consulted."

The generals, in turn, relied on a staff. In 1920 Schlesinger created an informal group, referred to simply as the Neighbors. Drawing astronomers from as far south as Charlottesville, as far west as Albany, and northward to Amherst, the Neighbors assembled quarterly at some convenient location (most often New Haven). Typically, the meeting would begin on Friday afternoon. A formal dinner (sometimes honoring a special guest) took place in the evening, and Saturday would see a continuation of discussions and informal conversations between members of the group. The object was not to hear papers, but to discuss astronomy-related topics, either in small groups or one-on-one. The Neighbors had all the characteristics of an exclusive male social club.

There was no fixed membership. A 1930 list included fifteen names, of which six were members of the National Academy of Sciences and nine directors of observatories ranging from Amherst to the Smithsonian Astrophysical. A decade later the list included seventeen names drawn from the same geographical area. The number of NAS members stood at three and observatory directors at seven.[29] The issue of opening the group to women astronomers was debated, but resolved in favor of keeping the organization a male preserve. The result of excluding women from informal professional gatherings such as the Neighbors meant, as historian Margaret Rossiter (1982: 92–4) points out, their isolation from discussions of the latest advances in science, as well as from the real sources of power within a scientific community. In response, women astronomers formed their own version of the Neighbors.

Control over those invited to meetings of the Neighbors gave Schlesinger and the generals great power. As Shapley indicated in 1927, "I understand, of course, that some meetings we intend to keep restricted in

29. "List of Neighbors" ("Probably 1930" in pencil at top) and "Neighbors 1940." YUA, Astronomy Department Papers. I have not been able to locate the archives of the Neighbors. References are scattered in the Astronomy Department Papers at Yale. Surviving documents suggest the range of activities and indicate that the Neighbors were tightly controlled by Schlesinger, with advice from Shapley and Russell.

attendance."[30] Shapley often pressed Schlesinger to secure invitations for younger members of his staff and, "when the Neighborhood meeting is about to turn out unexpectedly thin," to use the occasion to invite special guests; for example, a Brown University faculty member in order to kindle his interest in research. Such an invitation, Shapley remarked, might make the individual "feel flattered to be recognized a little."[31]

In addition to women, apparently the line was drawn at Roman Catholics. Schlesinger indicated that he had no objection to Shapley inviting Ernst Opik (b. 1893), an Estonian astronomer visiting Harvard, but objected to an Irish priest, one Father O'Connell. As Schlesinger enjoined Shapley, "if you have to choose between the two . . . the farther East the better."[32] There was no recorded objection from any quarter to inviting the young Martin Schwarzschild, a Jewish refugee from Nazi Germany.

Considering the informal nature of the group, it is not surprising that surviving letters give little indication of agendas. It is clear that Schlesinger wanted to use the October 1930 get-together to discuss the 1932 meeting of the International Astronomical Union at Harvard. Perhaps, this Neighbors gathering discussed the politically sensitive issue of who in the American astronomical community should be invited to the IAU.[33]

In 1935, Russell arranged a meeting at Princeton at which Einstein was the guest of honor. Visiting European cosmologist Abbe Georges Lemaître (1894–1966) and Princeton mathematical physicist H. P. Robertson (1903–61) also joined the astronomers. After dinner, the men retired to the lounge of the Engineering Building. "Einstein was in good form and started in telling a story to Russell. Russell came back with one of his. One chair after another was pulled into the circle until finally all sat together swapping stories, but mostly listening to Einstein as he touched on one subject after another."[34] An opportunity like this was available only to the inner circle.

Writing of the Neighbors, Frank Schlesinger (1926: 596) underscored the social and political importance of the organization. "In a small group like this . . . our knowledge of each other has gotten to be intimate and has given rise to a network of friendships, which in themselves, aside from any questions of scientific results, justify the existence of the Neighbors."

30. Harlow Shapley to Frank Schlesinger 17 March 1927. YUA, Astronomy Department Papers.

31. Shapley to Schlesinger, 22 September 1925. YUA, Astronomy Department Papers.

32. Schlesinger to Shapley, 5 February 1932. YUA, Astronomy Department Papers.

33. Frank Schlesinger to the Neighbors, 16 September 1930. YUA, Astronomy Department Papers.

34. Raymond S. Dugan, "A Meeting of the Neighbors, Princeton, N.J., 3–4 May 1935." YUA, Astronomy Department Papers.

It is an axiom of modern social life that social networks can be converted into political networks on short notice, as was the case in the struggle over the appointment of the director of the Nautical Almanac Office, discussed later in this chapter.

MOBILIZING RESOURCES: COMPETITION FOR FUNDING

Before World War II, American astronomy usually depended on private patronage for infusions of capital that would permit construction of new facilities. Patrons ranged from private individuals, for example, Charles T. Yerkes, who financed the 40-inch telescope and original buildings at the observatory which bears his name, to the Carnegie Institution of Washington. In addition, major gifts from individuals allowed observatories to undertake large, expensive projects (Miller 1970: chap. 5). A good example is the Henry Draper Memorial at Harvard which permitted Pickering to embark on the great spectroscopic *durchmusterung*, financed by Draper's widow.

The following discussion examines two aspects of funding. The first deals with small grants made to individuals, the second with the highly competitive process of securing large sums to be used for the creation of new institutions.

In the age of Little Science, securing a grant was, it seems, ridiculously simple. There were neither forms nor rules, or any need to negotiate indirect overhead or institutional cost-sharing; neither was there a formal system of peer review. In the epoch before the advent of massive federal support for scientific research, the process was informal.

Small grants often made the difference between success and failure for the research programs of individual astronomers. Sums, ranging from $500 to $2,000 made it possible to embark on new projects. Auxiliary equipment could be purchased or existing instrumentation modified; assistants and computers could be hired and the collection and reduction of data get underway. For some projects, one grant was enough; for others it was necessary to seek aid over a period of years. In 1890, Catherine Wolfe Bruce (1816–1900), whose benefactions were important for several of the great factory observatories, offered to aid astronomical research worldwide with the sum of $6,000. Harvard director Edward C. Pickering was the executive officer for the fund (Jones and Boyd 1971: 278–79). He immediately sent a circular to astronomers in both Europe and America. Attached to the copy sent to director Edward S. Holden at the Lick Observatory was a note. Pickering asked, "Would an appropriation of five hundred ($500) dollars to the Lick Observatory for a stellar spectroscope be a useful and acceptable application of the gift described in the enclosed circular?" Pickering hoped that aid from the Bruce Fund would make clear to the administration of the University of California and the state legislature how highly leaders of the American astronomical community regarded the Lick. Pickering suggested that national recognition might make Californians "more ready to provide for your current expenses if . . . they realize how necessary such aid seems to other astronomers."[35]

Pickering used his power over research funds in other ways. He congratulated William Wallace Campbell, third director of the Lick, on the latter's appointment to office in 1900. Stressing cooperation in the astronomical community, Pickering pledged "this Observatory will always be ready to render any assistance in its power to advance the work of the Lick Observatory." Pickering went on to report that he had just been appointed chair of the Draper Fund of the National Academy. However, there was a problem. The fund had "over two thousand dollars available for research in astronomical physics and no applicants." Pickering offered Campbell $500 from the Draper Fund, renewable if preliminary results were satisfactory, for a project using the Crossley reflector to photograph the spectra of faint stars.[36]

For Campbell, the offer meant not only recognition and acceptance into the power elite, but the chance to move into a new research area. However, there was some loss of autonomy; individuals external to the Lick Observatory defined a portion of its research program. From Pickering's perspective, the offer was, indeed, a gesture of welcome, but it also permitted the Harvard director to press his own research agenda on Campbell. In this case, however, the interests of both parties coincided.

Informal procedures guided the dispersal of funds controlled by the National Academy of Sciences.[37] There is no history of the trust funds of the National Academy (Lankford 1987), but it is clear that between the 1880s and World War II, they provided a significant number of modest grants to astronomers. Small grants were important for several reasons. They stimulated research and the growth of astronomical knowledge, thus allowing successful applicants to improve their visibility and standing within the community. Beyond this, the distribution of NAS funds provided elite astronomers with a means of exercising power. Members of the elite used NAS trust funds to promote research they favored, thus shaping the content of American astronomy. This provides an example of the way

35. Pickering to Holden, 15 July 1890. The circular is attached. SALO, E. C. Pickering Papers.

36. Pickering to Campbell, 17 December 1900, pp. 1–2. SALO, E. C. Pickering Papers.

37. Benjamin A. Gould to Asaph Hall, 2 December 1891. LC, Newcomb Papers. Charles L. Doolittle to Seth Carlo Chandler, 9 August 1909. DOA, Papers Relating to the *Astronomical Journal*. Again, readers must remember that given the current condition of the Dudley Observatory Archives, any location is at best approximate.

in which the generic community influenced events at the level of the specialist research community.

The creation of the Carnegie Institution of Washington sorely tested cooperation in American astronomy. It was relatively easy for the power elite to hand out modest grants from the trust funds of the National Academy of Sciences; there were sufficient resources to support almost all who applied. However, the possibility of CIW funding for large projects brought the elite into conflict. The situation was exacerbated by the fact that members of the power elite served as advisors to the CIW trustees. In the end, the distribution of CIW funds for astronomy turned on personalities, shifting political alliances, and competing visions of the future of American astronomy, as well as motives dictated by institutional affiliation. As an exercise in planning for the future of astronomy, these experiences were a far cry from the system of priorities and community-wide consensus developed in the 1960s (Greenstein 1972).

Soon after it was chartered, the CIW appointed a number of committees to advise the trustees. Edward C. Pickering chaired the astronomy advisory panel. Before accepting, the Harvard College Observatory director requested clarification. Could advisory committee members also apply for CIW funds? Only when the question was answered in the affirmative did Pickering agree to serve (Jones and Boyd 1971: 422).

The other members of the panel included George Ellery Hale, director of the Yerkes Observatory; Samuel P. Langley, director of the Smithsonian Astrophysical Observatory and secretary of the Institution; Lewis Boss, director of the Dudley Observatory; and Simon Newcomb, retired superintendent of the Nautical Almanac Office and President of the Astronomical and Astrophysical Society of America. While the group seemed to be weighted in favor of the new astronomy (Hale, Pickering, and Langley), it soon became apparent that Langley's administrative duties and interest in aeronautics left him little time for active participation. Further, Pickering, whose previous involvement in funding astronomical research earned the distrust of many American astronomers, did not have the full confidence of the CIW administration (Jones and Boyd 1971: 424). Newcomb, retired and working hard to finish several major projects, did not play a significant role. Before the end of the year, Lick director Campbell was added to the group. It soon became clear that Boss and Hale would be the antagonists in a major power struggle, while Campbell found himself caught between the two, tempted by Boss to join the battle against Hale.

Without formal guidelines or traditions to define behavior and limit conflict, the early years of the CIW were fraught with difficulty. As the first CIW president, Daniel Coit Gilman, remarked to the president of the University of California, "college professors composing the [Advisory] Committees were almost invariably inclined to favor the particular things in which they are individually interested, and make extravagant estimates for that purpose; and, further, that it was almost impossible to get them to compromise or admit the wisdom of doing anything else." William Wallace Campbell reported these remarks to Lewis Boss at the Dudley. The Lick director went on to suggest that perhaps Gilman's strictures do "not apply to the Directors of Observatories."[38] In fact, members of the power elite were at least as difficult to deal with as the professors Gilman caricatured. The CIW provided members of the power elite with a chance both to aid their own institutions and to define the direction of astronomical research. The situation was ripe for conflict.

Boss and Hale (Wright 1966: 159–78) represented conflicting views on the organization of scientific institutions and the direction of astronomical research. For Hale, the scientific enterprise was organized on a large scale. Teams of researchers attacked problems in solar physics and stellar spectroscopy. Improved instrumentation was essential to solving scientific problems. Observatory directors were, in Hale's view, CEOs who supervised the operations of factory observatories.

Boss held very different views. Committed to micro-management, Boss, as his son remembered, "considered . . . that unless he did a thing personally it could not be done at all."[39] Boss believed that a director must take an active part in all aspects of the work of his institution. He was skeptical of large-scale science and held up the productive but highly specialized small observatory as the ideal form of scientific organization. To be sure, there was a growing conflict between rhetoric and practice. In 1904 the Dudley Observatory became a department of the CIW and its staff grew rapidly. Boss criticized astronomical research on a large scale, but did not scruple, when he had an opportunity, to improve the fortunes of his own institution.

As the debate over CIW support for astronomical research developed, the protagonists mobilized resources in order to enhance their power. As directors of major research institutions, each had high standing in the American astronomical community. Boss had long been a member of the NAS, while both Hale and Campbell had been recently elected. Hale, although younger than Campbell, had already garnered important international honors including election as foreign associate of the Royal Astronomical Society and an award from the French Academy of Sciences. He was also well known as the founder of the *Astrophysical Journal*. Campbell, on the other hand, could point to his Nobel Prize nomination as well as similar European honors.

Perhaps the greatest difference between Hale and Campbell was so-

38. William Wallace Campbell to Lewis Boss, 9 September 1903, p. 3. SALO, Campbell Papers.

39. Benjamin Boss to William Wallace Campbell, 24 November 1911, p. 2. SALO, Benjamin Boss Papers.

cial. Hale knew virtually all the leading members of the American astronomical community at first-hand. He was a great traveler and took advantage of every opportunity to meet other scientists. Campbell, isolated on a California mountain top, had not developed a comparable network. Both Hale and Boss had access to resources that were denied Campbell. They either knew influential members of the CIW governing board or had social connections that could be used as informal means of communicating with the trustees.

In the end it was Hale who could mobilize a vital resource that seemed inaccessible to Boss and Campbell. Hale was the most entrepreneurial of the group. Ever alert to new opportunities, he seized the main chance whenever he saw it. Within two weeks of reading about Carnegie's gift to endow the CIW, Hale fired off a proposal asking for funds to complete a 60-inch reflector.

As the advisory panel worked to achieve consensus and present the CIW trustees with its report, Boss applied pressure on Hale. He urged a unanimous report in order "to avoid the unseemly personal struggle over spoils which is sure to ensue if we leave the Trustees in doubt as to what astronomers want." Boss then quipped, "I am quite sure you would fare well in such a struggle, without unseemly exertion on your part." And he added, parenthetically "I wish I were as well off."[40]

Boss emphasized the need to organize the report around clearly defined priorities. He did not support the construction of large new telescopes, but rather the work of existing observatories. This line of argument was, obviously, to the advantage of the Dudley. Boss had no need for new instrumentation. What he sought were funds for more staff and for an extension of his work in the southern hemisphere. CIW support for a new astrophysical research institution would conflict with the interests of Boss.

The director of the Dudley Observatory went out of his way to make sure Hale appreciated his connections. Speaking of CIW trustee Charles D. Walcott, America's leading paleontologist, Boss wrote, "Dr. Walcott is an old friend of mine. I have known him ever since you [Hale] were a small boy." Then, in response to Hale's report of Walcott's visit to Yerkes and his interest in CIW financing the 60-inch reflector, Boss curtly dismissed the geologist's views by questioning his "judgment upon a technical matter in astronomy."[41]

Hale, pushing for a grant of a quarter of a million dollars in order to establish a solar observatory (which would, incidentally, include the 60-inch reflector to be used for stellar spectroscopy), found Boss an im-

40. Lewis Boss to George Ellery Hale, 27 August 1902, pp. 1–2. HPME.

41. Boss to Hale, 27 August 1902, pp. 2–3. HPME.

placable foe. In a veiled attempt to denigrate Hale's abilities as a scientist and administrator, Boss praised Campbell for organizing the Lick Observatory for research in a few carefully selected fields. Boss went on to criticize astronomers who spread themselves too thinly. He would support Hale only if the younger man promised to give up all other interests and concentrate on solar physics.[42]

Boss went out of his way to disparage astrophysics and, by implication, Hale and others who might seek funding for research in the field. "There are astronomers accustomed to the severe aspect of some of the more exacting forms of mathematical research in astronomy who find it difficult to feel much sympathy with astrophysics." This attitude Boss explained as follows. He cited the "tentative character" of early research in astrophysics but suggested that some recent work did meet the standards of exact science. Here, the reference was not to Hale's research in solar physics but to Campbell's work on stellar radial velocities, investigations that were of particular interest for astrometry. Unable to leave well enough alone, Boss continued: "Astrophysics also has to labor under the disadvantage that a number of inferior investigators are attracted to it by its novelty." Then, administering the coup de grâce, Boss observed "Even among its better known men there has been a great deal of discussion which reminds one of art criticism in its inconclusiveness and general futility."[43]

Boss devoted the rest of this letter to a review of his own career. At times, the discussion bordered on the pathological. The older man revealed the driving force of his ambition and how often he had met with setbacks. At some level, the letter involved an implicit comparison between his career and that of Hale, the *wunderkind*. "Do you wonder that the demon of envy is strongly developed in my character?" Moving from thinly veiled self-pity to bitter criticism, Boss lamented his fate. "What is the use? There are other astronomers anxious to do things. Why should I have it all?"[44] While he hoped to get some support from the CIW, Boss feared his application might be ignored completely.

Then, in a truly mean-spirited paragraph, he consoled Hale, who was having one of his periodic bouts with ill-health. As Boss understood it, Hale was losing his eyesight and could no longer take an active part in the work of the Yerkes Observatory. This, to Boss, was tantamount to leaving science. Extolling a hands-on approach to the management of a research institution, he concluded that "Nothing tends so much to ripen judgment" as the active participation of the observatory director in all aspects of re-

42. Ibid., pp. 4–5.

43. Boss to Hale, 3 September 1902, p. l. HPME.

44. Ibid., p. 2.

search. Boss went on to suggest that work carried out without the participation of the director was bound to be inferior.[45]

A week later, without any warning, the older scientist changed his tactics. Hale's proposal for a solar observatory including a 60-inch reflector for astrophysical work had competition. Lick director Campbell intended to request funds for a 48-inch reflector to be dedicated to the study of variable star spectra. Sensing an opportunity to make mischief by pitting Hale against Campbell, Boss urged Hale *not* to withdraw his proposal. By this point Hale must have been confused! In a series of sometimes daily letters, Boss had exerted great pressure on Hale, requesting him to abandon his proposal for a new observatory. Then, after an unexpected about-face, Boss now appeared willing to support the project. Warning Hale not to abandon the field to Campbell, who, with the aid of the "U. of Cal. lobby might pull off" his attempt to secure funding for a large telescope, Boss assured Hale that "I am dead sure yours is the most deserving of the two on many grounds."[46]

Hale's response was to withdraw his proposal. As Hale confessed to Campbell, he found the dual role of advisor to the CIW and suitor for its funds difficult. At no point in the letter did Hale refer to the pressure applied by Boss. It must, however, have been crucial to his decision. Opposition from a respected senior member of the advisory panel would be taken seriously by the CIW Trustees.[47]

In December 1902, the CIW Trustees provided funds for site selection in the event they might later decide to fund a new astronomical research facility. Campbell was to direct the project, and it was to be carried out by Lick staff member William J. Hussey, who was to test observing conditions at various locations in southern California and then press on to Africa and Australia. Hale, who kept in close touch with Campbell and Hussey, was not optimistic. "This whole scheme for a new observatory . . . is contingent upon the remote possibility of an additional gift from Mr. Carnegie, and has nothing to do with the existing funds of the Institution."[48]

In 1903, both Hale and Boss changed tactics. Hale divorced himself from the advisory committee. By the end of the year he was packing to go

45. Ibid., pp. 4–5.

46. Boss to Hale, 11 September 1902, p. 1. HPME.

47. George Ellery Hale to William Wallace Campbell, 11 September 1902, pp. 1–2. See also Hale to Campbell, 8 September 1902. SALO, Hale Papers.

48. Hale to Campbell, 8 October 1902, p. 2. SALO, Hale Papers. See also Hale to Campbell, 7 January 1903. SALO, Hale Papers. The Hussey Papers in the SALO provide a wealth of information on the process of site selection. The scientific, economic, and political history of site selection from Charles Piazzi Smyth's 1856 expedition to Teneriffe to the location of the Mauna Kea Observatory would make a valuable contribution to the history of modern science.

to California, virtually without financial backing, except for his personal credit and aid from his family, to lay the foundations for a new institution on Mount Wilson. Boss, meanwhile, shifted his attention to Campbell. For Boss, it was now a question of forming an alliance with Campbell and pushing for an observatory in the southern hemisphere. Campbell, who was mounting his own expedition to South America in order to collect data on the radial velocities of southern stars, was clearly open to the idea of an observatory below the equator. He confessed to Boss that his interests were "equally great along both lines [the old and new astronomy]."[49] Campbell seem to suggest that a southern observatory would be devoted to both the old and new astronomy.

In the late summer of 1903, Boss moved to gain Campbell's assistance in cutting Hale down to size. Boss viewed the Yerkes director as an upstart who represented a style of doing science and a field of scientific research for which he had nothing but contempt. From Campbell's perspective, it may have been a question of protecting his institution. Campbell and the Lick *were* California astronomy. The addition of another major observatory, even though privately funded, and hundreds of miles to the south, might lead to divided loyalties among Californians, and thus lessen support for the Lick. Campbell did not want to loose his monopoly. He had every reason to support Boss.

Boss pressed Campbell to join in forcing Hale to choose between a solar observatory and one dedicated to the all facets of astrophysics. Further, he wanted CIW support for a solar observatory to be temporary, limited to one sun-spot cycle, about eleven years. Boss reminded Campbell that "The Lick Observatory is an important vested interest, which has fairly earned consideration from those who have the management of scientific enterprises."[50]

By September, Campbell was ready to make common cause with Boss. He apparently agreed that Hale's ambition had to be curbed. The advisory group should recommend a sum, perhaps $400,000, to the trustees for *both* a southern station and a solar observatory in southern California. There was an unspoken quid pro quo: Boss would support Campbell in his request for a large reflector for the Lick.[51]

While Boss and Campbell were maneuvering, Hale was implementing a dramatic plan. With limited resources and no immediate prospect of

49. Campbell to Boss, 22 May 1903, p. 2. SALO, L. Boss Papers. See also Boss to Campbell, 1 May 1903. SALO, L. Boss Papers.

50. Boss to Campbell, 6 August 1903, p. 10. SALO, L. Boss Papers. This is an eleven-page document, organized and written with great skill and cunning. For a discussion of Campbell's negative reactions to Hale's involvement in the site selection process, see Wright (1966: 168–9).

51. Campbell to Boss, 9 September 1903, pp. 2–4. SALO, Campbell Papers.

assistance from the CIW, he was preparing to leave for California. Once there, he would establish a modest observatory on Mount Wilson. "It is not only important to do the solar work for its own sake, but also in the hope of convincing Mr. Carnegie that means should be provided for carrying out the whole scheme later." Hale was not without hope. He had family connections that provided access to Carnegie. He informed Campbell that "an intimate friend in New York, who is very close to Mr. Carnegie, says the latter told him the other day during a golf game that he was very much interested in the solar observatory and that it would probably be the next thing provided for."[52]

The fact that the CIW trustees took no action at their December 1903 meeting did not deter Hale. Once in California, he constructed quarters on top of Mount Wilson. By April 1904, Hale and his assistant were making photographs of the sun, using equipment borrowed from Yerkes. Luck was on his side. In January, Hale was awarded the Gold Medal of the Royal Astronomical Society, one of the most prestigious honors available to astronomers, and in April Hale became the recipient of the Henry Draper Medal of the National Academy of Science. These honors can be seen as strengthening Hale's position with Carnegie and the CIW, thus increasing his power and improving his chances for funding. Whatever Boss might say, these evaluations of Hale by peers in the national and international scientific communities demonstrated the high regard in which the young solar physicist was held. Further, it may be that Carnegie himself, recognized a kindred spirit. Both men were, after all, entrepreneurs in the fullest sense of the word. After many delays the CIW trustees agreed to fund the Mount Wilson Observatory at their December 1904 meeting. Hale, the risk taker, captured the prize

More than twenty years passed before astronomers had another chance to compete for large amounts of research funding. When they did, the process of selecting projects for support was tightly controlled and conflict held to a minimum.

In May 1925, George Ellery Hale called a conference in New York to discuss ways of securing private sector support for basic research (Wright 1966: 365–70). He hoped to convince leaders of American industry to make substantial contributions to a fund that would be divided between leading workers in various scientific fields. Secretary of Commerce Herbert C. Hoover, an engineer elected to the National Academy in 1922, agreed to become chairman of the fund. The National Research Council and the Academy were actively involved. In addition to Hoover, a number of prominent business and political leaders served on the board (Tobey 1971: 206–17).

52. Hale to Campbell, 16 December 1903, pp. 1–2. SALO, Hale Papers.

In December 1925 the National Academy of Sciences announced the creation of a National Research Fund and a campaign to solicit pledges got underway. The goal was twenty million dollars. However, even in the flush days of the mid-1920s, captains of industry were slow to respond. In 1928 the goal was cut in half. While giants like AT&T and leaders of the steel and electric-power industries made pledges, many American businessmen apparently did not see great value in the fund (Davis and Kevles 1974). By 1930 the NRF had collected no money and was able to reach its reduced goal of ten million only with pledges from the Rockefeller Foundation and private philanthropists. By that time the great Depression was underway and the Fund, in fact, was dead.

However, all this was in the future, In 1928, the president of the Academy requested Henry Norris Russell, one of the generals who exercised great power in the American astronomical community, to recommend projects worthy of NRF support.[53] The fund's administrators anticipated that about two hundred thousand dollars would be available in 1929. Russell had only six weeks to submit a report before sailing for the Leiden meeting of the International Astronomical Union. Since he could not convene a special conference, the Princeton astrophysicist solicited written responses from sixteen leading astronomers. Russell was especially interested in significant work in progress that could be completed quickly with NRF support. As if seeking to avoid the confusion associated with the CIW advisory committee a generation earlier, he assured each astronomer that "It is especially requested that you will give full consideration to the claims of your own institution and your own work."[54]

The list of astronomers who received the circular from Russell is revealing. Only Ernest W. Brown did not occupy a directorship. He was, however, the leading researcher in celestial mechanics in the United States. Fifteen were members of the National Academy. Seven of the group (44 percent) were members of the east coast Neighbors. In short, Russell sought advice from the power elite in American astronomy.

Two surviving responses indicate that directors spent little time on the work of other institutions. Indeed, Robert G. Aitken, who took over as acting director of the Lick after Campbell became president of the University of California, said frankly that he would not go into detail about research at other institutions. Their directors could do that. Aitken concen-

53. Henry Norris Russell to Vesto M. Slipher, 30 April 1928. LOA, V. M. Slipher Papers.

54. Russell to Slipher, 30 April 1926, p. 2. LOA, V. M. Slipher Papers. The following astronomers received identical letters from Russell dated ca. 30 April 1928: Shapley (HCO), Schlesinger and Brown (Yale), Curtis (Allegheny), Curtiss (Michigan), Frost (Yerkes), Stebbins (Wisconsin), Slipher (Lowell), Adams (Mount Wilson), Aitken (Lick), Abbot (Smithsonian Astrophysical), Mitchell (Virginia), Leuschner (Berkeley), Miller (Swarthmore), and Hale (retired, Mount Wilson).

trated on the needs of the Lick. These included funds for the study of spectroscopic parallaxes, research on star clusters, and Aitken's revision of the Burnham double star catalogue.[55]

Lowell director Vesto M. Slipher's response was even more revealing. In the 1920s Henry Norris Russell developed a special relationship with the Lowell Observatory. He advised both the trustee and director on political (improving the image and reputation of the observatory) and scientific matters, and encouraged Slipher and his staff to undertake various lines of astrophysical research. Russell often visited Mars Hill, and on one occasion he and his family took advantage of the opportunity to tour the scenic wonders of northern Arizona, guided by members of the staff. Russell's relationship with the Lowell Observatory differed from his position as research associate at Mount Wilson. Russell played the role of patron at Lowell, while in Pasadena he was a scientific advisor.

Slipher acknowledged Russell's status as patron in responding to the circular letter. "It was so good of you to remember your good friends here at Lowell Observatory. . . . We are glad that we have a good friend at court . . . but we shall not let that fact make us grasping." Slipher went on to outline three projects for which he desired support. All were in progress, but they would be helped by an infusion of cash. The projects included radiometric observations of the planets, planetary photometry, and the study of the spectra of the night sky. In addition, Slipher mentioned a grand scheme for a photographic survey of nebulae and later added the search for a trans-Neptunian planet.[56]

In the fall of 1928, the NAS committee supervising the National Research Fund appointed subcommittees to advise in the distribution of funds among the various sciences. Russell was to chair the astronomy panel which included Shapley, Schlesinger, Abbot, Hale (with Adams serving as alternate), and Campbell (with Aitken as alternate). All eight were members of the Academy and directors of leading observatories. Half the group were active in the east coast Neighbors. The astronomy panel was to assemble just before the fall meeting of the National Academy and prepare its recommendations. Russell anticipated that $25,000 would be available. Small in comparison to expenditures of the CIW, this amount was far in excess of what could be secured from the trust funds of the Academy.

For consideration at the November meeting, Russell developed a list of projects arranged in three categories.[57] Presumably, this list reflects the

55. Robert Grant Aitken to Henry Norris Russell, 10 May 1928, SALO, Russell Papers.

56. Slipher to Russell, 6 May 1928, pp. 1–2, and Slipher to Russell, 12 May 1928. LOA, V. M. Slipher Papers.

57. Prior to the November 1928 meeting, the chairman offered an important suggestion. Russell indicated that in judging applications, members should consider (a) the importance

suggestions made earlier in the year by sixteen leading astronomers. The first, "Projects of Exceptional Merit," included four proposals for which a total of $27,300 was requested from the NRF. The successful applicants included Harlow Shapley, who requested $15,000 per year for five years to support research in galactic astronomy; Robert G. Aitken for a total of $4,500 to support spectroscopic research at Lick and his own studies in double star astronomy; Samuel A. Mitchell at Virginia asked for $4,800 for research in stellar proper motions and Frank Schlesinger requested $3,000 to expand the staff at the new Yale South African station. Each of these applicants was the director of a factory observatory. They were all members of the Academy and three of the four active in the Neighbors.[58] Apparently, the bulk of NRF money for astronomy was to be kept within the power elite.

A second class of proposals was labeled "Projects of Great Merit," totalling $11,000. They included $3,000 to the Lowell Observatory for a photographic survey of nebulae, studies of sky glow and the search for a trans-Neptunian planet. Heber D. Curtis at Allegheny would receive $3,900 for research in spectroscopy and the study of stellar parallaxes. Joel Stebbins at Wisconsin asked for $1,000 to aid his work in photoelectric photometry and Ralph H. Curtiss at Ann Arbor requested $3100 to support research in stellar spectroscopy.[59]

A final category, "Projects Worthy of Consideration If Funds Are Available," included two items that totalled $13,500 and listed two other projects, but did not provide detailed cost estimates. Charles G. Abbot requested $12,000 to support an African station to measure the solar constant and Frederick Slocum of Connecticut Wesleyan University asked for $1,500 to support his work on stellar parallaxes. Edwin B. Frost, Director of Yerkes, requested a very large amount of money (at least $53,500 for the first year) for five projects. Blind and long past his prime as a scientist, Frost's wish list of expensive projects received little attention from the subcommittee. Leuschner at Berkeley submitted a scheme for a central computing bureau, where the orbits of newly discovered solar system objects could be calculated. There was no price tag attached to the proposal.[60] At the end of the letter, Russell struck a note of caution. Neither Mount Wil-

---

of the results to be produced by each project; (b) the competence of the individual scientist; (c) and whether adequate instrumentation was available. Henry Norris Russell to Members of the Sub-Committee, 31 October 1928. Copy in SALO, Aitken Papers. See also Russell to Schlesinger, 31 October 1928. YUA, Astronomy Department Papers.

58. The list of successful applicants is appended to Russell's letter of 31 October to those members of the subcommittee who did not make it to the NAS meeting. See Henry Norris Russell to "Dear Sir," 31 October 1928, pp. 1–4. MTWA, Adams Papers.

59. Henry Norris Russell to "Dear Sir," 31 October 1928, pp. 4–6. MTWA, Adams Papers.

60. Ibid., pp. 6–8.

son nor Princeton submitted proposals in 1928. This should not be taken as an indication of future behavior.[61]

Since Walter S. Adams of Mount Wilson did not plan to attend the November NAS meeting, he sent his evaluations to Russell. Many of the applicants requested funds for additional staff members for three to five years. Adams opposed grants for more than two years and would have preferred one year with the possibility of an extension for another. He was especially concerned about the amount of money going to Shapley, for his research in galactic astronomy. "Harvard ought not to be obliged to apply to every fund in the country, as seems to be pretty much the case at present." Nor did Adams wish to see other applicants dropped from the list in order to fund Shapley.[62] Adams proposed a revised budget that reduced the Harvard request by two-thirds. The question of support for Shapley became one of the few issues that divided members of the advisory committee, and may have been an outgrowth of the east/west split in the American astronomical community.

In November, the committee reviewed Russell's original document and made changes. On the whole, these involved reducing requests to fit the announced level of funding available for astronomy ($25,000). Shapley took the largest cut (33.3 percent), while Schlesinger received more than 80 percent ($2,500) of his original request. Leuschner was added to the list with $1,000 for his work on the orbits of asteroids. Stebbins and Curtiss were dropped, but Hale agreed to assist the young Wisconsin astronomer with discretionary funds connected with the 200-inch telescope project. *If* the NRF cut the amount for astronomy to $20,000, Russell proposed a contingency plan. Shapley still received the largest share of the pie and Schlesinger's portion remained unchanged.[63]

In early December, when the subcommittee report was almost ready to be sent to the NRF Board, Russell learned that Curtis was planning to be away from Allegheny on an eclipse expedition for much of 1929. The chairman confided to Schlesinger that had he known this he would "not have made the recommendation that I did." The money allotted Curtis was promptly divided between Lick and Virginia.[64]

Apparently Adams made a last-minute effort to block the Shapley

61. Ibid., p. 8.

62. Adams to Russell, 15 November 1928, pp. 1–2. MTWA, Adams Papers.

63. Russell to "Members of the Sub-committee on Astronomy of the NRF," 26 November 1928. SALO, Russell Papers. A few late proposals had been received by Russell, but apparently the committee declined to consider them. Russell to "Members of the sub-committee on Astronomy of the NRF." no date, SALO, Russell Papers.

64. Russell to Schlesinger, 3 December 1928. YUA, Astronomy Department Papers. The final division is contained in Russell to "Members of the Sub-committee on Astronomy of the NRF," 5 December 1928. SALO, Russell Papers.

appropriation of $8,000. In response to his telegram, Russell wrote that Shapley's request had been cut to the bone. He then went on to characterize the Harvard project as "the most important which we have been asked to aid." The final report to the NRF listed six projects totalling $20,000. Four (Shapley, Aitken, Mitchell, and Schlesinger) were from the "exceptional merit" category, in Russell's October memorandum. One, aid to the Lowell Observatory, was drawn from the "great merit category," and one (Leuschner) was from the last category, to be funded if money was available. Shapley received 53 percent of his original request while Schlesinger suffered least and was awarded 83 percent of his requested budget. Aitken of the Lick came third with 78 percent of his budget while Mitchell and Slipher each got approximately 65 percent. Since Leuschner's original proposal contained no budget, his funding can not be compared in this way.[65]

The final report went to the NRF in late December. There was one last hitch. Russell had been informed that the general committee would probably be making "relatively few grants at the present time and deferring consideration of the others until its next meeting." It seemed wise, therefore, for the astronomy advisory panel to rank-order its requests. After a flurry of letters and phone calls, Russell achieved consensus and informed the NRF general committee that Shapley and Aitken were to be given highest priority.[66]

Why did Mount Wilson *not* make application to the NRF? Perhaps the CIW administration had forbidden Adams to do so. There is, however, another possibility. It may be that Hale, in discussion with Adams, urged his successor to stay out of the fray, and thus avoid the bruising embarrassment he suffered as a member of the CIW advisory committee in 1902–3. I can find no documents to substantiate this supposition, but it has a certain plausibility.

In retrospect, Russell and his colleagues were engaged in an exercise in futility. The National Research Fund never paid out a dollar to support scientific research. Of course, in 1928, none of the participants had any idea that the NRF would fail. Russell exercised great power in selecting recipients of NRF support. He did so with the assistance of other members of the power elite. There was a tendency to favor East Coast institutions. The Midwest was ignored, and the last minute inclusion of Leuschner may have been a sop to California astronomers. In the process of seeking NRF funding for astronomy, it paid to be a member of the Neighbors and a friend of Henry Norris Russell.

65. Russell to Adams, 5 December 1928. MTWA, Adams Papers. These figures are contained in Russell to "Members of the Sub-committee on Astronomy of the NRF," 5 December 1928. MTWA, Adams Papers.

66. The final report is contained in Russell to Campbell, 22 December 1928. SALO, Russell Papers.

THE POWER ELITE VERSUS THE NAVY

From the 1870s to the eve of World War II, leaders of the American astronomical community fought a series of skirmishes with the federal government, first over the Naval Observatory and then the Nautical Almanac Office. Although the naval bureaucracy proved too strong for even the most powerful astronomers, the long struggle to rid the community of anachronistic and inefficient institutions throws light on the nature and use of power. It also provides an opportunity to compare conceptions of power and its uses across two cohorts of elite astronomers.

Beginning in the 1870s, leading members of the American astronomical community sought to wrest control of the U.S. Naval Observatory from the government and place the reorganized institution under civilian control as a national observatory rather than a military installation.[67] Historian of science Howard Plotkin (1978) has provided a detailed discussion of these complex events. This section interprets the struggle between the navy and the power elite in light of the ideas developed in this chapter.

In the last quarter of the nineteenth century, the Naval Observatory fell on hard times. Its scientific output declined, and the caliber of the line officers who served as superintendents fell off. By the 1880s leading members of the American astronomical community were becoming increasingly critical of this state of affairs. Even the move to a better location and the acquisition of advanced instrumentation in the 1890s could not still criticism. In 1899, the secretary of the navy, appointed three leading civilian astronomers to canvass the American astronomical community concerning the USNO and its future.

The committee consisted of HCO director Edward C. Pickering as chairman; George C. Comstock, director of the Washburn Observatory at the University of Wisconsin and a leading practitioner of the old astronomy; and George Ellery Hale, director of the Yerkes Observatory and editor of the *Astrophysical Journal*. Pickering and Comstock represented the power elite of the second cohort while Hale was on his way to becoming one of the most powerful astronomers of the early twentieth century. Pickering and Hale ran large-scale factory observatories organized along corporate lines. Comstock's operation at Madison represented the smaller type of observatory whose financial base and organizational structure reflected limited resources available for astronomical research at most state universities. Comstock's research output and political skill earned him

67. The motives which impelled astronomers in their criticism of the USNO and NAO varied (Plotkin 1978: 387). Some saw it as a chance to advance their own careers. Others, distressed by waste and inefficiency, wanted to bring these institutions into line with models provided by industrial capitalism.

membership in the elite. The members of the committee found conditions at the USNO unacceptable. Its annual income was larger than that of any other comparable institution (Greenwich, Paris, or Pulkovo), but the USNO was almost a decade in arrears publishing transit-circle observations and more than a quarter century behind in completing the analysis of the transits of Venus (1874/1882). Such inefficiency ran counter to the entrepreneurial values of industrial America. The leaders of the American astronomical community wished to take over an unproductive but affluent scientific institution and reorganize it.

While leaders of the American astronomical community made clear their displeasure with the organization and performance of the USNO, and the secretary of the navy appointed a Board of Visitors to examine conditions and make recommendations, the naval bureaucracy and a few entrenched civilian scientists were able to block substantive reforms. In 1902 the board was disbanded (Plotkin 1978: 394–7). Members of the power elite, acting with widespread support from the astronomical community, were unable to "denavalize" the observatory. As late as 1910, Lewis Boss, still hoping for reform, drafted a long letter to President William Howard Taft, suggesting ways of getting sound (that is, nonmilitary) advice concerning candidates for a civilian director of the observatory and urging the importance of the National Academy as Taft's proper scientific advisor.[68]

In 1903, the superintendent of the Naval Observatory used his annual *Report* to attack the defunct Board of Visitors and enhance the reputation of the observatory. "History will show," wrote Admiral C. M. Chester (1903: 13), that among "astronomers who have become eminent in this branch of work [astrometry] the Naval Observatory has contributed quite its share of able men who first learned their profession here, or . . . gained largely the experience which enabled them to take a prominent stand in scientific research." The quantitative discussion of employment patterns in chapter 5 suggests otherwise. After the first cohort, neither the Naval Observatory nor the Nautical Almanac Office were viewed as desirable locations for employment. Those who moved from them to a second position often made a lateral move or sacrificed institutional potential. Neither the USNO nor the NAO were able to hold many of the elite astronomers who began their careers there.

Motivated perhaps by a lingering sense of inferiority, the naval bureaucracy viewed attempts by civilian astronomers to privatize the USNO

68. Simon Newcomb coined the word in a letter to Edward S. Holden, 4 December 1896. SALO, Newcomb Papers. Draft letter, Lewis Boss to President William Howard Taft, 23 April 1910. DOA. There is no indication in the Dudley Observatory Archives that this letter was actually sent.

as hostile takeover bids. Retiring observatory superintendent Davis (Plotkin 1978: 398) wrote to his successor, "I am fully persuaded that these men [members of the power elite] will never relax their efforts nor will they ever have any compunction as to the means which they employ until they have accomplished their end, which is to gain full control of the [Naval] Observatory." He urged the line officer who filled his place to make friends in Congress, as well as in the naval bureaucracy, and to keep close watch on Pickering and Newcomb, the astronomers he considered leaders of the movement.

In the midst of the mounting tide of criticism by civilian scientists, the navy transferred the Nautical Almanac Office to the observatory. This was, in part, a measure to clip the wings of its distinguished superintendent, Simon Newcomb, an implacable foe of the naval bureaucracy, and in part, to reward one of the staunch supporters of the navy within the observatory, William Harkness. Newcomb and Harkness had long been political enemies. To the dismay of many civilian scientists, Harkness succeeded Newcomb as director of the NAO in 1897.

Writing from the field of battle, longtime NAO staff member Eben Loomis (b. 1828) confided to David P. Todd, director of the Amherst College Observatory, "The [Nautical Almanac] Office has become too common and too undignified a place for a gentleman to remain in." With considerable hyperbole, Loomis continued, "If I could find some crank to turn [at] 10 [cents] per hour, or a place to act as motor man on an electric car line, I would resign tomorrow."[69] However much Loomis may have exaggerated, the future of the NAO after Newcomb was, indeed, bleak. A succession of mediocre astronomers directed the Almanac Office. Only in 1910, when William S. Eichelberger took command, did the NAO regain some of its lost prestige. Eichelberger's retirement in 1929 would spark another round of activity, as the power elite tried to block the appointment of a third-rate scientist with first-rate political connections.

Before considering that episode, it is important to look at the ways in which the nature and exercise of power in the American astronomical community changed between the 1890s and the 1920s. Until the early twentieth century, the power elite in American astronomy lacked cohesion. In the Gilded Age, powerful directors of the great observatories tended to be loners. They came together only when their vital interests appeared to require it, but otherwise operated in a relatively independent manner. This behavior rests in part on structural foundations. The American astronomical community shared few institutional resources. The first community-wide organization (the Astronomical and Astrophysical So-

69. Eben Loomis to David P. Todd, 12 December 1894, p. 2. YUA, Todd Family Papers.

ciety of America) was organized in 1899. Before that date, astronomers had only the American Association for the Advancement of Science and the National Academy as institutional foci which, of course, had to be shared with other scientists. Further, only in the mid-1890s was the *Astrophysical Journal* founded, and Gould revived the *Astronomical Journal* a scant decade earlier. Lacking community-wide institutions, the very nature and scope of power was different from what it would become after World War I. For much of the nineteenth century, the locus of power was the individual observatory: its creation, funding, development, and preservation in the face of competition. The development of observatories and conflict between them characterized the politics of American astronomy.

After World War I, a new generation assumed power in a very different structural context. American astronomy was better endowed with community-wide institutions. Its leaders where more inclined to cooperate in the exercise of power than to make the first priority their own institutions. To be sure, the post–World War I elite was guilty of exercising power for individual and collective advantage, but often they acted from higher motives. Schlesinger, Shapley, and Russell remained in constant contact and were active in the affairs of the American Astronomical Society. Further, they made conscious use of institutions such as the Neighbors. These men behaved as if they had a mandate to guide the development of American astronomy. In both cohesion and motive, they differed from the more isolated and individualistic power elite of an earlier day.

In the 1920s personal and professional ties developed between the generals and staff astronomers at the USNO. These ties were strengthened when Schlesinger was asked to advise the USNO on developing photographic astrometry at Washington.[70] NAO director Eichelberger, whose stature grew during these years, was a member of the Neighbors considered to be an astronomer of recognized ability. He was awarded a star in the first (1906) edition of *American Men of Science* and held office in both the AASA and the AAAS. Eichelberger was elected a member of the *Bureau des Longitudes* in 1919 and a foreign associate of the Royal Astronomical Society in 1925. He was also active in the International Astronomical Union. During the 1920s, Eichelberger was considered for membership in the National Academy of Sciences, but never achieved enough support to become the nominee of the astronomy section. Eichelberger was, in short, a highly visible research scientist, well connected and active in the inner circle of the American astronomical community.

In early 1929, as Eichelberger's official retirement date approached, leading members of the power elite became concerned about the fate of

70. Frank Schlesinger to Frank B. Littell, 8 November 1927. YUA, Astronomy Department Papers.

the NAO. Unless Eichelberger was permitted to remain beyond the mandatory retirement age of sixty-four, "it appears all but certain that a Mr. James Robertson will get the appointment." Robertson, who had served in various positions at the NAO since 1893, was described as a man of "considerable political influence" both on Capitol Hill and within the naval bureaucracy. In the years since his appointment, Robertson published only a few papers and had made little impression on peers outside government service. Schlesinger wrote to William Wallace Campbell, then president of the University of California, "In my opinion, and in that of every one else who is conversant with the facts, it would be in the nature of a scientific calamity if he would direct the Nautical Almanac."[71]

Schlesinger, who must have relied on information supplied by USNO staff members active in the Neighbors, discussed Robertson in some detail. He was characterized as a "good computer in the sense of being able to apply a formula that is presented to him" but "would be totally unable to set up a formula in any new case . . . or in any new phase of a known application." Further, Schlesinger reported that at the Leiden meeting of the IAU in 1928, and a subsequent meeting of the *Astronomisches Gesellschaft*, Robertson, an official delegate from the NAO, was a source of great embarrassment to other astronomers. "His lack of technical equipment is all too apparent," Schlesinger observed. The letter ended with a request that Campbell discuss the matter with Armin O. Leuschner, head of the Berkeley astronomy department, and then, if Campbell agreed, "it would be a considerable public service on your part to do what you can to prevent this appointment."[72]

In March, Schlesinger passed on to Campbell a copy of Robertson's testimony before a congressional committee considering the 1930 naval appropriations bill. Robertson's statement indicated that he lacked a firm grasp of optical theory and telescope construction and had been deeply influenced by the ideas of former Mount Wilson staff member George Willis Ritchey, who figured the great 60 and 100-inch mirrors. Ritchey (Osterbrock 1993) had been dismissed from Mount Wilson and was living in Paris, where Robertson met him. Robertson's testimony also demonstrated a limited knowledge of both the goals and methods of astrometry. It is clear, however, that no matter how restricted his technical knowledge, Robertson and Observatory superintendent Charles S. Freeman were seeking increased appropriations, so that the USNO could begin research in photographic astrometry. In writing to Leuschner, Campbell characterized Robertson's testimony as "fearful and wonderful" and

71. Frank Schlesinger to William Wallace Campbell, 15 January 1929, p. 1. YUA, Astronomy Department Papers.

72. Ibid., p. 2.

alerted Leuschner to the "incorrect or grotesque points" made by the NAO astronomer.[73]

As the date of his mandatory retirement drew near, Eichelberger wrote Walter S. Adams at Mount Wilson enclosing a copy of Robertson's testimony. Adams, in turn, forwarded the material to Hale, suggesting he help Eichelberger. Specifically, Adams urged Hale to write President Herbert Hoover, with whom Hale had been associated on the National Research Fund. In Adams's view, Robertson was "little more than a fool, both personally and as regards his knowledge of astronomy."[74]

As with earlier attempts, this skirmish ended in defeat for the power elite. Robertson assumed the directorship of the Nautical Almanac Office and Eichelberger accepted a position with Eastman Kodak, his great catalogue of zodiacal stars unfinished, even though the former director volunteered to stay on without pay until it was complete.

Adams reported to Charles Abbot at the Smithsonian that "Mr. Hale and I attempted to stir things up at the Naval Observatory in connection with Eichelberger's retirement. Unfortunately we got nowhere." Robertson was more than a match for the elite. Apparently he had powerful friends in the Senate as well as close connections with the naval bureaucracy.[75]

Robertson won the office, but elite astronomers sought to block honors that might be expected to come to a scientist occupying such an important position. When asked to evaluate Robertson for an honorary degree, Leuschner (class of 1888) replied that the director of the NAO was hardly among the distinguished graduates of the University of Michigan. "It is the general impression of American astronomers that Robertson has maintained and advanced himself . . . purely by political activities. His standing among American astronomers is not very high."[76]

Leading scientists like Leuschner found the large appropriations for the Naval Observatory galling. The Berkeley celestial mechanician encouraged Jan Schilt, chair of the Columbia University astronomy depart-

73. Campbell to Leuschner, 26 March 1919. UCBL, Astronomy Department Papers. There are a number of letters from USNO superintendent C. S. Freeman outlining his plans for the institution in the Astronomy Department Papers in the Yale University Archives. They begin in 1928.

74. Walter S. Adams to George Ellery Hale, 11 September 1929, p. 2. MTWA, Adams Papers.

75. Walter S. Adams to Charles G. Abbot, 10 October 1929. MTWA, Adams Papers. For a discussion of Robertson's political connections see Schlesinger to Campbell, 15 January 1929, p. 1. YUA, Astronomy Department Papers.

76. Armin O. Leuschner to Thomas P. Hayden, 5 January 1937, p. 2. BLUCB, Astronomy Department Papers. Leuschner sent a copy of this letter to the Chairman of the Michigan astronomy department, Heber D. Curtis.

ment, to find ways to link the Hollerith Computing Bureau at Columbia with the Naval Observatory. Columbia astronomers had made giant strides in applying machine technology to the reduction and analysis of astronomical data but were short of funds. If USNO astronomers were associated with the project, federal funds might become available to support the Hollerith Bureau.[77]

Many astronomers feared that Robertson might remain in office for a long time. Leuschner believed that Robertson and USNO superintendent J. F. Hellweg were working to extend "their tenure after expiration of their terms of office."[78] Given Robertson's political connections, this seemed a genuine possibility. Events, however, took a different course.

In early 1939, Virginia astronomer and active Neighbors member Samuel A. Mitchell learned that Robertson would soon retire and alerted Schlesinger, who immediately took charge of the situation. He asked the cooperation of Russell and Shapley in selecting "the best candidate and then to unite in pushing him."[79] Schlesinger favored either Dirk Brouwer, a Dutch-born astronomer on the staff of the Yale Observatory, or Wallace Eckert, a recent Yale Ph.D. working at Columbia. Eckert was the driving force behind the Hollerith Bureau, and one of the pioneers in the application of new computing technologies to astronomy. The Yale director feared that because of his foreign birth, Brouwer might be objectionable to politicians and the Navy. On the whole, Schlesinger preferred Eckert. His experience with computing machines would be of great value to the NAO. Leaving nothing to chance, Schlesinger reported that he had talked with Eckert, who "would be willing to push his candidacy and to accept the appointment if it should be offered to him."[80] While Schlesinger supported Eckert, he was willing to discuss other options with Shapley and Russell.

From this point the trail becomes difficult to follow. By November a total of five individuals were being considered by the generals. The list included Carl L. Stearns and Fred L. Whipple, both members of the Harvard staff, former Harvard astronomer Willem Luyten, who had moved to the University of Minnesota, and Brown University astronomer Charles Smiley (1903–77). Brouwer had been dropped, but Eckert remained on the list. These individuals were ranked according to twenty-three criteria, but unfortunately, there apparently is no key in the Schlesinger papers explaining the criteria. We can see the generals at work but have no clear

77. Armin O. Leuschner to Jan Schilt, 27 October 1938. BLUCB, Astronomy Department Papers.

78. Leuschner to Hayden, 5 January 1937, p. 2. BLUCB, Astronomy Department Papers.

79. Frank Schlesinger to "Dear Colleague" (Shapley, Russell and Mitchell), 7 March 1939. YUA, Astronomy Department Papers.

80. Ibid.

idea of their thinking or the negotiations that went on between them. We know that Whipple actively sought the post and asked his mentor, Lick director William H. Wight, for a recommendation.[81]

The 1939 success of the generals in securing the NAO directorship for Wallace Eckert remains shrouded in mystery. Shapley, Russell, and Schlesinger played key roles. Members of the Neighbors such as Mitchell were of great help. In all probability the generals received important information from Neighbors who worked at the Naval Observatory, but until more documentation is discovered we can not reconstruct the episode in detail.

The contest over the USNO and the NAO suggests the gap between the appetite of the power elite and their political strength. When thrust on a larger stage, beyond the confines of the astronomical community, these leaders lacked both the knowledge and resources available to politicians and bureaucrats. Only in 1939 were they able to achieve their long-sought goal.

GEORGE ELLERY HALE AND *THE ASTROPHYSICAL JOURNAL*

George Ellery Hale was always in a hurry. At twenty-three, he undertook, in cooperation with W. W. Payne (1837–1928), professor of astronomy at Carleton College, and editor of the *Sidereal Messenger*, to publish a new journal, *Astronomy and Astro-Physics*. In 1894, at twenty-six, he was appointed associate professor of astrophysics at the new University of Chicago and, in the same year, received his first major honor, a prize from the French Academy of Sciences in recognition of his research in solar physics. Once launched, Hale's career trajectory continued to arc upward: election as foreign associate of the Royal Astronomical Society at the age of thirty-one and three years later (1902) membership in the National Academy of Sciences.

Hale broached plans for a new journal in letters to leading astronomers in October 1891. He was very careful to legitimate these activities by involving senior astrophysicists in Europe and America. As the young Chicago astronomer wrote to Holden at the Lick, the decision to publish a new journal was reached with the blessings of European scientists as well as encouragement from Young at Princeton.[82]

At this early stage in Hale's thinking, the journal would take the place

81. Unsigned sheet of graph paper titled "Director Nautical Almanac," November 1939. YUA, Astronomy Department Papers. Fred L. Whipple to William H. Wright, 9 October 1939, SALO, Whipple Papers.

82. George Ellery Hale to James E. Keeler, 12 October 1891. UPAIS, Allegheny Observatory Papers, and George Ellery Hale to Edward S. Holden, 16 October 1891, p. 1. SALO, Hale Papers.

of the publications of his Kenwood Physical Observatory. He envisioned the *Astro-Physical Journal* as devoted primarily to solar and stellar spectroscopy.[83] Hale asked leaders of the American and international astronomical communities to evaluate his plans, commit themselves to support the journal, and respond with comments he could print.

At the beginning of his career and still unschooled in the ways of powerful observatory directors, Hale probably did not expect the response he got from Holden, who was always on the lookout for ways to augment his power and improve the image and reputation of the Lick. Holden suggested a journal that would appeal both to research scientists and interested laypersons. His vision rested an undifferentiated view of the scientific enterprise, more appropriate to the mid rather than late nineteenth century. In an era of increasing specialization Holden's views were a throwback to an earlier epoch.[84]

But there was also a hidden agenda. How could Holden coopt Hale and his project and turn them to his own ends? Under the guise of providing a firm financial foundation for the journal, the Lick Director suggested Hale throw his lot with the Astronomical Society of the Pacific. Here was an assured pool of subscribers. All Hale had to do was make his journal a department of the *Publications of the ASP*. Holden controlled the ASP and used it as a support group for the Lick Observatory (Bracher 1989). Since the membership of the ASP was composed primarily of amateurs and wealthy Californians who deemed membership a social privilege or civic duty, it is difficult to see how Hale would have benefited.

There were, however, certain practical questions. Was there an audience? Were there contributors? Could such a journal survive without subsidies? Hale did not have the answers. Perhaps for these reasons, he abandoned his original scheme and joined forces with Payne (Wright 1966: 88). The *Sidereal Messenger* was an established journal, read by amateurs and professionals; there would be a ready-made subscription list. During the fall of 1891 Hale and Payne discussed the partnership. While Payne and his staff would take care of routine matters, the editorial decision making would be kept separate, with Hale in complete charge of the astrophysical section.[85] *Astronomy and Astro-Physics* appeared in January 1892.

Hale, whose mind was always on the future, confided to James E. Keeler that "One of these days I want to get the whole journal away from

83. Hale to Holden, 16 October 1891, p. 1. SALO, Hale Papers.

84. Holden to Hale, 22 October 1891, p. 1. UCYOA, Directors Papers.

85. The negotiations are documented in the Hale papers at Williams Bay. See, for example, W. W. Payne to George Ellery Hale, 26 October and 6 November 1891. UCYOA, Directors Papers.

Payne, and have some first-class mathematical astronomer take his place in editing the general astronomy part, and have three or four associate editors with him" and an equal number for astrophysics. Hale indicated that the University of Chicago would act as publisher. Hale went on to invite Keeler to become an associate editor of *A&A-P*, with the implicit promise of a position on its successor when the time came. At first Keeler demurred, but at length joined Hale. This fruitful collaboration, first editing *A&A-P* and then the *ApJ*, ended only with Keeler's death in 1900.[86]

In planning the *ApJ*, Hale was especially careful to create an editorial board that was international in scope. This insured the legitimacy of the venture, provided a source of contributions and guaranteed subscribers. Foreign scientists serving on the *ApJ* board included physicist M. A. Cornu (1841–1902) of the École Polytechnique, and astrophysicists H. C. Vogel (1841–1907), Director of the Potsdam Astrophysikalisches Observatorium, Nils C. Dunér (1839–1914) of the Astronomiska Observatorium, Upsala, Father Pietro Tacchini (1838–1905) of the Osservatorio del Collegio Romano, and Sir William Huggins. The American board members were of equal stature. Physicists Albert A. Michelson (Chicago), Charles S. Hastings (Yale), and Henry R. Rowland (Johns Hopkins) were joined by astrophysicists Young of Princeton and Pickering of Harvard. That three of the five senior American board members were located in physics departments suggests how deeply Hale felt about the cross-disciplinary nature of the new journal. These ten individuals would not participate in day-to-day work on the *ApJ*. They served to legitimate the venture with American and European scientists and provide conduits for papers. The real work was done by Hale and Keeler as coeditors, and a staff of five assistant editors drawn from the ranks of younger American astronomers and physicists. Physicists Henry Crew of Northwestern and Joseph S. Ames of Johns Hopkins were carried over from *A&A-P*. They were joined by astrophysicists Edwin B. Frost of Dartmouth and William Wallace Campbell of the Lick. University of Chicago astronomer and instrument designer Frank L. O. Wadsworth (b. 1867) completed the staff of assistant editors.

Only Campbell declined Hale's offer of a position on the *ApJ*; however, he was a key element in Hale's strategy. Hale, who had so carefully included associate editors representing the major astrophysical research institutions in Europe, and sought a balance between physics and astrophysics in selecting Americans, needed to achieved geographic balance as well. Campbell of the Lick represented West Coast astronomy.

86. Hale to Keeler, 13 September 1892. UPAIS, Allegheny Observatory Papers. Keeler to Hale, 19 September 1892. UCYOA, Directors Papers. Hale to Keeler, 16 October 1893. UPAIS, Allegheny Observatory Papers.

Yet, Hale seemed ambivalent toward the Californian. With the publication date for the first number of the *ApJ* just three months away, he still had not invited Campbell to serve. As Hale told his coeditor, Keeler, "Campbell is all right except in his rather sharp replies to men who should be treated with great respect." Hale was firm in insisting the new journal do nothing to offend leading European spectroscopists. In writing to Keeler, he concluded that "it may be that a word to Campbell on this point would be sufficient to restrain him." However, when Hale included this caution in his letter of invitation, Campbell took offense. The Lick astronomer declined to serve as an assistant editor of the *ApJ*.[87]

A month later, Campbell changed his mind. However bruised his ego, Campbell could not afford to ignore the political implications of Hale's invitation. A position on the *ApJ* board would provide enhanced visibility and status within the astronomical community. To be sure, Campbell would have to moderate his language, especially since some of his critics, including Huggins and Vogel, were also members of the board, but in terms of increased visibility and power the trade-off must have seemed worth it.

The last number of *Astronomy and Astro-Physics* (December 1894) and the first issue of the *Astrophysical Journal* (January 1895) carried identical notices. Written by Hale, the statement reviewed reasons for creating a new journal, listed its editorial board, and outlined editorial policy. For Hale, there was nothing radical in creating the *ApJ* (Hale 1895:80). "In returning to the original idea of a purely astrophysical journal we are simply following out a long-cherished plan. Few who appreciate the true scope of astrophysics, and have its best interests at heart, will deny the advisability of devoting an entire journal to this, the most fascinating and at the same time the most rapidly advancing department of astronomical research." After this rhetorical flourish, Hale moved to the heart of the matter. There were no cross-disciplinary journals read by physicists and astronomers (Hale 1895: 81). "The astronomer and physicist should be able to meet on common ground, and this only an astrophysical journal can supply."

Hale continued to invoke the names of leaders in the field. They would provide legitimation and increased visibility for a journal founded by a young man at the beginning of his career. He made reference to a recent European visit and conversations with leading astronomers. In-

87. Hale to Keeler, 29 September 1894, pp. 1–2. UPAIS, Allegheny Observatory Papers. Hale to Campbell, 4 October 1894, pp. 1–2. SALO, Hale Papers. Campbell to Hale, 9 October 1894. UCYOA, Directors Papers. Campbell spent the better part of five pages defending himself against the charge that he had used sharp and disrepectful language in replying to critics.

deed, he suggested that the final form of the editorial board emerged from discussions held with astronomers in Berlin. Of course, all members of the editorial board would gain in proportion to the success of the journal. It was a situation in which participants could enhance their reputations, improve their standing within the discipline, and garner resources useful in the future exercise of power. Hale stood to benefit most: his career would gain immeasurably if the new journal proved successful. By the same token, if the journal failed, he would bear most of the blame.

Shaping the editorial policy proved challenging. What should the scope of the new journal be? This was an economic as well as a political problem. A narrowly based journal might not attract subscribers or contributors, and be doomed to a short and impoverished existence. In describing the editorial policy of the *ApJ*, Hale made it clear that spectroscopy would take pride of place. Space would be available for a wide range of papers, from wave-length determinations through spectro-photometry, and theoretical studies in related areas of physics. The *ApJ* also would be open to discussions of innovative instrumentation and reports on new observatories and physical laboratories. Hale (1895: 82–83) emphasized that the journal would reproduce photographs using the latest engraving and printing techniques.

Toward the end of the announcement, Hale (1895: 83) reported that the American editors had gathered in New York in November 1894 to work out details of the editorial policy and that the board voted to meet annually. Out of these meetings would come policy decisions affecting fundamental practice in astrophysical research.

The *ApJ* was not simply a vehicle for the publication of research. Hale and the board sought to define technical standards in the field. The actions of the *ApJ* board in setting standards must be understood in light of the discussion of entrepreneurship and factory observatories. The board was doing nothing less than standardizing products of the great factory observatories by specifying the technical language and format in which their products were to be presented. A common technical language was fundamental to the process of rationalization that transformed production in the private sector. Hale and the *ApJ* Board were simply taking the next step in the complex process of institutional isomorphism.

The American members of the *ApJ* board held their first meeting on 2 November 1894 at the Fifth Avenue Hotel in New York. All the American associate editors (Young, Hastings, Rowland, Michelson, and Pickering) were present, as were coeditors Hale and Keeler. None of the assistant editors attended. Annual gatherings of American members of the editorial board continued until at least 1901. Generally, meetings took place after the fall gathering of the National Academy of Sciences. All of the associate editors were members of the Academy.

At the first meeting, Pickering was elected chairman of the editorial board and Keeler secretary. The morning was devoted to a variety of issues ranging from the title of the journal through format and frequency of publication. After lunch the editors considered Hale's preliminary discussion of the scope of the journal and charged Keeler and Hale with perfecting the statement for publication. The subscription price was set at $4.00.

Up to this point, the discussion had been routine, devoted largely to housekeeping matters. Hale, however, made a motion that turned the discussion in a very different direction. "Professor Hale moved that in printing maps of spectra, the red end should be placed on the right, and in tables, the small wave-lengths should be placed at the top." As Keeler noted in the minutes, "An interesting discussion followed."[88]

Several important issues were involved, including national styles of doing science, individual perception, and mental habits of scientists. Many British and German spectroscopists were in the habit of printing spectra with the red end to the left and reporting measurements of wavelengths in tables that started with the longer wavelengths. Beyond the question of national styles, there was individual perception. In his mind, how did a scientist envision the spectrum ? Both Hastings, the Yale optical theorist and Hale, the student of solar spectroscopy, admitted that they pictured red to the left, "an illogical habit which he [Hale] had no wish to defend." Hastings, who raised the question of conflict with European scientists, indicated that he "regarded the other method [red on the right] as preferable."[89] After a brief physics lecture from Rowland, the board voted on Hale's motion. It carried, apparently without opposition.

From that point, the board acted quickly. Michelson moved, and the other members agreed without discussion, to adopt Rowland's scale of wavelengths as standard for the *ApJ*. Hastings then moved the Angstrom unit (one ten-millionth of a millimeter) be adopted as the *ApJ* standard for expressing wavelengths. Hale, as if in deference to European colleagues, suggested that Vogel's method of designating hydrogen lines be used. The kilometer was declared the unit for expressing radial velocities measured by spectroscopists.[90]

Then Pickering suggested another area in which he felt uniform standards were needed: the designation of stars when two catalogues overlapped. He was concerned about stars south of the celestial equator listed in both the Cordoba and Bonn *durchmusterungen*. The issue was related to the Harvard studies of stellar magnitudes and spectra. Thus it was, in

88. Minutes of a meeting of the editorial board of the *Astrophysical Journal* at the Fifth Avenue Hotel, New York, on November 2, 1894, p. 4.

89. Ibid.

90. Ibid., pp. 4–5.

part, an astrophysical question, but it was also an issue for astrometry. The meeting agreed to Pickering's nomenclature, but this decision, more than any other, would open the board to criticism. The *ApJ* board proceeded, apparently without dissent, in adopting each of these standards. They did, however, decide to poll the foreign editors. The board could not afford to appear divided on these critical issues.

Albert A. Michelson, soon to become America's first Noble laureate, ended the day with a benediction. "The meeting has been so pleasant in every respect that he [Michelson] should be glad to look forward to another one of the same kind, and he therefore moved that the Editorial Board" assemble at the same time in 1895.[91]

Decisions concerning standards were not announced by the editorial board of the *ApJ* until January 1896. More than a year was devoted to canvassing European members and soliciting comments from assistant editors. At this stage, Hale and his American colleagues introduced another issue into the discussion. The printed questionnaire that went out to all assistant and associate editors included an item that was not considered in November 1894. Each editor was asked to vote on the following proposition. "I am [am not] in favor of adopting the standards named above, *not only in The Astrophysical Journal, but also in all publications in which they might properly find a place.*"[92] An affirmative vote on this proposition would, in effect, define technical standards for astrophysical observatories in the Western world.

All of the American and four of the five European editors responded to the questionnaire. There was considerable disagreement over the Angstrom unit as a measure of wavelength, centering on what the unit should be called. Only Vogel of Berlin objected to printing spectra with red on the right and tabular data beginning with the shorter wavelengths. *No one* challenged the supplementary proposition concerning the use of these standards for all of their publications, not just those appearing in the *ApJ*. The board concluded, "It is greatly to be hoped that the action of the majority . . . will be concurred in by all astronomers, astrophysicists and physicists, and adopted in their publications." Never again did the board of editors of the *ApJ* take such dramatic action.[93]

By far the most heated discussion arose in relation to Pickering's desire to standardize stellar nomenclature in zones where major catalogues overlapped. Frost at Dartmouth and Campbell of the Lick felt strongly about this proposal. Frost suggested that the dean of American astrome-

91. Ibid., p. 5.

92. Copies of the questionnaire are in the Directors Papers, UCYOA. The emphasis is mine.

93. A copy of the call and agenda for the 1895 meeting is in the Allegheny Observatory Papers, UPAIS. See also Hale to Campbell, 25 November 1895, pp. 1–2. SALO Papers.

try, Benjamin Gould, retired director of the Argentine National Observatory at Cordoba, be brought into the discussion. Campbell agreed. In late November and again in December, Campbell wrote Hale concerning the actions taken at the New York meeting. He objected to calling the standard of wavelength an Angstrom unit, but devoted most of his energy to the problem of stellar nomenclature. For Campbell, the issue involved the limits of power. "This is not a question for astrophysicists to settle: it is outside of our jurisdiction, and should be left for the catalogue-makers and meridian-circle men to decide."[94]

The published standards contained no discussion of stellar nomenclature. In moving beyond astrophysics, the board entered a domain in which it had little power or influence. Had the *ApJ* tried to impose Pickering's nomenclature, "the *Journal* thereby [would loose] . . . influence with men in the other departments [of astronomy]."[95]

Initial reaction to the standards selected by the *ApJ* Board seemed favorable. Neither the Royal Astronomical Society (Anon. 1896: 222) nor individual British scientists opposed the plan. Indeed, Keeler reported that "Many compliments were paid to the Journal while I was in England [late summer 1896]. It seems to be regarded there as pretty nearly the right thing."[96]

Within a year, however, opposition did surface. Greenwich astronomer E. Walter Maunder (1851–1928) announced at a meeting of the RAS (Anon. 1897: 86–87) that he was a heretic, vigorously objecting to the actions of the *ApJ* with respect to printing spectra and wavelength tables. He was joined by Alexander Herschel (1836–1907), son of the great Sir John. Maunder used the analogy of the book, which a reader begins on the left-hand side of the page, while Herschel suggested the proper analogy was to a piano with the high notes and short wavelengths of the spectrum both placed on the right. Several speakers expressed the hope that the editors would reconsider. The distinguished German spectroscopist Heinrich J. G. Kayser (1853–1940) added his voice to those appealing the Board's decision. Hale urged that the policy on standards remain unchanged and, apparently, other members of the board agreed.[97]

These actions of the *ApJ* Board were successful for several reasons. The standards were formulated by a group of leading physicists and astro-

94. Responses from Campbell and Frost are in the Directors Papers, UCYOA. See also Campbell to Hale, 28 November 1894 and 3 December 1894.

95. Campbell to Hale, 3 December 1894, p. 1. UCYOA, Directors Papers.

96. Keeler to Hale, 6 November 1896. UCYOA, Directors Papers.

97. Hale to Keeler, 9 November 1896, p. 2. UPAIS, Allegheny Observatory Papers. Sir William Huggins supported printing spectra with the red on the right, but argued that wavelength tables should begin with the longer values and continue in descending order (red to violet).

physicists drawn from both the European and American scientific communities, whose stature lent weight to the actions of the board. Further, the choice of specific standards reflected current practice in some (but by no means all) laboratories and observatories on both sides of the Atlantic. Nonetheless, it was a daring move on the part of Hale and his colleagues.

The material discussed in this chapter suggests that power can most fruitfully be understood as a complex social process. Developments at the level of specialist research or generic communities, as well as in the larger society and culture, helped to define and direct the exercise of power. Neither the generic nor research communities occupied a privileged position in this process. Resources acquired at one level helped define the uses of power at the other. A deep understanding of power in a modern scientific community involves knowledge of these dynamic relationships.

8 ✦

THE REWARD SYSTEM IN A MODERN SCIENTIFIC COMMUNITY

The following discussion locates the reward system in the context of the scientific community. How was the reward system defined and how did it function? The chapter opens with a quantitative profile of a scientific elite, and then considers the social and political processes that provided access to the reward system in American astronomy, beginning with an examination of the means by which astronomers won stars in successive volumes of the *American Men of Science* series. The focus then shifts to the National Academy of Sciences. Beyond the anecdotal level, historians of science know very little about the politics of nomination and election to the Academy.

The chapter concludes with an empirical model of the reward system. The acquisition of honors and awards is viewed as contingent across the career. Far from being unidirectional and cumulative, the process involved alternative starting points that entail well defined developmental paths. For some scientists, terminal honors marked the end of the process, while for others the acquisition of honors and awards proceeded to higher levels. For the first fourteen years of the career, patronage rather than merit seemed to play a central role in determining the allocation of honors. Later in the career, merit tended to assume an increasingly important role. Both literary and quantitative data were employed in reconstructing the reward system across the careers of three cohorts of astronomers.

PROFILE OF A SCIENTIFIC ELITE

Elite status is defined as the award of a star in the *American Men of Science* series or election to the National Academy of Sciences. A total of 170 astronomers (166 males and 4 females) belong to the elite group. This is just over fourteen percent of the total population active between 1859 and 1940. Members of the elite captured the bulk of honors and awards available to astronomers and occupied most of the positions of power in the American astronomical community. Within the elite, members of the National Academy garnered more honors and power than did astronomers who earned only an *AMS* star. Differentials between the elite and rank and file or women are several orders of magnitude greater.

For eighty-nine of the elite, an *AMS* star was the highest honor (see table 8.1). Obviously, not many of the cohort that entered in 1859 or before survived to earn a star in the first edition of *American Men of Science* in 1906. George Washington Hough, director of the Dearborn Observatory at Northwestern University, is the sole representative of this cohort.

The vast majority of astronomers who earned stars in cohort two worked in astrometry or celestial mechanics. This reflects the dominant research interests of astronomers who entered the community between 1860 and the end of the century. In the third cohort the situation is reversed. Practitioners of the new astronomy gained more stars than did those who worked in the old astronomy.

Table 8.2 provides a breakdown of the population elected to the Academy but never starred in *AMS*. The majority are found in the cohort that entered in 1859 or before. Most were dead or had retired by the time the first stars were awarded. What is striking about table 8.2 are the second and third lines. In cohort two, only six astronomers who were elected to NAS failed to achieve stars in *AMS*. For the post-1900 cohort, there are no data to report; *all* astronomers elected to the Academy also won stars in *AMS*. This evidence provides the first indication that the reward system in American astronomy was a complex process that involved clearly defined patterns.

Table 8.3 deals with the population that earned both an *AMS* star and election to the Academy. For cohort one, only three individuals survived to achieve this distinction. Representing the old astronomy were longtime director of the Nautical Almanac Office and doyen of American astronomy, Simon Newcomb, and Asaph Hall, who retired from the Naval Observatory to lecture on celestial mechanics at Harvard. Pioneer solar physicist, Charles A. Young of Princeton, represented the new astronomy. While the old and the new astronomy reach parity in the second cohort, the most interesting information in table 8.3 concerns cohort three. For those who entered after 1900, only workers in the new astronomy earned both an *AMS* star and election to the Academy. So completely had the intellectual climate changed, that cohort three did not see any workers in the old astronomy both starred and elected to the Academy.

The socioeconomic class of the fathers of elite astronomers and the highest level of educational attainment of these scientists are two of the factors associated with the attainment of elite status. The class standing of the fathers of elite astronomers was estimated using obituaries and biographical materials (Elliott 1982: 84). Categories employed were blue col-

Table 8.1 The Elite: *AMS* Star Only

Cohort	Old Astronomy	New Astronomy	Mixed Astronomy	Total N=89
1859 or before	1	0	0	**1**
1860–1899	33	9	3	**45**
1900–1940	10	30	3	**43**

Table 8.2 The Elite: NAS Only

Cohort	Old Astronomy	New Astronomy	Mixed Astronomy	Total N=34
1859 or before	26	2	0	**28**
1860–1899	3	3	0	**6**
1900–1940	0	0	0	**0**

Table 8.3 The Elite: *AMS* Star and NAS

Cohort	Old Astronomy	New Astronomy	Mixed Astronomy	Total N=47
1859 or before	2	1	0	**3**
1860–1899	13	13	2	**28**
1900–1940	0	16	0	**16**

lar (manual laborer), white collar (nonmanual) or professional (minister, professor, lawyer, or doctor).

Table 8.4 reports data on the socioeconomic class of the fathers of elite astronomers. For both groups (*AMS* star and NAS), fathers whose socioeconomic status was white collar or professional predominate. As the notes to Table 8.4 indicate, these findings are statistically significant. While his findings are not differentiated by scientific community, Visher (1947: 533) reports in his study of the *AMS* star population that fathers of 46 percent of all starred scientists were professionals, while 23 percent were white-collar workers and 31 percent were in blue-collar occupations. *AMS*- starred astronomers closely reflect the pattern of the larger community, but this is *not* the case for those who earned a star and then went on to membership in the Academy. Here the percentage of those whose fathers were from blue-collar backgrounds was much *smaller* than the total *AMS*-starred population, while the percentage in the white-collar or professional category was greater. The socioeconomic standing of fathers apparently provided a significant advantage for those who entered the elite.

Table 8.4 The Elite: Father's Socioeconomic Class

Socioeconomic Class	*AMS* Star N = 36, data missing for 53 cases.	NAS N=66, data missing for 15 cases.
Blue Collar	11 (31%)	12 (18%)
White Collar and Professional	25 (69%)	54 (82%)

NOTE: $\chi^2 = 4.60$. Significant at the 0.5 level for two-tailed test.

Table 8.5 Highest Level of Education for the Elite

Educational Level	*AMS* Star N=77 Missing Data=12	NAS N=74 Missing Data=7
Secondary Only	2 (3%)	6 (8%)
Collegiate	15 (20%)	34 (46%)
Graduate (Includes three medical degrees.)	48 (62%)	27 (37%)

NOTE: $\chi^2 = 13.61$. Significant at the .01 level.

Only a few of the NAS-bound or those who would one day win *AMS* stars came from homes in which fathers were scientists. Five *AMS* star scientists (15 percent) and five future NAS members (9 percent) came from families where the father was a scientist. Astronomers (5) and medical doctors (3) predominated.

Fathers' highest level of educational attainment provides an interesting point of comparison between the two elite groups. Future NAS members came from homes in which fathers had more formal education than did the *AMS* star group. For almost all (98 percent) fathers of the *AMS* starred group, the highest level of educational attainment was no more than a secondary education. Sixty-five percent of the fathers of the NAS-bound population had no more than a secondary education. The father of one *AMS*- starred astronomer had a college education. Twenty-five fathers of NAS astronomers earned undergraduate degrees, and three did graduate work. Clearly, NAS-bound youngsters had an advantage in comparison to whose who earned no more than an *AMS* star. Many came from homes in which fathers had a relatively high level of educational attainment and their fathers tended to be white- collar or professional workers.

When the levels of educational attainment of the two groups of elite astronomers are compared, interesting patterns emerge (table 8.5). There are differences by educational level. For the NAS-bound, the advantage came at the level of secondary education. Some of the data in table 8.5 appear counter-intuitive. On balance, the *AMS*- star group had more formal education than those who one day would be elected to the National Academy. Conditioned by late twentieth century emphasis on educational credentials, we may find these data difficult to explain. Considered in light of the discussion on education and career development (chapters 4–6), this is not the case. Pluralism and diversity are key to understanding both education and career development. Nor was study abroad a significant factor in shaping the careers of elite scientists. Ten *AMS*-starred astronomers studied in Europe and three returned with degrees. This compares to eighteen NAS-bound scientists, of whom three earned European degrees. The collective biography provides information on the distribution of honors, awards, and offices within the American astronomical community. The differences between the two components of the elite as well as between the elite, rank and file, and women suggest the shape of the stratification system.

Table 8.6 indicates the participation of astronomers in two of America's most prestigious learned societies, the American Philosophical Society and the American Academy of Arts and Sciences. NAS astronomers outnumber those whose highest honor was an *AMS* star by four to one in the American Philosophical Society and three to one for the American Academy. The differentials are even more striking when the elite are compared with nonelite astronomers.

The distribution of major American prizes (table 8.7) for research in astronomy suggests similar patterns. NAS astronomers led those members of the elite who attained only an *AMS* star by ten to one. The skewed distribution of prizes is even more evident when the rank and file are considered.

American astronomers looked to four elite foreign academies and learned societies for special recognition (table 8.8). The *Bureau des Longitudes* was established to supervise the Paris Observatory and the French Nautical Almanac. The bureau elected a few distinguished foreign astronomers in recognition of their contributions to science. Only one non-NAS astronomer ever earned election to the bureau, Nautical Almanac Office director, William S. Eichelberger, who was a perennial, if unsuccessful, candidate for Academy membership.

Election as foreign associate of the Royal Astronomical Society of London was an important step on the way to membership in the National Academy of Sciences. Of the forty-five Americans so honored before World War II, 80 percent were also elected to the NAS. Honorary memberships in the RAS were created specifically for women, and two female members of the elite (Williamina P. Fleming and Annie J. Cannon) were so honored. *Only* NAS members were honored with foreign membership in the French Academy of Sciences or the Royal Society of London. Four rank-and-file males were elected foreign associates of the Royal Astronomical Society. Women never won any of these honors.

Table 8.6 Membership in Other Major American Academies

	Elite		Non-Elite	
	AMS Star	NAS	Rank and File	Women
American Philosophical Society	13	58	7	1
American Academy of Arts and Sciences	19	56	12	1

Table 8.7 Major American Prizes

AMS Star N=4		NAS N=40	
American Astronomical Society	1	NAS	18
Astronomical Society of the Pacific (ASP)	2	ASP	14
National Academy of Sciences (NAS)	1	American Academy of Arts and Sciences	7
		American Philosophical Society	1

NOTE: The rank and file received a total of twenty-three prizes, eighteen of which came from the ASP (primarily medals for discovering comets) or from the Franklin Institute in Philadelphia. Women were not awarded any prizes by major American academies or learned societies.

Table 8.8 Membership in Elite Foreign Academies

Academy	***AMS* Star**	**NAS**
Bureau des Longitudes	1	7
Foreign Associate of the Royal Astronomical Society	9	36
Honorary Member of the Royal Astronomical Society	2 (Both Women)	0
Foreign Member of the French Academy of Sciences	0	15
Foreign Associate of the Royal Society of London	0	14

Elite astronomers garnered most of the prizes offered by elite foreign academies. These include the Gold Medal of the Royal Astronomical Society, and prizes from the French Academy of Sciences and the Royal Society of London. American astronomers won sixty-five of these awards with the largest amount (82 percent) going to those elite astronomers who were also members of the National Academy. *AMS*-star awardees garnered just over 9 percent of these honors while rank-and-file males earned five (8 percent) and one woman was so honored.

Thirty-three Americans (thirty NAS members and three whose highest honor was an *AMS* star) were awarded honorary degrees by major European Universities (Oxford, Cambridge, Paris, and Berlin). No members of the rank and file or women were honored in this way.

Sixty-nine elite astronomers were awarded honorary degrees by major American institutions (Harvard, Chicago, Johns Hopkins). These included fourteen (20 percent) holders of *AMS* stars and fifty-five (80 percent) NAS members. Neither rank-and-file males nor women astronomers were given honorary degrees by these institutions.

Election to office, editorial appointments, and membership on commissions of the International Astronomical Union (IAU) represent measures of both status and power. Among those so honored were the directors of all the factory observatories and most of the major university observatories. As argued in chapter 7, these individuals constituted an interlocking network that controlled major institutions in the American astronomical community.

Editorial appointments and membership on IAU commissions can be defined as *gatekeeper* positions. Gatekeepers are involved in assessing the performance of other scientists. They both establish and enforce standards within the generic and specialist research communities (Merton and Zuckerman 1972: 522). Election to office, especially the presidency of a major scientific organization, tends to come later in the career. It represents a complex amalgam of both power and recognition by peers. We must not forget that before the creation of permanent bureaucracies, elected officers of professional scientific societies exercised a wide range of power. Thus elected office can be seen as both a form of gatekeeping as well as recognition.

Of the 352 offices filled by members of the American astronomical community between 1859 and the coming of World War II, 80 percent were occupied by members of the elite. A comparison of the performance of the two elite groups is instructive. *AMS*-star awardees held a third of these offices. Those whose highest honor was membership in the Academy held twice as many. Here is further evidence of differentiation within the elite as measured by power and recognition.

Twice as many NAS astronomers (50) held office in the American Association for the Advancement of Science as did those with only *AMS* stars. The two groups achieved parity in the American Astronomical Society, but sharp differentials show up when we compare offices held in the IAU, where NAS astronomers led those with only an *AMS* star by almost six to one. Equally significant differentials can be observed when office holding in the American Philosophical Society or the Astronomical Society of the Pacific are examined. Altogether, NAS members held sixty-nine presidencies and thirty-seven vice-presidencies. *AMS*-starred astronomers won fourteen presidencies and seven vice presidencies.

For comparative purposes, it should be noted that the rank and file held only about 18 percent of the offices and women less than 2 percent. Rank-and-file males held only two presidencies and four vice presidential positions. Women held three presidencies and one vice presidency, all of which were in the American Association of Variable Star Observers.

Editorial positions provide another index of location within the stratification system of a scientific community. Of course, the administrative tasks of an editor have, since World War II, become much more complex and demanding than was the case earlier (Wali 1991: 216–27). However, the social and political dimensions of the role remain virtually the same. As John Ziman (Merton and Zuckerman 1971: 461) put the matter, "An article in a reputable journal does not merely represent the opinions of its author; it bears the *imprimatur* of scientific authenticity, as given to it by the editor and the referees." Indeed, Ziman sees the gatekeeping role of editors as "the linchpin about which the whole business of science is pivoted."

Of the sixty-eight editorial positions held by American astronomers, almost two-thirds (65 percent) were occupied by members of the NAS. Those with only *AMS* stars held a mere 13 percent. This is, surprisingly, less than the percentage of editorial positions (19 percent) held by members of the rank and file. Women held 3 percent of these positions.

Examining the distribution in detail, we find that NAS members held eighteen editorships and twenty-four associate editorships. These included sixteen positions associated with the *Astrophysical Journal* and ten connected with the *Astronomical Journal.* No astronomer with only an *AMS* star served on the board of the *ApJ* and only two were associated with the *AJ.* At the *AMS* star level, we find five editors and four associate editors. Among the rank and file and women, none held editorships but they did have positions as associate editors or members of editorial boards. The combined total for the rank and file and women is fifteen editorial positions. For both the rank and file and women, editorial activities connected with *Popular Astronomy* bulk large.

A final indicator of power is membership on IAU commissions. Founded after World War I, the International Astronomical Union set fundamental technical standards for the international astronomical community. For example, IAU commissions established standards for spectroscopy and photometry as well as the study of stellar parallaxes. Of the 227 IAU commission seats filled by American astronomers, almost half (47 percent) went to members of the Academy. *AMS*-starred astronomers followed closely with 43 percent of the seats. Rank and file occupied just over 8 percent of the IAU commission seats given to American astronomers and women less than 2 percent.

These data suggest that members of the elite controlled most of the positions of power associated with gatekeeping functions. Almost without exception, the elite outdistanced the rank and file or women in every category. Within the elite, members of the Academy usually were far ahead of those whose honors peaked with an *AMS* star.

Honors and awards are, by their very nature, a scarce resource. They are valued in proportion to their scarcity (Merton 1957: 286–324; Hagstrom 1965: 71–72; Storer 1966: 20–26). Honors available to all members of a scientific community would be a contradiction in terms.

George Ellery Hale discussed honors as a scarce commodity when he informed Edward C. Pickering that the University of Chicago wanted to award the Harvard astronomer an honorary degree. In the first decade of its existence, Hale noted, Chicago had conferred only one honorary degree in the sciences. Hale emphasized that "the greatest care is taken in the selection of candidates" and since "very few are to be bestowed, the degree should have some value."[1] Hale went on to describe the steps through which a nomination for an honorary degree passed. The process was complex and guaranteed that the degree was, indeed, a scarce commodity.

Recognition in the form of honors and awards is a fundamental element in the scientific career. Writing to Walter S. Adams of the Mount Wilson Observatory, Carnegie Institution secretary Vannevar Bush remarked on the "yearning for recognition" among scientists and characterized it as "very natural."[2] Frank Schlesinger, one of the generals who directed American astronomy between the wars, was keenly aware of the significance of honors. In November 1928, the secretary of the Astronomical Society of the Pacific informed Schlesinger that he had been awarded the Bruce Medal. Established with funds from Catherine Wolfe Bruce, whose patronage was so important for astronomical research at the end of the nineteenth century, the award was highly coveted. To Schlesinger, the award of the Bruce was "a source of great satisfaction." In an age when astrophysics was in the ascendancy, it suggested astrometry was still an important field of scientific investigation.[3]

Schlesinger then sought to locate himself within the ranks of Bruce

1. George Ellery Hale to Edward C. Pickering, May 1901. Harvard College Observatory, HUA, Director's Papers. At Chicago, the following steps were involved in selecting candidates for honorary degrees. The candidate had to be recommended by the corresponding department and accepted by a committee representing allied fields (in this case, mathematics and physics). The recommendation then had to be approved by the university senate and finally the trustees. Only then was the nominee informed. Candidates were required to attend the ceremony in person. Under these circumstances, an honorary degree was, indeed, a scarce commodity and its value was enhanced accordingly.

2. Vannevar Bush to Walter S. Adams, 9 October 1939, p. 2. MTWA, Adams Papers.

3. Frank Schlesinger to William Eichelberger, 27 February 1922. See also C. H. Adams to Frank Schlesinger, 24 November 1928, and Schlesinger to Adams, 4 December 1928. YUA, Astronomy Department Papers.

medalists. At the bottom of the letter announcing the award and on the carbon copy of his reply, he carefully drew up a set of lists. These included living Bruce medalists, distribution of the Bruce by nationality, those who had been awarded both the Bruce and the Gold Medal of the Royal Astronomical Society and those who won the Bruce but never the RAS Gold. These lists enabled Schlesinger to place himself on a larger map representing the reward system of twentieth-century astronomy. Already the recipient of an award from the French Academy of Sciences (1926) and the RAS Gold Medal (1927), the Yale director joined a select company: the eight living astronomers who had won *both* the Bruce and the RAS Gold. American astronomers on the list included Mount Wilson founder George Ellery Hale, William Wallace Campbell of the Lick, Walter S. Adams of Mount Wilson, Yale celestial mechanician Ernest W. Brown, and Princeton's Henry Norris Russell. The names of the Astronomer Royal, Sir Frank Dyson, Meudon Observatory director Henri Deslandres and Cambridge astrophysical theorist Sir Arthur Eddington, suggest the status of European representation in this distinguished group. Schlesinger's private reflections underscore the centrality of recognition in the minds of astronomers, most of whom had a keen sense of the importance of honors and awards for their careers, understood the hierarchical nature of the reward system, and could locate themselves within it.

ON THE WAY TO THE TOP: THE *AMS* STAR SYSTEM

In 1903, Columbia University psychologist James McKeen Cattell (1860–1944) started work on what would become the *American Men of Science* series. Before World War II, Cattell published six editions of *AMS* (1906 [4,000 names], 1910 [5,500 names], 1921 [9,500 names], 1927 [13,500 names], 1933 [22,000 names], 1938 [34,000 names]). One of the most important aspects of the *AMS* series was a system of awarding stars to leading scientists. Beginning with the first edition, Cattell sought to identify the thousand most eminent scientists in America.[4] Cattell used a form of peer review to secure nominations and evaluations in awarding stars.

From 1906 through the 1944 edition of *AMS* (the last to include stars), the star system was an integral, if controversial, element in the reward system of American science. Because stars were awarded by discipline, and based on the judgment of peers within each of twelve major fields of American science, the starring system also took on an important

4. Professor Michael M. Sokal of Worcester Polytechnical Institute is at work on a biography of Cattell. I am indebted to Professor Sokal for information on Cattell and the starring system. Historians have not generally treated Cattell and the star system with all the care they deserve.

role within generic scientific communities. Writing to Cattell in 1936, Berkeley physicist L. B. Loeb pointed out that "while I recognize the possible fallibility of such listings as the starred men, I am convinced that the starring of men in *American Men of Science* furnishes an exceedingly valuable incentive . . . There are all too few rewards for proper scientific achievement in a crowded field such as physics." Loeb argued that since only a small number of physicists could be elected to the National Academy, the star system provided an effective means of identifying and rewarding successful younger scientists.[5] It is significant that Loeb ranked the *AMS* star just behind membership in the National Academy in the hierarchy of honors and awards available to American scientists.

In June 1903, Edwin B. Frost of the Yerkes Observatory, received an invitation from Cattell to take part in selecting astronomers to be starred in the first edition of *AMS*. Cattell enclosed a packet of slips listing the names and addresses of the most prominent members of the astronomical community. Frost was to arrange the slips in order of merit. It is not clear how many other astronomers received similar requests. The group probably included the directors and senior scientists at the leading observatories and university departments of astronomy.

Cattell was at pains to explain his conception of merit. In evaluating colleagues, contributions to research were to be given first priority. However, Cattell went on to suggest that "teaching, administration, editing, [and] the compilation of text-books" should also be taken into consideration. He urged referees to pay careful attention to the mix of activities that made "a man efficient in advancing science."[6] In ranking scientists, Cattell cautioned that contributions to only one discipline should be considered. Cattell also asked those on whom he called for evaluations to avoid being negatively influenced by age and to reward performance, not promise.

With the 1910 edition of *AMS*, Cattell opened himself to considerable criticism. Once individuals had been awarded a star (in this case in the 1906 edition) they would retain the honor for life. The star system did not involve a reevaluation of past awardees for each new edition. Peers voted only on new nominees to the ranks of starred scientists. Despite phenomenal growth in the number of scientists between 1900 and World War II, Cattell remained committed to the magic number of one thousand starred scientists. This meant that stars would remain a scarce commodity and their value would be enhanced as the total size of the American scientific community increased. For the second edition, the system of peer review was more clearly defined. In each of the dozen fields of science recognized

5. L. B. Loeb to James McKeen Cattell, 5 October 1936. LC, Cattell Papers.

6. "Memorandum," pp 1–2, attached to James McKeen Cattell to Edwin B. Frost, 24 June 1903. UCYOA, Frost Papers.

by Cattell, ten leaders were asked to nominate scientists who have "done the most valuable work."[7]

When Cattell solicited nominations of scientists to be starred in the third edition of *AMS*, some astronomers offered criticisms of the list of previously starred colleagues. Campbell of the Lick objected to fully a third of the astronomers who received stars in previous editions. Critics insisted that the list of starred astronomers had to be revised. Some individuals should never have been awarded stars. They must be stripped of this honor in the 1921 edition. A number of leading astronomers not starred earlier must be added. Speaking bluntly, Campbell told Cattell "that unless the list is mercilessly cut I shall not be able to regard it as representative of the best that there is in American astronomy."[8] These criticisms did not prevent Campbell from nominating candidates to receive stars in the next edition of *AMS*.

In 1926 Cattell asked all living starred astronomers to nominate colleagues for the fourth edition of *AMS*. The number of starred scientists would be capped at fifty for the American astronomical community. This figure was roughly twice the size of the astronomy section of the National Academy during the 1920s. In his 1926 covering memorandum, Cattell admitted that while "there may be room for question as to the desirability of such a list . . . the preponderance of argument appears to be in its favor."[9] Cattell did make one major concession. He dropped all starred astronomers who were retired.

The selection of astronomers to receive stars in the sixth edition (1938) of *AMS* makes an interesting case study. Cattell expanded the system of nomination. Star holders were encouraged to nominate in their fields and to nominate, as well, deserving scientists, even if not in their disciplines. Also, self-nomination was permitted for the first time.

Cattell planned to award a total of 250 new stars in the sixth edition. Astronomers were permitted twenty-five nominations. Of astronomers requested to vote, 73 percent returned ballots. This figure was down slightly from 81 percent in 1926.

The return rate for balloting on stars for the sixth edition ran from a high of 86 percent in anthropology to a low of 55 percent in botany.

As the work sheets in the Cattell papers at the Library of Congress

7. A copy of the 1909 instructions to evaluators is found in the Lick Observatory Archives. SALO, Cattell Papers.

8. William Wallace Campbell to James McKeen Cattell, 3 September 1919, pp. 1–2. SALO, Cattell Papers. See also the "List from Harlow Shapley," 16 February 1920. LC, Cattell Papers.

9. "Memorandum," from James McKeen Cattell, 7 July 1926. YUA, Astronomy Department Papers.

indicate, thirteen astronomers were awarded stars in 1938. The cutoff point for election was drawn at thirty votes. Three individuals were tied with thirty votes (Dirk Brouwer of Yale, Gerard Kuiper of Yerkes-McDonald, and Jan Schilt of Columbia). Cattell and his staff drew the line between Kuiper and Schilt; the Columbia professor did not receive a star. Solar spectroscopist Charlotte Moore-Sitterly led the field with a total of fifty-four votes. She was followed by Theodore Dunham, Jr., and Walter Baade of Mount Wilson Observatory.[10]

The director of the Yale South-African station, Harold Alden, was awarded a star in the 1938 edition of *AMS*. He stood eighth in the balloting, tied with John C. Duncan of Wellesley College. Each man had thirty-four votes. Alden felt that his long residence abroad had cut him off from the reward system of American science and was gratified to be honored with a star.[11]

PROGRESS TOWARD ELITE STATUS

The honors and awards available to astronomers represent a series of grades or statuses that make up the stratification system in the American astronomical community. Any discussion of elite astronomers should also examine the ways in which they move through the stratification system. The process is not random. Some awards reliably precede others in the course of a career. In this process, the award of an *AMS* star marked an important career milestone.

Table 8.9 reports data on forty-seven individuals who earned *both* an *AMS* star and election to the National Academy. For the whole group (far right-hand column) the *AMS* star preceded election to the Academy in over half (53.2 percent) the cases. If we add the two individuals who won an *AMS* star and NAS election in the same year, the figure rises to 57 percent. Chronological disparities between the founding of the NAS in 1863 and the awarding of the first *AMS* star in 1906 distort relationships for the first two cohorts. Relying on data for the third cohort (1900–1940), it is clear that the *AMS* star was a harbinger of things to come. The award of a star preceded election to the Academy in 79 percent of the cases.

Table 8.10 suggests another way to look at the sequencing of honors preliminary to NAS election. Here the data are for the third-cohort elite (both *AMS* star holders and members of the Academy). Major American honors (election to the American Philosophical Society, the American

10. These data are taken from pages 2a and 3 of the file containing a quantitative analysis of the voting for the sixth edition of *American Men of Science*. It is located in the Cattell papers in the Library of Congress. The organization of the Cattell papers was, at the time I used them, minimal. It will take a little detective work to locate the file.

11. Harold L. Alden to James McKeen Cattell, 20 October 1937. LC, Cattell Papers.

Table 8.9 Sequencing of *AMS* Stars and Election to the NAS

	1859 or Before			1860–1899			1900–1940			
	Old Ast.	New Ast.	Mixed Ast.	Old Ast.	New Ast.	Mixed Ast.	Old Ast.	New Ast.	Mixed Ast.	**Totals**
AMS★ first				5	7	2		11		**25**
NAS first	2	1		8	8			3		**20**
AMS★ and NAS same year								2		**2**
Mean years between *AMS*★ and NAS				9.2	9.9	12.5		7.3		

NOTE: *AMS*★ = *AMS* Star

Table 8.10 Third Cohort Elite: Sequencing of Major American Honors

Honors	**Mean Professional Age**
American Academy of Arts and Sciences	13.3 Years
AMS Star	13.8 Years
American Philosophical Society	15.5 Years
National Academy of Sciences	18.1 Years

Academy of Arts and Sciences, the National Academy, and the award of an *AMS* star) are arranged according to mean professional age at the time of the award (calculated from date of first entry into astronomy).

Election to the American Academy began the process and was closely followed by the award of an *AMS* star. Then came election to the American Philosophical Society. Almost five years separate the first honor in this sequence from election to the NAS. While not all third-cohort elite astronomers were elected to the American Philosophical Society or the American Academy, two-thirds of those who were rewarded by peers with election to the NAS first received an *AMS* star. The percentage jumps to over 80 if we add the two individuals were elected to the Academy in the same year they earned *AMS* stars. Thus, in a very real sense, the *AMS* star served as a discrete status, marking the progress of individuals through the hierarchy of honors and awards, toward membership in the National

Academy of Sciences, the ne plus ultra of American recognition for outstanding scientists.

THE NATIONAL ACADEMY OF SCIENCES

From its creation, astronomers played a central role in the affairs of the National Academy of Sciences. Benjamin Gould was a member of the inner circle that founded the Academy.[12] Members of the astronomy, geography, and geodesy section of the mathematics and physical sciences class were the largest single membership block (30 percent) when the NAS was organized in 1863. As the Academy grew, this percentage declined, but astronomers made their presence felt in a variety of ways. Before World War II, one astronomer served as president of the Academy, five occupied the vice presidential chair, three served as home secretary, two as foreign secretary and one filled the office of treasurer. Down to World War I, there was almost always one astronomer on the council and often up to three. George Ellery Hale led the drive to create the National Research Council and to construct a permanent Academy building in Washington in the 1920s. Between the wars astronomers became the focus of controversy. They were accused by other scientists of seeking to elect too many of their number to the NAS.

Fundamental to understanding nomination and election to the Academy are the rules and procedures that guided the process. These changed over time and were not always clear to all members. For nine years after its founding in 1863, the Academy was divided into two classes, mathematics and physics, and natural history. Each had five sections (e.g., within mathematics and physics there were sections for mathematics, physics, the combined group of astronomy, geography and geodesy, mechanics, and chemistry). During this period Academy membership was limited to fifty. In 1872 the system of classes and sections was abolished. For more than a quarter century, the Academy debated plans for reorganization, but no resolution was reached until 1899, when the constitution was amended to provide for six permanent sections: mathematics and astronomy, physics and engineering, chemistry, geology and paleontology, biology, and anthropology. In 1914, the mathematics and astronomy section divided into separate units.

In 1870 the membership limit of fifty was removed and from 1872 through 1907 up to a total of five new members could be elected each

12. Cochrane (1978) is the official history of the Academy. Dupree (1957a, 1979) provides important correctives to the official history of the Academy. In comparison to the Royal Society of London or the French Academy of Sciences, the National Academy has been, in Dupree's words, "true to the principle of not viewing anything, even itself, historically." Dupree (1979: 348).

year. In 1907 the ceiling was raised to ten new members per year and in 1915 the figure increased to fifteen. In any given year, the Academy did not necessarily elect as many scientists as the constitution permitted.

From 1872 through the reforms of 1899, the Academy nominated and elected new members as a whole. Nominations were generally brought before the fall meeting and voted on the following April. This system placed a premium on patronage. At least five Academy members had to nominate an individual for membership, and one of them had to take responsibility for orchestrating the campaign. Supporters were required to pay for printing the candidate's bibliography and mailing it to all members, thus incurring liabilities measured not only in time and political capital, but cash as well. Patrons with parliamentary skills probably had more luck in electing their clients than did those who knew little about the complex electoral process. Visibility within the Academy may also have helped patrons swing votes for their clients.

Beginning in 1907, section chairs were required to canvass their members for nominations. This process involved two ballots. The first was a preliminary or informal vote that often involved a large field of candidates. The second or formal ballot selected one or more nominees, by majority vote, from those who stood highest on the preliminary ballot. One or, at most, two names from each section went to the home secretary. The next step was an Academy-wide preferential ballot. The names of the nominees from each section, *including the number of votes candidates received in their respective sections,* were printed on the preferential ballot. A bibliography and biographical sketch for each individual on the preferential ballot was also made available. The final vote was by Academy members attending the April meeting. The merits of candidates were discussed openly, and, on occasion, the Academy rejected a nominee. In 1924 electoral procedures were changed to require a two-thirds vote of the section in order to present a name to the home secretary for inclusion on the preferential ballot.[13]

Archival materials provide information on astronomers nominated and voted on by the astronomy section of the Academy. From the formation of the mathematics and astronomy section in 1899 through World War II there are only a few gaps in the record. Before 1899, it is much more difficult to develop lists of nominees, and virtually impossible to locate information on votes cast for individual candidates under the old system.

Readers should understand that data on nomination and election to the National Academy are not drawn from the archives of the NAS. If such

13. This summary is based on a close reading of the annual reports of the Academy. The file I used is in the Archives of the NAS in Washington.

data exist in Washington, they are very much off limits to historians or anyone else. Over the years, I found that certain astronomers were pack rats, who carefully filed away copies of ballots or memos from section chairs or made notes on how they voted on the preferential ballot and actions taken on the floor at the April meeting. If formal minutes of the April meetings were kept, they must not have been circulated to members, or, at least, I have never located copies. The following collections are rich in materials concerning NAS nominations and elections: the astronomy department papers of the University of California at Berkeley; the Comstock and Stebbins papers at Madison; the astronomy department papers at Yale; and the Struve papers at Yerkes. A number of collections at the Library of Congress contain information on NAS elections. I found the J. C. Merriam papers most useful; there I learned what to look for in the papers of other scientists.

Individuals considered by the astronomy section (or mathematics and astronomy between 1899 and 1914) can be grouped into three classes. The first are *outsiders,* individuals nominated on the informal ballot by at least one member of the section, *but who received no votes on any formal ballot.* Available records provide no indication of who nominated these individuals or why. The group of outsiders includes four individuals. Sherburne W. Burnham was nominated in 1907. Possessed of an angular and independent personality, Burnham ranked as the premier double-star observer in the United States, but stood outside the major power configurations in American astronomy. In 1911 both Thomas Jefferson Jackson See, superintendent of the observatory at the Mare Island Naval base in California, and the Reverend Joel Metcalf, optical craftsman and astronomical photographer par excellence, were nominated for NAS membership. The 1925 nomination of Harvard spectroscopist Annie Jump Cannon rounds out the list. All shared one characteristic: they earned stars in *American Men of Science.* Three of the four (Burnham, See, and Cannon) were affiliated with observatories. Metcalf had a well-equipped private observatory and optical shop in which he produced state-of-the-art optics for the Harvard Observatory. The data do little more than point to the status of the group as marginal. Cannon as a woman was not seen as eligible for the Academy. See was marginal in the sense that he was both institutionally and intellectually isolated from the main stream of astronomy. Burnham never quite lost his amateur status, even after appointments at two major research institutions, and Metcalf clearly remained an amateur.

A second group, numbering six individuals, received *courtesy nominations. They received one vote on the informal section and one vote on the formal section ballot.* While it is impossible to be certain, my presumption is that the same individual voted for a candidate on both ballots.

These names appeared only once on the informal and formal ballots of the astronomy section.

This group included University of Pennsylvania Observatory director Eric Doolittle, Yerkes photometric expert John A. Parkhurst, and Harvard solar system astronomer William H. Pickering, all of whom were considered in 1918–19. Yerkes comet and doubl- star specialist George Van Biesbroeck and Mount Wilson solar astronomer Francis G. Pease were nominated in 1924–25. Edward S. King, Harvard astronomical photographer, was considered in 1929. These individuals won stars in *American Men of Science* and were affiliated with astronomical research institutions. Although evidence is lacking, it may be that these astronomers were nominated by close associates or superiors as a form of recognition or courtesy, even though supporters knew full well that they were not likely to be endorsed by other members of the section.

Detailed information is available on twenty-six individuals who were nominated but never elected to the Academy. This group includes those in the categories discussed above as well as more serious contenders. Perhaps the most fruitful way of looking at this group is in terms of the length of time an individual name remained before the astronomy section. This measure is simply the difference in years between the date of first appearance before the astronomy section and the last time an individual was voted on.

In the second cohort, the data permit discussion of twelve individuals who were nominated but never elected. The six courtesy nominations were before the astronomy section for only one year. But, perhaps surprisingly, four names were considered by the astronomy section for periods ranging from eleven to twenty years. This group includes Charles L. Doolittle, who preceded his son Eric as professor of astronomy and director of the observatory at the University of Pennsylvania, Michigan Observatory director William J. Hussey, John A. Miller, director of the Indiana University and then Swarthmore College observatories, and Charles D. Perrine, who left Lick to direct the Argentine National Observatory at Cordoba. These individuals had enough support to keep their names before the section for a considerable time, but never commanded the votes needed to win a place on the preferential (Academy-wide) ballot.

For the third cohort, fourteen individuals were nominated but never elected to the NAS. Four were considered by the section for only one year; six, however, remained under consideration for periods ranging from eleven to twenty years. This latter group includes Philip Fox, longtime director of the astronomy program at Northwestern University, Berkeley astronomer Russell T. Crawford, University of Chicago mathematical astronomer William Macmillan, Lowell planetary astronomer Carl O. Lampland, Keivin Burns of the Allegheny Observatory and Cecilia Payne-

Gaposchkin of Harvard. All of these astronomers won stars in *AMS*, and Fox was director of an observatory.

Another way of looking at the group of astronomers nominated but never elected, is to examine patterns of balloting for selected individuals. Berkeley celestial mechanician Russell T. Crawford remained before the section for nineteen years (1921 through 1940). This is one of the longest candidacies on record. Crawford's name appeared on a total of ten informal ballots and eleven formal ballots. In 1936 and again in 1938 Crawford peaked in the informal balloting with nine votes. The first time his name appeared on the formal ballot (1925), he got seven votes. He secured ten votes on the formal ballot for 1935 and peaked with fifteen the next year. In the period 1938–40, support for Crawford on the formal ballot declined rapidly. By 1940 he had the support of only two members of the section.

William Eichelberger, director of the Nautical Almanac Office, represents a mid-range case. Eichelberger's name was before the astronomy section for a total of seven years. He first appeared on the informal ballot in 1924 with only one supporter, but later that year, on the formal ballot, received seven votes. In 1927 five NAS astronomers gave Eichelberger their support on the informal ballot, and he won seventeen votes on the formal ballot. This would be the high-water mark of his campaign. In 1928 fourteen astronomers supported the NAO director in the informal balloting while in 1929 he received thirteen on the formal ballot. In 1929 a total of sixteen votes were cast for Eichelberger on the informal ballot, but this translated into only ten votes in the formal vote a year later. In 1930 support for Eichelberger, now retired from the NAO and working for Eastman Kodak, fell to thirteen on the informal ballot, and he received only one vote on the formal ballot of 1931.

It appears as if candidates were considered from one perspective in the informal balloting but from a quite different perspective when the formal vote was taken. Probably, the first or informal vote was an evaluation of individuals in isolation, while the formal vote was comparative, involving the ranking of candidates based on their perceived merit and, perhaps, the degree of support within the section.

It may be that the longer an individual was before the astronomy section, the less chance there was of gaining enough votes to be placed on the preferential ballot for consideration by the whole Academy. As William Wallace Campbell remarked to Joel Stebbins, the longer the election of a senior astronomer was postponed, the less probable his chances for election. "There will be other and younger men coming up each year who I suspect will appeal more strongly to a majority of the members of the astronomical section."[14]

14. Joel Stebbins to Henry Norris Russell, 11 August 1933. UWMA, Stebbins Papers.

One final characteristic of those nominated but never elected deserves consideration. Of these twenty-six individuals, fully half were either observatory directors or chairmen of astronomy departments. The concentration of directors and chairmen was highest in the second cohort, where two-thirds of the nominees occupied directorships or chaired departments. In cohort three, the figure fell to 36 percent. These data suggest that while the position of observatory director or departmental chairman may have played a part in gaining consideration by the NAS astronomy section, it was by no means enough to guarantee sufficient support to send a name forward for consideration by the Academy.

Before the creation of an independent astronomy section in 1914, nomination and election to the Academy depended on the activities of patrons and brokers. Late-nineteenth-century attempts to elect double star observer Sherburne W. Burnham provide an example of patrons and brokers in action. In this case, Lick director Edward S. Holden assumed the role of patron for Burnham, but Asaph Hall also played an important part. As Academy home secretary, he acted as a broker, passing on information and advice. A patron like Holden was at a disadvantage; he lived too far away to attend East Coast meetings of the Academy on a regular basis. Brokers like Hall supplied information that could be known only to those who took part in meetings, as well as definitive interpretations of the by-laws governing the nomination and election process. Soon after the Lick opened, Director Holden nominated Burnham for NAS membership. Holden, however, had secured only four cosponsors. Hall pointed out that Academy regulations required five members to support a nomination.[15] Once Burnham's nomination was in order, Hall requested Holden to prepare a list of the candidate's publications and send copies to all members of the Academy. Hall reminded Holden that the expenses for printing and mailing the list were the responsibility of those making the nomination. He also reported that the name of Dudley director Lewis Boss would be brought to the Academy for a vote at the April meeting.[16]

Boss was elected to the Academy at its April 1889 meeting; Burnham was not. Undaunted, Holden initiated the process again. Hall indicated that Burnham's name would be presented at the November 1889 meeting and voted on in April 1890, and that it was not necessary to do a second mailing of biographical and bibliographical material on the candidate.[17] In the summer of 1889, Holden sought to strengthen Burnham's nomination by enlisting the support of Edward C. Pickering, director of the Harvard College Observatory. All was not, however, smooth sailing. "You

15. Asaph Hall to Edward S. Holden, 9 July 1888. SALO, Hall Papers, and C. H. F. Peters to Edward S. Holden, 26 July 1888. HCA, Peters Papers.

16. Hall to Holden, 2 November 1888. SALO, Hall Papers.

17. Hall to Holden, 28 May 1889. SALO, Hall Papers.

ought to know," Asaph Hall reported to Holden, "the objection that is raised to Burnham . . . [is that] he knows very little of theory."[18] For workers in the old astronomy, this was a serious criticism. Burnham was a qualitative rather than a quantitative scientist. He knew little about celestial mechanics, and did not analyze his observations in order to produce mathematically exact orbits for binary stars. Burnham failed of election at the April 1890 meeting, and Holden again renewed the nomination. Hall, who may have had a difficult time maintaining neutrality in Burnham's case, reported that he had lost a key letter listing Burnham's sponsors. Holden secured new supporters, but found some senior astronomers, like Charles A. Young of Princeton, unwilling to cooperate. Hall candidly advised Holden that he ought to secure the most powerful sponsors available and suggested California astronomer and geodesist George Davidson.[19] For whatever reasons, Burnham's nomination did not receive the required number of signatures and could not be presented at the fall meeting. Hall later reported that the April 1891 meeting was so deeply divided that no one was elected.[20]

Holden continued his role as patron during the last decade of the century. In 1896 he asked George C. Comstock, his successor as director of the Washburn Observatory, for 100 copies of his bibliography. On receipt of the material, Holden reported that he would "take pleasure in proposing your name for the N.A.S." and, further, that Simon Newcomb concurred in the nomination. "Strictly speaking you should not know of my proposal to nominate you. . . . Even the best names have to wait two or, at times, three years," but Holden assured the younger man "this is no discredit." Holden indicated that he had written four or five of "the leaders to ask them to second your name" and pledged to "keep things going till you wake up and find yourself elected." But, Holden counseled, "In the mean time, forget it!" Comstock did not demur when Holden shifted the financial responsibility for printing the bibliography to the candidate.[21]

Holden again turned to Asaph Hall for advice. The Naval Observatory astronomer urged Hall to secure the support of Benjamin Gould and Simon Newcomb. "With either one opposed the election would fail." Hall went on to remind Holden that Boss had been elected by only one vote. Isolated on Mount Hamilton, Holden may not have appreciated how dif-

18. Hall to Holden, 25 July 1889. SALO, Hall Papers.

19. Hall to Holden, 9 December and 26 December 1890. SALO, Hall Papers.

20. Hall to Holden, 21 May 1891. SALO, Hall Papers. Apparently, Holden had not been able to secure Davidson's support for Burnham.

21. Holden to Comstock, 27 October 1896; Holden to George C. Comstock, 10 October 1896; Comstock to Holden, 20 October 1896. UWMA, Comstock Papers.

ficult it was to elect astronomers under the old system. As Hall pointed out, "A good astronomer is not generally known well enough to command a large vote outside his own profession."[22] It was necessary for astronomers to close ranks behind their nominees. Hall believed astronomers should "unite on one or two" candidates if they wanted to elect one of their number to the Academy.[23]

In the end something went wrong with the campaign. Not enough astronomers were willing to back Comstock's nomination. Without going into detail, Holden wrote the Wisconsin astronomer "I'm afraid some one has blundered about your N.A.S. nomination so as to require a year's delay. It was not I—but it is provoking."[24] It is difficult to see where the blame should be placed, if not on the shoulders of Holden, Comstock's patron.

Efforts to elect Comstock ran into serious difficulties when Edward S. Holden resigned as Lick director and fled Mount Hamilton. This effectively ended Holden's scientific career; he remained a member of the Academy, but he lost virtually all standing and credibility. Seth Carlo Chandler, editor of the *Astronomical Journal,* stepped in as Comstock's patron. In requesting a fresh list of major publications, Chandler wryly assured the younger man that he had no "nefarious" schemes in mind.[25]

By the end of the century the role of broker was being redefined. A growing number of talented and highly visible young astronomers were pressing for admission to an Academy whose rules and procedures for nomination and election were geared to a much smaller scientific community. Asaph Hall suggested to Newcomb that "It seems to me we ought to consult in Washington, and agree on two names, for if we nominate three or four astronomers, no one will be elected."[26] After 1914, the astronomy section would provide an arena for collective brokering between contending candidates and their patrons. Early the new century, however, a variety of ad hoc solutions emerged.

Chicago mathematical astronomer Forest Ray Moulton was elected to the Academy in 1910 and soon became an active participant in the affairs of the mathematics and astronomy section. Moulton was, by his own admission, in a unique position: "The mathematicians of this part of the country naturally rely on me somewhat for advice as to astronomers

22. Hall to Holden, 4 November 1896. SALO, Hall Papers.

23. Hall to Holden, 31 March 1897. SALO, Hall Papers.

24. Holden to Comstock, 1 April 1897. UWMA, Comstock Papers. See also Hall to Holden, 1 April 1897. SALO, Hall Papers.

25. Seth Carlo Chandler to George C. Comstock, 11 August 1897. UWMA, Comstock Papers.

26. Hall to Newcomb, 10 March 1897, p. 1. LC, Newcomb Papers.

who would be suitable candidates for our section." Moulton opened a correspondence with William Wallace Campbell. The major topic of discussion was nominating and electing astronomers to the NAS. The Chicago astronomer rejected the older idea that the "heads of certain departments of the government, and men who occupied important administrative positions elsewhere" should be elected without regard to their scientific accomplishments. He then listed three possible candidates: Abbot of the Smithsonian Astrophysical Observatory, Adams of Mount Wilson, and Schlesinger of Allegheny. Moulton asked Campbell for his views on these potential nominees.[27] Moulton concluded with an invitation for Campbell to lecture at Chicago. By this invitation, Moulton indicated a proper sense of deference toward the senior scientist.

Campbell expressed complete agreement with Moulton concerning qualifications for membership in the Academy: "No one should be elected by virtue of the office which he may be filling; and success or failure as a teacher or administrator should not be considered seriously." The Lick director then went on to evaluate the three astronomers suggested by Moulton. Charles G. Abbot ranked first. "His work is marked by originality, value and sustained effort, and I should be glad if we could agree to present this point of view to such other members of our division as we shall have the chance of addressing orally or by letter between now and October." Campbell was not enthusiastic about Moulton's other suggestions. Campbell was critical of Schlesinger's spectroscopic investigations, rating them as derivative, but indicated that he did look forward to voting for Schlesinger "in the near future." When it came to dealing with California astronomy, Campbell judged the competition harshly. He rejected Adams, damning him with very faint praise. While Adams was, "no doubt a tremendous worker and perhaps an excellent administrator" he worked along lines developed by others and was hardly an original scientist.[28] Campbell took the opportunity to suggest his colleague, double star observer, Robert G. Aitken as more deserving of consideration. If we use the same criteria, it is difficult to understand how Aitken's case differed from that of Adams. There was nothing original about his great survey of double stars. It was simply a continuation of work done by Burnham and others. Perhaps the difference had more to do with institutional location than scientific activities.

Moulton agreed with Campbell's ranking but suggested that he had, perhaps, undervalued Adams. However, the younger man made this sug-

27. Forest Ray Moulton to William Wallace Campbell, 5 May 1913. SALO, Moulton Papers.

28. Campbell to Moulton, 9 May 1913, pp. 1–2. SALO, Moulton Papers.

gestion only after complimenting Campbell on his handling of Leuschner's election to the Academy and indicating agreement with Campbell on the question of the standards that should guide the nomination and election process. Moulton urged further discussion between astronomers and volunteered to lobby mathematicians.[29]

Campbell took the initiative and in June wrote to Comstock concerning the need for astronomers to achieve consensus on a single candidate. There was a tendency for the mathematics and astronomy section to divide, thus precluding the nomination of any candidate. He asked Comstock's opinion of Abbot as the astronomers' candidate. The Lick director then went on to suggest that his colleague Aitken should follow Abbot as the next astronomer elected to the Academy. Neither Schlesinger nor Adams were mentioned as possible nominees.[30]

By mid-October, Moulton was able to report the probable choice of the mathematicians and urged Campbell to work for consensus on a single astronomer (Abbot was his preference), so that the section could agree on one nominee from each group. Apparently, there were some hitches, but Abbot and the mathematician with whom he was paired were elected in 1915. At length, Campbell made it clear to Moulton that he was ready to support Schlesinger as the next astronomer for Academy membership. Campbell also wrote to the newest astronomer in the Academy, his colleague Leuschner at Berkeley, advising him how to vote. It was done with finesse and diplomacy, but the intent was clear enough.[31]

After 1914 the need for brokers declined. The astronomy section provided the primary arena in which patrons struggled to elect their clients. Home secretaries never again played the part of powerful brokers. Individuals like Campbell and Moulton concentrated on specific campaigns without the need to balance their candidates against those of the mathematicians.

Patronage was indispensable in achieving election to the Academy, and patrons made sure that clients were aware of their actions. In part, this was inescapable, since patrons needed a bibliography and biographical information, but, in a larger sense, patrons wanted to make sure that

29. Moulton to Campbell, 31 May 1913, p. 2. SALO, Moulton Papers.

30. William Wallace Campbell to George C. Comstock, 12 June 1913, pp. 1–2. UWMA, Comstock Papers

31. Moulton to Campbell, 17 October 1913. SALO, Moulton Papers. See also B. Osgood Peirce to William Wallace Campbell, 19 October 1913. SALO, Peirce Papers. For a view of the Lick Director pressuring a new member of the Academy see William Wallace Campbell to Armin O. Leuschner, 20 November 1913. He enclosed Moulton's letter of 6 November 1913. See also Leuschner to Campbell, 2 February 1914, p. 2. SALO, Leuschner Papers.

clients appreciated the exertions undertaken in their behalf. In response to a telegram from Edward C. Pickering announcing Campbell's election to the Academy, the Lick director wrote: "This is an honor that could not well be surpassed, and to all my friends who have taken an interest in the subject, I return my cordial thanks." [32] The record suggests that Campbell himself took considerable interest in the subject. He paid for the printing of a twelve-page pamphlet containing a listing of his scientific papers.[33]

The creation of an astronomy section did not necessarily make the electoral process any less political. As Frank E. Ross wrote Vesto M. Slipher, concerning their perennial struggle to elect John A. Miller to the Academy, "I would really like to see him get in, but you can't buck the big interests. Politics in astronomy as in anything else."[34] The "big interests" were always on the lookout for an opportunity to secure Academy membership for one of their own. Director Adams of Mount Wilson wrote to emeritus Lick director Aitken that he hoped soon to secure the election of his colleague Alfred Joy. It appeared, however, that Adams saw Joy's election as a reward for "self-sacrificing work as Secretary of the Observatory" rather than recognition of his contributions to stellar spectroscopy.[35]

Harvard biometrician Edwin B. Wilson criticized the tendency of astronomers to exaggerate the abilities of their nominees when speaking from the floor at the April meeting. Why, he asked Leuschner, did they need to employ hyperbole in describing candidates? Wilson reported that "I tackled [Henry Norris] Russell after a recent election . . . [concerning the claims he made about one of the astronomers] and I told him that he knew just as well as I did and everybody else that this astronomer was no first-class scientist, that he had been on the list of the astronomy section for years and years and if he had been first-class would have been taken in before he was 45 . . . and that he [Russell] knew perfectly well . . . that there were outside of the Academy men of much higher distinction." Wilson called the nominee "a border line case" who "would lower rather than raise the standard of the Academy." [36]

During the interwar years, astronomers pushed hard to elect as many of their number as possible. Led by the directors of the great factory observatories, the campaigns were well orchestrated and often intense. Lob-

32. William Wallace Campbell to Edward C. Pickering, 19 April 1902. HUA, Harvard College Observatory Director's Papers.

33. William Wallace Campbell to Simon Newcomb, 27 February 1902. LC, Newcomb Papers.

34. Frank E. Ross to Vesto M. Slipher, 4 December 1930. LOA, V. M. Slipher Papers.

35. Walter S. Adams to Robert G. Aitken, 31 December 1940. MTWA, Adams Papers.

36. E. B. Wilson to Armin O. Leuschner, 27 March 1934, pp. 2–3. BLUCB, Astronomy Department Papers.

bying went on right up to the final discussion of candidates on the floor of the April meeting. Walter S. Adams promised William Hammond Wright that a Mount Wilson staff member would "do a little missionary work among members of the Academy" on behalf of Lick spectroscopist Joseph H. Moore as soon as he reached Washington. There were two astronomers to be considered in April 1930 and, as Adams pointed out, critics like J. McKeen Cattell were likely to make "sarcastic remarks" about the proportion of astronomers elected to the NAS.[37]

Perhaps nothing demonstrates the political nature of election to the Academy more clearly than a congratulatory note from Mount Wilson director Walter Adams to Frank E. Ross, on the occasion of the election of the Yerkes astronomer to the NAS. "It is hardly necessary to say that you received a unit vote from Mount Wilson."[38] The implication is clear. Adams discussed the election with other Academy members at Mount Wilson and they all voted for Ross on the preferential ballot.

The April meeting of the National Academy of Sciences, traditionally held in Washington, D.C., was frequently the scene of high drama as various factions jockeyed to secure the election of favored candidates. Sometimes these dramas provoked astronomers to write their allies concerning the tactics employed or the merits of candidates. Before the 1899 reforms, the April meeting often broke up in rancorous confusion. William C. Winlock, a staff member at the Smithsonian Institution, where the April meeting was usually held, reported to his old friend Edward S. Holden, "I understand they couldn't come to any agreement [at the April 1886 meeting], on the elections and so they go over to next spring." A decade later Winlock reported "I wish you were there at the Academy meeting today—they have just been having their squabble over election. . . . Don't you wish you could have taken a hand in the scrimmage!"[39]

Astronomers were aggressive, putting forward as many of their kind as possible. Sometimes this provoked other Academy members. In 1896 Benjamin A. Gould suggested to Simon Newcomb "that it might be wiser to wait a little, before electing more astronomers, until the other branches of science have a more equable representation."[40] However, astronomers continued to work for the election of more of their number. Nearly a generation later, Harlow Shapley remarked that he had been lucky in getting

37. Walter S. Adams to William H. Wright, 14 April 1930. MTWA, Adams Papers.

38. Walter S. Adams to Frank E. Ross, 12 May 1930. MTWA, Adams Papers.

39. William C. Winlock to Edward S. Holden, 23 April 1886 and April [?] 1896. SALO, Winlock Papers.

40. Benjamin A. Gould to Simon Newcomb, 8 November 1896. DOA, Gould letter book II, p. 66.

into the Academy, "especially as I hear from different quarters that some people think astronomers are overconspicuous in national science."[41]

If other NAS scientists felt that astronomers were "overconspicuous" in the Academy, within the astronomy section there were strong feelings concerning what might be called the California lobby. Samuel A. Mitchell of the University of Virginia expressed this point of view. According to Mitchell, in 1935 fifteen of the twenty-five members of the NAS astronomy section lived in California, two others either had lived or were trained at the University of California, while four others frequently spent time at Mount Wilson or Mount Hamilton doing research. Complaining to Armin O. Leuschner at Berkeley, Mitchell continued: "To one who does not live in California, but in a section of the country fairly isolated as far as Academy members are concerned, I wonder if you realize how much weight the California influence seems to have in the elections of the Academy?" Mitchell continued, "You remember that W. W. Campbell for many a long year has thought that the Academy was not a *national* organization on account of the concentration of members either in New York, New England or in California. This concentration is especially shown in astronomy where on account of the two great observatories, California is really over represented."[42] Geography was as important as scientific merit. A well-oiled California machine worked for the election of candidates from Lick and Mount Wilson. Some Academy astronomers felt strongly about the power of the California lobby, but could not afford to alienate powerful directors. Clear skies, large telescopes, and entrepreneurial directors gave California astronomy virtual hegemony within the astronomy section between the wars.

Support for this interpretation comes from an unlikely source. In 1920, when Campbell's protégé, Joel Stebbins, was being considered for membership, the Lick director was upset to find that Stebbins had been listed as a Lick staff member. Campbell urged the secretary of the Astronomy Section, Charles G. Abbot, to correct the mistake before the April meeting. Otherwise, Campbell wrote, "this may militate against his chances," because Lick astronomers had been elected at the two previous April meetings. "Too much of a good thing is pretty sure to be mildly resented by numbers of Academy members."[43]

After he left Mount Wilson for Harvard, Harlow Shapley became increasingly critical of his former California colleagues, and protective of New England science. Writing to Schlesinger at Yale, Shapley reflected on

41. Harlow Shapley to Robert G. Aitken, 14 May 1924. SALO, Shapley Papers.

42. Samuel A. Mitchell to Armin O. Leuschner, 25 November 1935. BLUCB, Astronomy Department Papers.

43. William Wallace Campbell to Charles G. Abott, 11 March 1920. SALO, Abbot Papers.

the power of the California lobby. "I wish I could feel that there were not a (unconscious?) systematic exchange of elections in operation between the northern [Lick Observatory] and southern [Mount Wilson Observatory] parts of California."[44] The evidence suggests that there was, if not a systematic exchange, at least very close cooperation between Lick and Mount Wilson in Academy elections. For example, in April 1930, William Hammond Wright wrote Adams that no Lick NAS members would attend the April meeting. He and Aitken wanted to know if a Mount Wilson astronomer would act as sponsor for Joseph H. Moore, a Lick spectroscopist, whose name would be voted on at Washington. The Lick senior staff would be grateful "for a few kind words in his [Moore's] favor—whatever might conscientiously and appropriately be said. I realize this is a delicate request, and if it would be embarrassing to meet it please do not attempt to do so—Only let me know in time . . . so I can try to make other arrangements."[45]

It took another year to secure Moore's election. Either he was voted down by the Academy in 1930 or the elections were closed before his name was reached. Schlesinger confided to Shapley that "If I were present at the meeting [April 1931] it would take very little to induce me to speak against his [Moore's] election. Aside from any such questions as geographical or institutional bias, I feel strongly that this election will perceptibly lower the standard of membership in our section."[46]

In 1931, when the Academy elected a new president to replace California Institute of Technology geneticist Thomas Hunt Morgan, Shapley heard that either Cal Tech president Robert A. Millikan or University of California president William Wallace Campbell would be the nominee. Good men, but Californians both, the Harvard Observatory director lamented. "It is too bad that the weaker institutions in the New England area cannot produce men of this calibre, he said, ironically."[47]

Within the astronomy section, the question of the over representation of West Coast astronomy was a simmering issue in the late 1920s and 30s. Matters were exacerbated by the lengths to which the leaders of the California lobby would go to secure election of their candidates. A classic case was the election of Swiss-born Robert J. Trumpler, a staff member at the Lick since 1919. Trumpler did important work in stellar statistics and in 1932 his name had advanced through various stages to the floor of the

44. Harlow Shapley to Frank Schlesinger, 16 April 1931. YUA, Astronomy Department Papers.

45. William Hammond Wright to Walter S. Adams, 5 April 1930, p. 1. MTWA, Adams Papers.

46. Frank Schlesinger to Harlow Shapley, 21 April 1931. YUA, Astronomy Department Papers.

47. Shapley to Schlesinger, 16 April 1931. YUA, Astronomy Department Papers.

April meeting. Academy president and former Lick director, William Wallace Campbell, was prepared to take extreme measures to ensure Trumpler's election. Lick astronomers were pressing Trumpler's claims, primarily on the basis of an unpublished manuscript that resolved problems connected with the use of stellar positions photographed at the time of a total solar eclipse to test General Relativity. When it appeared that the paper would not be in print until after the April meeting, Campbell wrote Schlesinger that, if necessary, he would turn over the meeting to the vice president and take the floor to support Trumpler's nomination. Campbell hoped to avoid such an unusual situation and asked Schlesinger, Trumpler's former boss at Allegheny, to attend the meeting and speak on the candidate's behalf. Campbell then went on to describe his ace in the hole. Trumpler's unpublished manuscript had been read by Einstein and pronounced "completely convincing." Campbell promised to "take all essential precautions" to avoid bringing Einstein into the debate, "but I shall take his letter to Washington with me to show personally to you and a few others."[48]

Schlesinger, who, left for the Philadelphia meeting of the American Philosophical Society before the vote on Trumpler, was later briefed by Shapley. The Harvard director had mixed emotions concerning the election of Trumpler: "Happy for Campbell and Trumpler; unhappy on the Academy's account . . . because we are not being fair." Shapley described the order of battle (doubtless orchestrated by Campbell). Walter S. Adams of Mount Wilson "generously" spoke first. Henry Norris Russell of Princeton "spoke second in [a] most unrestrained and enthusiastic manner. . . . He very much overstated the case, and so did Leuschner." No one spoke against Trumpler, but, Shapley continued, "the success was narrow as you see by the vote." The Harvard director felt that undue pressure had been used to secure Trumpler's election; the California lobby had pushed too hard. When a candidate is elected by a very narrow margin, and the victory hinged on "a statement such as Leuschner's that Trumpler had, in an unpublished manuscript, done one of the grandest things in recent science . . . Well, I don't feel comfortable." Further, Shapley was disturbed that in each of the last three or four years, "the astronomers have added one member to the Academy by the scarce margin of two or three votes."[49] Little wonder that other Academy members were critical.

48. William Wallace Campbell to Frank Schlesinger, 4 April 1932. YUA, Astronomy Department Papers.

49. Harlow Shapley to Frank Schlesinger, 28 April 1932, p. 1. YUA, Astronomy Department Papers. The degree to which Shapley was hostile to Trumpler because of the latter's work on interstellar absorption is not clear. The Shapley model of galactic distances assumed that the absorption of light by the interstellar medium was negligible. Trumpler demonstrated that this assumption was incorrect. See Smith (1982: 158–60)

Schlesinger, perhaps one of the few senior astronomers who could stand up to William Wallace Campbell, wrote the retired Lick director concerning the Trumpler election. While he sought to keep the tone cordial, he did insist that it was time to find ways "to expedite the election of new members and especially to keep the discussion down to reasonable lengths." Schlesinger insisted that "a detailed defense of a candidate would seem to be in order only when some one has spoken against him for it is invariably the case that a candidate is elected unless some member speaks in opposition."[50] Apparently, these muted criticisms had little impact on the California imperialist.

LESSER HONORS

The American Philosophical Society and the American Academy of Arts and Sciences are the other major honorific bodies to which scientists sought election. Very little material has survived concerning campaigns to elect astronomers to the Philosophical Society or the American Academy. The Lowell Observatory archives prove a remarkable exception. Materials on Mars Hill provide a candid view of a reward system that has little to do with merit. For example, the struggle between the Lick and Lowell observatories discussed in chapter 7, spilled over into the election of William Wallace Campbell to the American Academy of Arts and Sciences. Percival Lowell used his influence to block Campbell's election.

In 1910, Lowell mounted a campaign to elect observatory staff member Carl O. Lampland to the American Academy of Arts and Sciences. Founded during the Revolution, the American Academy, while national in intent, long remained a New England institution, comprising primarily Harvard faculty and the Boston elite. In a real sense, Lowell may have believed he had a proprietary interest in the American Academy. Lampland's competition included Lick director William Wallace Campbell, a longtime critic of Lowell and his observatory. Lowell moved to oppose Campbell's claims to membership.[51]

Soon Lowell was joined by Sherburne W. Burnham, whose dislike of anything connected with the Lick Observatory seemed inexhaustible. Burnham urged Lowell to make sure members of the American Academy understood the importance of electing real workers rather than administrators.[52] Lowell set about collecting letters of opposition to the nomina-

50. Frank Schlesinger to William Wallace Campbell, 10 May 1932. YUA, Astronomy Department Papers. See also Campbell to Schlesinger, 30 April 1932, pp. 1–2. YUA, Astronomy Department Papers. This letter was written on the Golden State Limited as Campbell was returning from the NAS meeting.

51. Edwin H. Hall to Percival Lowell, 10 March 1910, p. 1. LOA, Lowell Papers.

52. Sherburne Wesley Burnham to Percival Lowell, 8 October 1910. LOA, Lowell Papers.

tion of Campbell. One of the first he solicited was from Vesto M. Slipher, whose animosity toward the Lick director stretched back a decade.[53] When Campbell was nominated, Lowell challenged the validity of the process, suggesting that Lampland's name had not appeared on all the lists from which members chose candidates for the preferential ballot. Lowell also enclosed letters from Barnard at Yerkes and the retired astronomer and geodesist George Davidson opposing Campbell. For good measure, Lowell supplied clippings from *Science* that were apparently unfavorable to the Lick director.[54]

Not content, Lowell pressured the secretary of the American Academy for a tally of the votes. Six candidates (Ernest W. Brown, Campbell, George C. Comstock, Edwin B. Frost, Lampland and Frank Schlesinger) were voted on by eleven of the twelve members of Class One, Section One. Campbell and Frost tied with nine votes each, followed by Brown with eight and Comstock six. Lampland and Schlesinger received two votes apiece.[55] All the nominees except Lampland and Brown were directors of factory observatories. Burnham reported to Lowell that he had voted for Lampland but had "turned down all applicants for the honor [of election to the American Academy] who counted on directorships of observatories to boost them up." Again, Burnham emphasized that "the real worker" should be considered.[56]

Only in 1915 did Lampland's friends succeed in securing his election. That Lowell could have exerted so much pressure provides dramatic testimony to his imperious nature. That the American Academy should elect the directors of the most important factory observatories before Lampland indicates the power and status of these institutions within American science.

Vesto M. Slipher was, in comparison to Lowell, modest and retiring. He preferred long hours at the telescope to the rough-and-tumble political activities associated with the reward system in American science. The second director of the Lowell Observatory, however, was aware of the importance of honors for his staff. Slipher worked to secure membership in the American Philosophical Society for Lampland.

Stretching back to Benjamin Franklin, the American Philosophical Society is the oldest learned society in the United States. Like the American Academy, the Philosophical Society was organized into classes, and both organizations employed a multistep system for nominating and electing members. While most commentators would rank the NAS above the Philosophical Society, there were some who felt differently. Writing to

53. Percival Lowell to Vesto M. Slipher, 26 October 1910. LOA, V. M. Slipher Papers.

54. Percival Lowell to Edwin H. Hall, 3 November 1910. LOA, Lowell Papers.

55. Hall to Lowell, 11 November 1910. LOA, Lowell Papers.

56. Burnham to Lowell, 30 November 1910. LOA, Lowell Papers.

congratulate Otto Struve on his virtually simultaneous election to the NAS and the Philosophical Society, Frank Schlesinger pointed out that "In some respects this [election to the Philosophical Society] is a greater honor than the election to the Academy, since many [of its members] are chosen [from] outside of the ranks of Science."[57] However that may be, there was careful coordination between the NAS and the Philosophical Society, which held its annual meeting in Philadelphia immediately after the Academy meeting. Many elite scientists made it a point to adjust their April schedules so they could attend both events.

Swarthmore Observatory director John A. Miller was a longtime friend and advisor to several Lowell astronomers. After moving to Swarthmore, Miller was elected to the American Philosophical Society (1915) and became active in its affairs. Thus it was only natural that in 1924 Slipher should turn to Miller for help in electing Lampland. In reviewing the situation, Miller was not sanguine. He doubted "the probability of his [Lampland's] election, not because of his lack of work, but because he has not published as widely as he could easily have done."[58] Slipher's response reflected the sense of isolation that marked the Lowell staff during the interwar years. "In view of the fact that the other circles are likely always to have friends to nominate, it appears that Lampland's case is up to us."[59] This statement, however oblique, indicates that Slipher was well aware of the competition between directors of the factory observatories ("the other circles") to secure honors for their staff, and that he was ready to assume responsibility for Lampland's election with, of course, help from Miller.

Four years later, Slipher reopened the issue. Miller, it seems, fumbled the Lampland nomination in 1928. In order, perhaps, to make amends he secured an invitation for the Lowell astronomer to deliver a lecture to the April meeting of the Society. The Swarthmore director, however, cautioned Slipher not to get his hopes up. There was a feeling in the Society that too many astronomers had been elected and that they were overrepresented in proportion to their numbers in the total scientific community.[60]

In 1929, Miller took the initiative in nominating Lampland. He prepared the necessary forms and together with Heber D. Curtis, director of Allegheny, signed them. Miller apologized for taking the initiative in the matter, but considering Slipher's "anxiety to have him come before the Society" hoped "this liberty has not been unwarranted."[61] For whatever reason, Lampland failed of election, but Miller indicated that his name would be considered again in 1930. However, he cautioned Slipher not be

57. Frank Schlesinger to Otto Struve, 6 May 1937. YUA, Astronomy Department Papers.

58. John A. Miller to Vesto M. Slipher, 21 January 1924. LOA, V. M. Slipher Papers.

59. Slipher to Miller, 27 January 1924. LOA, V. M. Slipher Papers.

60. Miller to Slipher, 13 February 1928. LOA, V. M. Slipher Papers.

61. Ibid., 16 January 1929.

become overly optimistic. Fifty-seven nominees were competing for sixteen places.[62] Again, Lampland did not succeed. Political considerations led to a number of "men of affairs" being elected over worthy scientific candidates. Miller assured Slipher that Lampland's name would be held over for consideration the next year and that "we may have better success" in 1931.[63] Miller was correct. After a campaign of seven years, the Lowell Observatory succeeded in getting one of its senior staff members elected to the American Philosophical Society.

AN EMPIRICAL ANALYSIS OF THE REWARD SYSTEM IN AMERICAN ASTRONOMY

In chapters 5 and 6 the scientific career was discussed as a complex social process. The following discussion directs attention to the career as movement through the reward system of a modern scientific community. The process involves both the stratification of the reward system and of careers as well. The analysis is based on the careers of elite astronomers who won stars in *American Men of Science* and/or were elected to the National Academy of Sciences. The findings suggest that previous investigators have emphasized merit at the expense of patronage (Cole and Cole 1973; Gaston 1978a, 1978b; Allison 1980), thus preventing scholars from seeing the complementary nature of merit and patronage in shaping the careers of scientists.[64]

In conceptualizing merit (universalism) and patronage (particularism) I have departed from traditional usage (Reskin 1979: 129). Since Robert K. Merton's classic 1942 paper, researchers have tended to view the norms that govern science as pattern variables. As sociological theorists Talcott Parsons and Edward Shils (1951: 77) define the concept, "a *pattern variable* is a dichotomy, one side of which must be chosen by an actor before he can act with respect to that situation." While this theoretical construct

62. Ibid., 14 March 1930, p. 2.

63. Ibid., 1 May 1930, p. 1.

64. Attentive readers will note that the discussion has abandoned traditional sociological language (universalism and particularism) and substituted merit and patronage. This decision was dictated in part by the analysis developed earlier in this book (especially chapters 5 and 6). Many historians of science and as well as practicing scientists are more comfortable with merit and patronage as categories for analysis and discussion. After World War II, the structural functional school in American sociology developed a language that, while scientific (neutral), tended to be sterile. This school lost sight of the complex nuances of language that evoke the complexity of human behavior which, after all, is supposed to be the central concern of sociology. In one other area, we have also sacrificed technical language. For some time it has been customary to use the phrase "status attainment" in place of the older word, "mobility." Like "merit" and "patronage," the older term is, for many readers, more familiar and comfortable.

may be useful in some areas of social research, it is not necessarily valuable for the historical study of collective behavior. Rather than a dichotomous situation, the available behavioral options range along a continuum. For the study of the collective activities of a scientific community, it is more fruitful to employ a continuum with perfect (if unattainable) patronage at one extreme and perfect (if equally unattainable) merit at the other. Most behaviors will reflect degrees of patronage or merit but not the pure type of either.

An emerging consensus in the sociological literature seems to suggest that a scientist's first job and perhaps later appointments as well, depend less on productivity (taken generally as a measure of merit) than on patronage. Reskin (1977: 502) suggests that "educators apparently overestimate the direct effects of graduate training on scientific achievement." Long and McGinnis (1981: 441) report findings that "do not support the idea that scientists are allocated to various contexts on the basis of contributions to the body of scientific knowledge." They point out that "recruitment operates independently of productivity," and conclude that the stratification system in science is not necessarily based on merit. Material presented below supports this position.

Central to the discussion that follows is a conception of the interplay between merit and patronage in determining the distribution of honors and awards. Merit and patronage *each* have a part to play and both factors must be taken seriously, if the reward system is to be understood. Merit and patronage are complementary rather than mutually exclusive. Both affect progress through the reward system.

A variety of constructs have been put forward to represent the stratification system in modern science (Merton and Zuckerman 1971; Cole and Cole 1973; Zuckerman 1977; Allison 1980; Hargens, Mullins and Hecht 1980; Hargens and Felmlee 1984). These range from relatively straightforward qualitative descriptions to complex theoretical models. The model of the stratification system in a modern scientific community proposed here is based on the allocation of honors and awards over the course of an entire career. The stratification system represents nothing less than the hierarchy of statuses reflecting the honors and awards that come to scientists at different points in the career. For these reasons the stratification system is conceived of as a series of discrete statuses hierarchically arranged. The allocation of honors and awards occurs along a well-defined time line. Certain honors generally precede others in the unfolding of a career. Later honors tend to be more prestigious than those that come earlier in the career.[65]

65. It is a source of amazement that sociologists of science have deliberately closed off this avenue of inquiry. Social historians find it difficult to understand why sociologists have not attempted to develop a model of the stratification system that takes into account the

The art of operationalizing theoretical constructs by converting them into measurable variables involves a number of steps. It is important for readers to understand the process so that they can evaluate our findings. Data used to construct the variables came from the collective biography. One of the first tasks was to control for the effects of background and demographic characteristics on career experiences. Background controls include region of birth, father's socioeconomic status, father's status as a scientist, and father's status as native- born or an immigrant. In addition, information was gathered concerning the highest earned degree and whether the subject's pre-career training included graduate study in Europe.

Background variables were measured in the following ways. Region of birth is defined according to standard Census Bureau practice (U.S. Department of Commerce 1973). Father's socioeconomic status is a hierarchical variable measured on a three-point scale (blue-collar, white-collar, and professional) with professional having the highest weight. Father's status as a scientist, father's status as native born or an immigrant, and graduate study in Europe are all dummy variables that measure the presence or absence of these background characteristics. Highest degree was operationalized as a dummy variable indicating whether or not the subject had earned a doctoral degree. Gender was not included as a background control. Only four of the 170 elite astronomers were women, too small a number for meaningful comparative analysis. Issues relating to gender and the reward system will be discussed in chapter 9.

Of the background variables examined, father's socioeconomic status proved the most interesting. Each of the other background variables has been examined by various researchers, but the findings have been inconsistent.[66] There were no significant effects at any stage of the model for region of birth, father's status as a scientist, father's immigrant status, or graduate study in Europe.

In order to examine mobility, the decision was made to divide careers into stages corresponding to seven-year increments (see figure 8.1). This procedure was mandated by the clustering of career experiences and mobility in cycles of approximately seven years' duration. Three primary variables were assessed at every stage of the model. These indicators mea-

---

temporal sequencing of honors and awards. Cole and Cole (1973: 61), for example, assert that "unlike society at large, science is difficult to describe as a series of discrete positions hierarchically arranged." Perhaps, in the case of Cole and Cole (1973: 211, 214–15), the decision reflects skepticism concerning the legitimacy of historical analysis applied to the study of modern science. Antihistorical attitudes reflect a break with an important element in the Mertonian (e.g. 1935; 1936; 1938; 1939; 1972; 1988) tradition and also limit the conceptual horizons of those who hold them.

66. The following have examined one or more of these background variables: Visher (1947), Knapp and Goodrich (1952), and for a historical overview Elliott (1982).

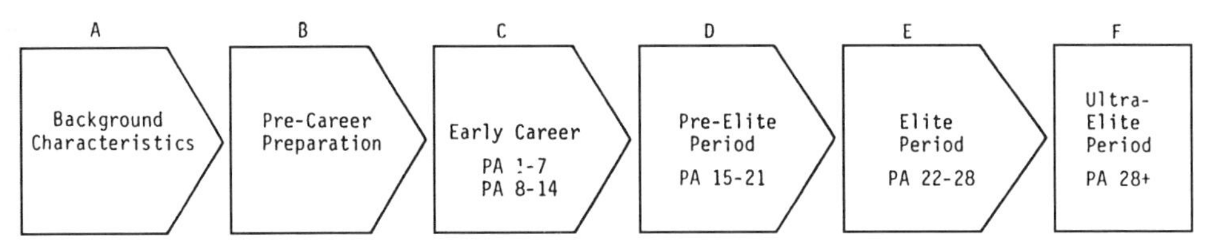

Figure 8.1. Conceptual model of the scientific career.

sure productivity, institutional potential and contacts with members of the National Academy of Sciences.

Merton and Zuckerman (1971) and Cole and Cole (1973: 28) view publication as a measure of productivity. Following this line of reasoning, we argue that productivity, as measured by publications, and merit are closely associated. Papers are published or rejected because of the contribution they make to a field. From another perspective, however, publications are not a pure measure of merit. In order to produce significant results, certain preconditions must be met. In a data driven science such as astronomy, scientists must have access to sophisticated instruments that can provide data with which to solve problems. This instrumentation is expensive and available only in a few observatories. In an age before guest investigator programs were common, and in which there were no national research facilities in astronomy, access was a critical limiting factor.

A second barrier involved access to resources that permitted the reduction and analysis of data. This, too, became increasingly technology-intensive as various kinds of plate- measuring devices were developed for use in both astrometry and astrophysics. In the early twentieth century mechanical computing devices made their appearance. Once purchased, these instruments demanded expansion of the observatory staff in order to carry out data reduction and analysis. Institutions lacking in capital were hard pressed to keep up.

In addition to large telescopes, advanced analytical technologies, and staff to operate them, colleagues provided resources of another kind. Workers at small institutions were not only handicapped by the lack of research instrumentation but were intellectually isolated as well. They could not discuss work-in-progress with colleagues who represented a variety of perspectives and a wide range of experience. For these reasons, it seems doubtful that publication can represent pure merit. On the continuum that has as its end points perfect merit and perfect patronage, publications should be seen as tending toward but not representing the perfect merit end of the scale.

Institutional potential is a measure of the ability of an institution to contribute to the advancement of an individual's career (see chapter 5). This measure involves aspects of both patronage and merit. Institutions were classified as public or private, government, college, or university, research-oriented and/or graduate teaching or undergraduate teaching only. The scale is arranged so that institutions providing maximum opportunity for original research rank highest. Private research institutions such as the Mount Wilson or Dudley observatories generally command advanced instrumentation, have ample budgets that are not tied to the whims of state politics, and offer research scientists an environment free from the demands of academic life. Opportunities tend to decrease as one moves down the scale.[67]

Institutional potential has received little attention from sociologists or historians of science. In contrast, researchers and theorists have frequently argued that reputational prestige (Hargens and Hagstrom 1967; Cole and Cole 1973; Long, Allison, and McGinnis 1979; Reskin 1979) of the institutions at which an individual obtains employment is an important indicator of subsequent mobility. We believe that institutional potential provides a better assessment, based on structurally determined advantage implicit in a position. Ideally, employment at an institution with adequate equipment, staff, and financing should lead to increased productivity (Long and McGinnis 1981). There are, however, factors that weaken the indicator as a measure of merit. Being located at a well-equipped observatory does not, in and of itself, guarantee access to the best instrumentation. Observatory directors control access to instruments. In some cases, this power has been abused (Frost 1926:4–5; Osterbrock 1984b). For example, allocation of time on large telescopes may be made on the basis of seniority rather than by any consideration of the merits of competing projects. For these reasons, we argue that the concept of institutional potential includes a mixture of both merit and patronage.

The purest indicator of patronage focuses on association with members of the National Academy of Sciences. This indicator represents an actual count of NAS members in the observatory or department at which the subject was employed during each career stage. The NAS contacts indicator measures a complex bundle of relationships. First and foremost, it is an index of interpersonal contact. Most of the observatories, government research facilities, and academic departments included in this study were relatively self-contained units, sometimes geographically isolated. Individuals of all statuses and ranks (from graduate assistants to the director

67. Astronomers were well aware of the hierarchy implied in this scale. For samples of their thinking, see Herbert R. Morgan to Seth Carlo Chandler, 16 October 1905. DOA. John M. Schaeberle to Armin O. Leuschner, 9 May 1892. BLUCB, Astronomy Department Papers. Robert S. Woodward to Ormond Stone, 2 June 1894. UWMA, Stone Papers. J. A. Hoogewerff to William Wallace Campbell, n.d. SALO, USNO Papers. Frank Schlesinger to Benjamin Boss, 2 January 1940. YUA, Astronomy Department Papers.

or chairman) were involved in a tight-knit web of personal contacts. Senior staff, who were most likely to be members of the Academy, were charged with a variety of day-to-day supervisory responsibilities. Virtually all research, from thesis projects to problems investigated by the staff, had to be approved by the observatory director or department chairman. In most instances, papers had to pass an in-house review conducted by senior astronomers before submission for publication. This context guaranteed a high degree of interpersonal contact between NAS members and those who had not yet attained Academy membership.

Given the institutional, social, and psychological environment that defined relations between Academy members and their junior colleagues, patronage rather than merit often played an important role in determining the life-chances for younger astronomers. This indicator tends further toward perfect patronage than either institutional potential or publication as a measure of merit.

One last item warrants comment. How do we conceptualize the scientific career as movement through the reward system? Figure 8.1 represents a model of the scientific career as movement through the reward system. The process begins with background characteristics (box A) and then proceeds to B, pre-career preparation. Boxes C through F represent the division of the career into seven-year intervals. The early career includes two seven-year intervals. Honors and awards begin to accrue after professional age fourteen (box D) and the process continues through the ultra-elite career phase represented by box F. This model reflects a conception of the career as movement through the reward system. While processual, the acquisition of honors and awards is neither unidirectional nor cumulative (Merton 1968). Neither, on this view, is the career to be understood as linear. Movement though the stratified reward system involves stages and tracks. Each stage is contingent and leads to potentially different outcomes.

The primary analytical task was to examine the causal linkages between various components of the model and to assess its explanatory power. Primary analyses employed ordinary least squares regression procedures and are presented in the form of path diagrams. The advantage of this form of analysis is that it holds constant other effects in the data, while examining the impact of a particular variable. Results are presented as path diagrams in which each line in the diagram represents a standard regression slope or beta weight, indicating the effect of one variable on another when controlling for all other variables in the model.

The action of variables arranged along the patronage-merit continuum have somewhat different etiologies and causata in the unfolding of careers. This suggests these indicators will vary independently in different contexts. The investigation of mobility within a processual stratification system dictates the continuous measurement of indicators throughout the entire scientific career. The isolation of one segment of the career for analysis would tend to reify that portion and render it nonprocessual. When the precursory etiologies are removed from the analysis, investigators are likely to arrive at erroneous causal schemata for the portion of the career examined. Further, to examine the independent effects of patronage and merit, the research design must employ multivariate analysis that permits investigators to examine the effects of various combinations of factors in different contexts. To study the stratification of scientific careers in this way requires a data base that provides continuous measurement of important indicators for the entire career of all subjects, coupled with techniques of analysis that afford the flexibility to examine mobility in a variety of contexts.[68]

The Early Career

The early career was broken into three stages as shown in figure 8.2. The first encompasses the career to professional age seven. The second stage covers the period from professional age eight through professional age fourteen. Stage three indicates the status of the indicators and the end of the early career. The three indicators on the patronage-merit continuum were assessed at each stage.

Analysis showed that, for the most part, background variables were insignificant in predicting mobility when controlling for the indicators of patronage and merit. One variable, however, deserves attention. As suggested earlier, Father's socioeconomic status (SES), plays an important role in later aspects of the scientific career. Figure 8.2 indicates that Father's SES is significantly related to productivity and contacts with NAS members during the pre-career training phase and the early career as well. Four paths connect Father's SES with variables at later stages of the career. Interestingly enough, Father's SES affects publications, the indicator we suggest tends most toward merit, *as well as* NAS contacts that are equated with a high degree of patronage. Father's SES does *not* affect institutional potential at all. Further, the effects of this background variable are felt at both the pre-career training stage and during the first seven years of the career as well.[69]

68. Readers unfamiliar with ordinary least squares regression analysis might wish to consult Sage University Paper Number 22, by Michael S. Lewis-Beck (1990) for an introduction to the subject. The definitive discussion is Cohen and Cohen (1975).

69. As sociologist Diana Crane (1970: 953) points out, "one of the principal concerns in the study of social mobility has been the extent to which sons inherit the social class status of their fathers." A key issue involves "the relative importance of achieved and ascribed characteristics in determining career patterns." Sociology of science, according to Crane, translates this concern into an examination of the relationships between "prestige of the doctorate, scholarly performance, and selection for a position in a leading department." However,

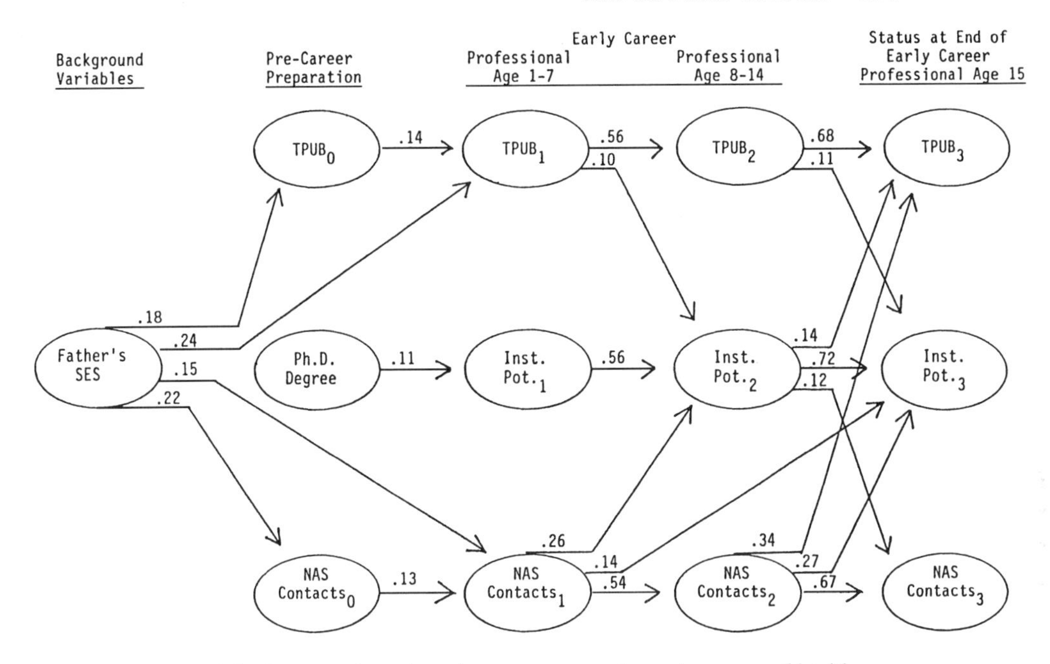

Figure 8.2. The impact of merit and patronage on the early career of highly mobile American astronomers.

Father's SES is often believed to affect career development of offspring by providing resources and opportunities for the acquisition of maximum educational advantages.[70] These advantages must be considered as both formal institutional experiences at the primary, secondary, and/or collegiate levels, *as well as* informal experiences in the context of the home. Fathers in relatively secure middle-class professional or business related occupations could afford to provide both formal and informal educational advantages to their children. In addition to formal opportunities, the existence of books in the home as well as an atmosphere that encouraged reading, discussion, and independent thinking were important. So was an

findings reported here suggest it is not necessary to translate father's SES in the way Crane proposes. There seem to be direct and significant relationships between father's SES and two key indicators at both the precareer training stage and during the first seven years of the scientific career. These findings underscore the importance of research design that takes background factors into consideration.

70. These comments are based on a reading of the biographical memoirs of all astronomers who were members of the NAS plus obituaries for as many other elite astronomers as could be located. DeVorkin (1982a: 347–85) provides a bibliographical introduction to biographical resources for the American astronomical community. Jencks et al. (1972) offers an introduction to the related sociological literature.

economic environment in which a young person's interest in science could be indulged through the purchase of equipment in order to set up a basement laboratory or a rooftop observatory. If this was the way in which Father's SES translated into educational advantage for children, it is not surprising to see that this variable directly affects publication during the pre-career training period and the first seven years of the career.

Connections between Father's SES and NAS contacts during pre-career training and the first seven years of the scientific career must be understood in terms of socialization provided by the family of origin. In this context, young persons interested in science developed identities and learned social skills that would assist them in successfully interacting with leaders in the profession. The historical record contains no indication that young persons actually encountered members of the Academy as visitors to the family of origin, although it may have occurred.[71]

Several indicators of pre-career training were examined. These include pre-career publication counts, highest earned degree, NAS contacts at the institution where highest degree was earned, and graduate study in Europe. The regression results indicate that studying in Europe was insignificant in the process of career development. As shown in figure 8.2, pre-career productivity and NAS contacts have significant positive relationships with both continued productivity and continued NAS contacts. The attainment of a Ph.D. is significantly and positively related to acquiring a first job with high institutional potential but becomes less important in the later career.

Productivity as measured by publication counts is often seen as the pivotal criterion for selecting scientists for positions and honors. If the stratification of scientific careers is based on merit, we would expect this indicator to produce the most significant effects on the other primary indicators. As shown in figure 8.2, publication counts in the pre-career training phase *have no significant effect* on the institutional potential of the first job or NAS contacts made at the first job. Publication counts during the first seven years have statistically significant but comparatively weak effects on institutional potential in later position acquisition. There is no significant effect of publication on NAS contacts. Thus, early productivity does not appear to increase the chances for attaining an advantageous first job and provides only minimal advantage in position acquisition during the first fourteen years of the career.

Institutional potential of the first job is not directly related either to publications or NAS contacts seven years into the career. At fourteen

71. Where fathers were Academy members (Lewis Boss or Asaph Hall, Sr.) the children did not achieve Academy membership. Benjamin Boss earned a star in *AMS*. Harold D. Babcock's son, Horace, did not become a member of the NAS until after World War II.

years, however, institutional potential is significantly and positively related to both publications and NAS contacts. Thus it appears that institutional potential becomes important after several years of working in a more or less potential-laden position.

Perhaps the most interesting results are those associated with NAS contacts. If, as is often argued, the stratification of scientific careers is based on merit (Merton 1960; Cole and Cole 1973), we would expect the indicator most strongly reflecting patronage to produce the least significant effects. Figure 8.2 indicates that this is not the case. Comparatively, the paths from NAS contacts at each stage of the early career are stronger than those from either publications or institutional potential. The results shown in figure 8.2 suggest that the indicator representing patronage is a better predictor of success than the variable reflecting merit. For this population, the stratification of early careers appears to be based more on patronage than merit. However, the results suggest that *both* types of advantage must be examined in order to fully understand the process of mobility.

Two additional findings demand attention. The first deals with the stability of the indicators. As shown in figure 8.2, each indicator in the model predicts subsequent states of the same indicator in a progressive manner. For example, prolific publishers at time one continue to be prolific publishers at time two and so on throughout the career, The same pattern holds for those who are successful in attaining jobs with high potential for making NAS contacts. This cumulative effect holds for each variable independent of any control variables employed.

Cohort had no effect on the three indicators at any stage in the career. The effects of cohort were tested using two approaches. First, a variable indicating cohort membership was entered into the model as a predictor variable along with the other background variables. It had no significant effects on any of the main variables in the model. Second, the model was tested independently with data obtained from each cohort. The independent results obtained for each cohort provided replications of the effects obtained from the amalgamated data. These results indicate the stability of the model across time. That is, the three indicators on the merit-patronage continuum operate in the fashion shown in figure 8.2 over an eighty-one-year period during which major technical and cognitive changes occurred in the American astronomical community.

The stability of the model can be demonstrated in one other way. Our findings indicating that early productivity is not related to position acquisition are in accord with those of sociologists (Crane 1970; Reskin 1977; Long, Allison, and McGinnis 1979; Long and McGinnis 1981) studying American science after World War II. Taken together, these data span more than 125 years of the history of American science. This suggests that, at a deep level, the social organization of science has remained remarkably stable in spite of dramatic structural, demographic, technical, and cognitive changes as well as changes in the way in which science was funded.

The Later Career

The pre-elite stage represents the career from professional age fifteen to twenty-two. On the basis of pairwise comparisons, the statuses were arranged as shown in figure 8.3. There are *two entry-level* statuses for this stage of the career: election to the American Academy of Arts and Sciences (AMADY) and election as a Foreign Associate of the Royal Astronomical Society of London (FARAS). They occur at approximately mean professional age fifteen and reliably preceded all other statuses at this stage. The succeeding three prestigious statuses occur at progressively higher mean professional ages, with receipt of an *AMS* star taking place at about mean professional age twenty-two. Pairwise comparisons indicate that first election to a non-presidential office (NPEL) in a professional society preceded the award of the Royal Astronomical Society's Gold Medal (RASGOLD) which in turn reliably preceded starring in *AMS*.

Let us turn now to the relationship between statuses. The path diagram indicates two patterns of mobility in the pre-elite stage. The first, which we call the *preferred pattern,* is shown by the single arrows and begins with election as a Foreign Associate of the Royal Astronomical Society (FARAS). Entering the pre-elite stage in this way has a moderately strong effect on acquiring a Gold Medal from the RAS, which in turn has a positive effect on being starred in *AMS*. The *alternate pattern,* shown in double arrows, begins with election to the American Academy of Arts and Sciences (AMADY) as its entry status and is positively linked to first election to a nonpresidential post in a professional organization. The nonpresidential post is a *terminal status* for the alternate pattern. It provides no significant advantage for the attainment of future mobility. Indeed, negative paths to future statuses from nonpresidential election and election to the American Academy suggest that the alternate pattern of mobility during this career stage provides significant *disadvantages* for future mobility. The discovery of a two-track system beginning with the pre-elite career phase, indicates the value of regression analysis in the social historical study of the reward system in a scientific community.

Astronomers were at least partially aware of patterns in the allocation of honors and awards. In a letter to J. McKeen Cattell, Henry Norris Russell commented that the Royal Astronomical Society was "pretty careful in picking its associates."[72] In effect, Russell told Cattell that *AMS* stars should not be awarded to astronomers unless they were already foreign

72. Henry Norris Russell to J. McKeen Cattell, 31 January 1919, p. 1. LC, Cattell Papers.

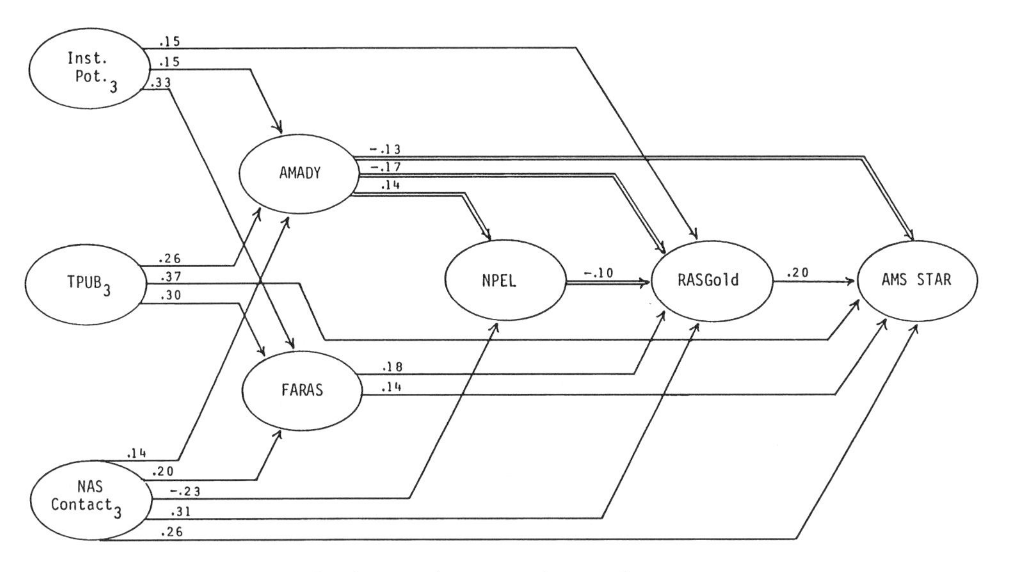

Figure 8.3. Findings for the pre-elite stage of scientific careers.

associates of the Royal Astronomical Society. In a sense, Russell was using the status of RAS foreign associate as a predictor.[73]

Figure 8.3 also suggests that merit and patronage continue to have important effects on mobility in the later stages of the career. Entry into the pre-elite stage is predicted by all three indicators on the merit/patronage continuum. Notice, however, that the measure of productivity (the indicator most closely related to merit) is strongly and positively related to both entry statuses (AMADY and FARAS) and, further, that productivity is the strongest predictor of the highest pre-elite status (*AMS* star).

The results also show that the indicator of patronage (NAS contacts) has a significant effect on the attainment of pre-elite statuses, that is, independent of productivity and institutional potential. Contact with NAS members favorably predisposes entry into the pre-elite stage through either entry status. The effect is more pronounced on the preferred entry status, and NAS contacts have a substantial direct influence on each status in the preferred pattern but are *negatively related* to non-presidential election, the terminal status for the alternate pattern at this career stage.

Institutional potential, the indicator located toward the middle of the merit/patronage continuum, is related to *both entry-level statuses.* As

73. In the 1990s, the scientific community (Holland 1990: 17) is well aware of "predictor prizes that traditionally anticipate Nobel committee selections." The hierarchical sequencing of honors and awards appears to be an accepted fact of life.

with the NAS contact variable, the effect is more significant for the preferred entry status (FARAS) than for the alternate (AMADY). Institutional potential is also related to the award of an RAS gold medal.

These results support my contention that both patronage and merit play a role in mobility. Each of the three indicators are positively related to entry into the pre-elite stage. Once having entered the pre-elite stage, these indicators help to explain movement through either the preferred or alternate pattern of mobility. However, an interesting shift in effect between the indicators on the merit/patronage continuum emerges in the pre-elite state. The indicator reflecting greater merit (productivity) actually has stronger effects on the entry statuses than does the patronage indicator (NAS contacts). Institutional potential also becomes more important at this point in the career. All three indicators of merit and patronage continue to affect the preferred pattern of mobility in a strong and relatively balanced manner throughout the pre-elite stage.

The next portion of the career is the elite stage (Figure 8.4). This stage represents the years from professional age twenty-three to twenty-eight. The elite stage forms a unique portion of the scientific career. Unlike any other stage, the statuses that occur during this phase show no clear evidence of *temporal hierarchy.* Each status is attained at about the same mean professional age. There was no evidence from pairwise comparisons that indicated the clear precedence of any status. There is, however, a *hierarchy effect.* As will be seen in the discussion of transitions between stages, there is a clear difference in patterns of effect between the two membership statuses (NAS and the American Philosophical Society [APS]) and the gatekeeper positions (first election to an International Astronomical Union [IAU] commission, first editorial post [EDPO] or appointment to an *Astrophysical Journal* editorial post [APJ]). In figure 8.4 paths leading to gatekeeper positions are indicated by double lines. But for now, let us view the elite stage as a plateau that follows the hierarchically structured pre-elite stage and during which competition is for the attainment of one or more of five virtually equal statuses.

As with the pre-elite stage, all of the indicators of patronage and merit are positively related to entry into the elite stage. In the elite stage, however, productivity (the indicator most closely reflecting merit) emerges as the more important element. Productivity alone is positively related to the attainment of every status at the elite stage, but the effect is more pronounced for gatekeeper positions. Institutional potential is related to the attainment of three statuses with stronger paths leading to NAS and APS election. The strong path (.34) from NAS contacts to NAS membership suggests at first glance that NAS contact is the key variable at this career stage. We must remember, however, that NAS contact is, by definition, a measure of interaction with electors to membership in the Academy. Therefore, one would expect a strong relationship between these two

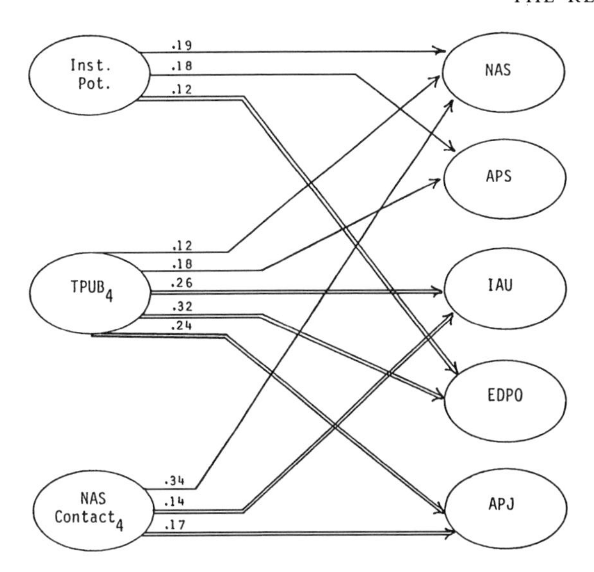

Figure 8.4. Findings for the elite stage of scientific careers.

variables. The real test of the importance of NAS contacts in the overall mobility process is the degree to which this indicator affects the acquisition of other independent statuses at each stage of the career. NAS contact does indeed, continue to affect mobility (other than NAS membership) in the elite stage, but its effects are moderate when compared to the effects of productivity. Taken together, these results indicate that merit emerges as the more important predictor of success in the elite stage of the career.

The arrangement of ultra-elite statuses is represented in figure 8.5. The ultra-elite stage also presents two patterns of mobility. The preferred route is shown by single arrows. Here scientists enter the ultra-elite stage by attaining either an award from the French Academy of Sciences (FRADY Award) or the Royal Society of London (RSL Award). Either of these awards predisposes astronomers to attain foreign membership in the Royal Society of London (RSL) or an honorary degree from a major foreign university (HDMF). These statuses are, in turn, positively related to election as a corresponding member of the French Academy of Sciences (FRADY). In addition, the French Academy award as an entry status directly affects election to the French Academy. The alternate pattern involves the entry status of first presidential election (PRES ELEC). This pattern is represented by double lines. First presidential election is positively related to two major American awards, the Bruce medal of the Astronomical Society of the Pacific and the Draper medal of the National Academy of Sciences. These two awards represent terminal statuses for the

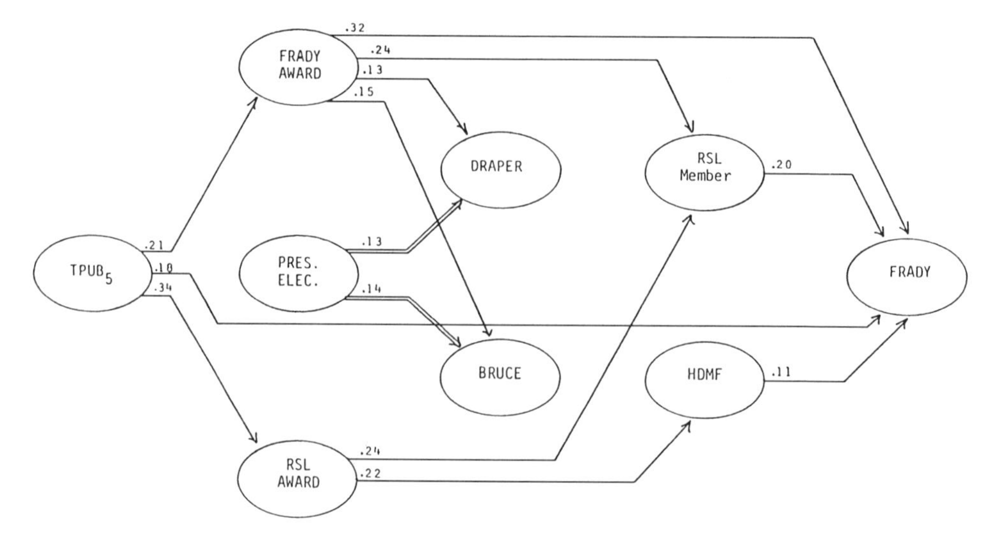

Figure 8.5. Findings for the ultra-elite stage of scientific careers.

alternate pattern. They provide no significant advantage for the attainment of future statuses. An interesting anomaly involves the paths from the French Academy award to the two American awards. These paths may be generated by a triggering effect. Committees administering the American awards may be favorably impressed by nominees who previously have earned European recognition.

No patronage effects exist for the ultra-elite stage. The findings related to productivity, however, warrant mention. The productivity of an astronomer positively and strongly predicts entry into and movement through the ultra-elite stage via the preferred pattern but does *not predict* the alternate pattern. Neither patronage nor merit affects patterns of mobility during the ultra-elite stage independently of effects they had at earlier stages of the career. Only productivity continues to produce *additional* effects on mobility. Those who continue to gain prominence are those who remain highly productive in the later stages of a career. The shift away from patronage begins in the pre-elite stage and ends in the ultra-elite stage with merit becoming the dominant factor in determining mobility.

Up to this point we have examined only intrastage relationships between statuses. We turn now to the relationships between the statuses from one stage to the next, that is, the transition between stages. In figures 8.6 and 8.7 beta weights have been omitted from internal paths for clarity. Readers who wish to remind themselves of these values can refer to the earlier figures. Figure 8.6 shows the paths of transition between the pre-elite stage and the elite stage. Paths leading to membership statuses (NAS, APS) are shown in single lines while paths to gatekeeper positions

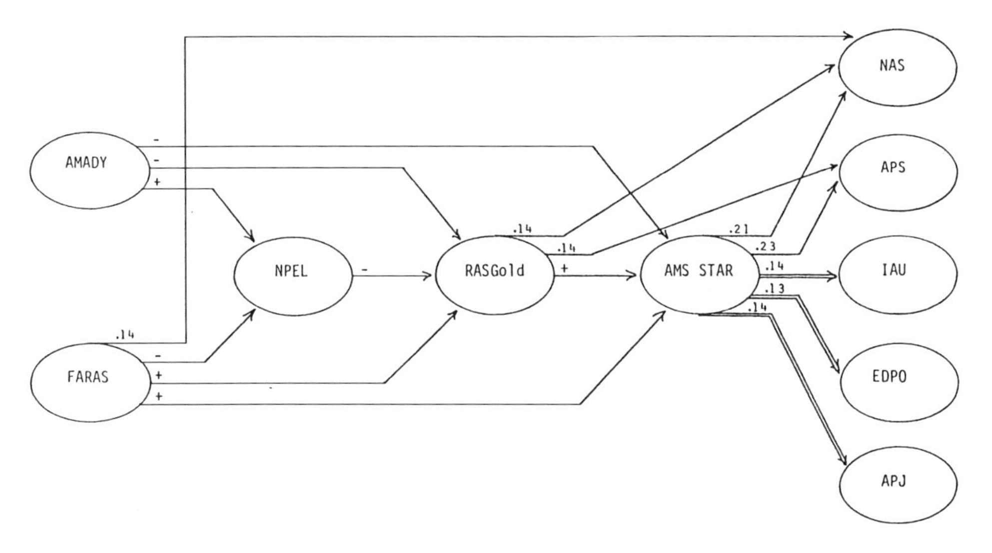

Figure 8.6. Transitions between pre-elite and elite stages in scientific careers.

(IAU, EDPO, APJ) are shown in double lines. Achieving an *AMS* star has a significant effect on entering the elite stage by obtaining any of these five statuses. Notice, however, that the paths are much stronger from *AMS* star to the membership statuses. In addition, direct paths exist from each of the statuses of the preferred pattern in the pre-elite stage to the membership statuses in the elite stage, but *not* to the gatekeeper positions. It is difficult to exaggerate the benefits gained by entering the stratification system via the *preferred status* of election as a Foreign Associate of the Royal Astronomical Society (FARAS). *This entry status appears to provide advantages for the rest of the career.*

Figure 8.7 shows the transition between the elite and ultra-elite career stages. The paths from membership statuses are shown in single lines and paths from gatekeeper positions are shown in double lines. Attainment of either type of elite status provides an advantage for entering the ultra-elite stage, as evidenced by paths from each status to at least one entry status in the ultra-elite stage. A sharp distinction exists, however, between those effects. As shown here, membership statuses predict entry into the ultra-elite stage via the *preferred entry statuses* while gatekeeper positions are positively associated with entry via the *alternate entry statuses.* Also, membership statuses are positively related to the most prestigious of the variables in the preferred pattern (FRADY membership). Gatekeeper positions, however, are directly related to the *terminal statuses* in the alternate pattern of status attainment. *Any associations between gatekeeper statuses and those of the preferred pattern are negative.*

From this evidence, it appears that there is a form of hierarchical dif-

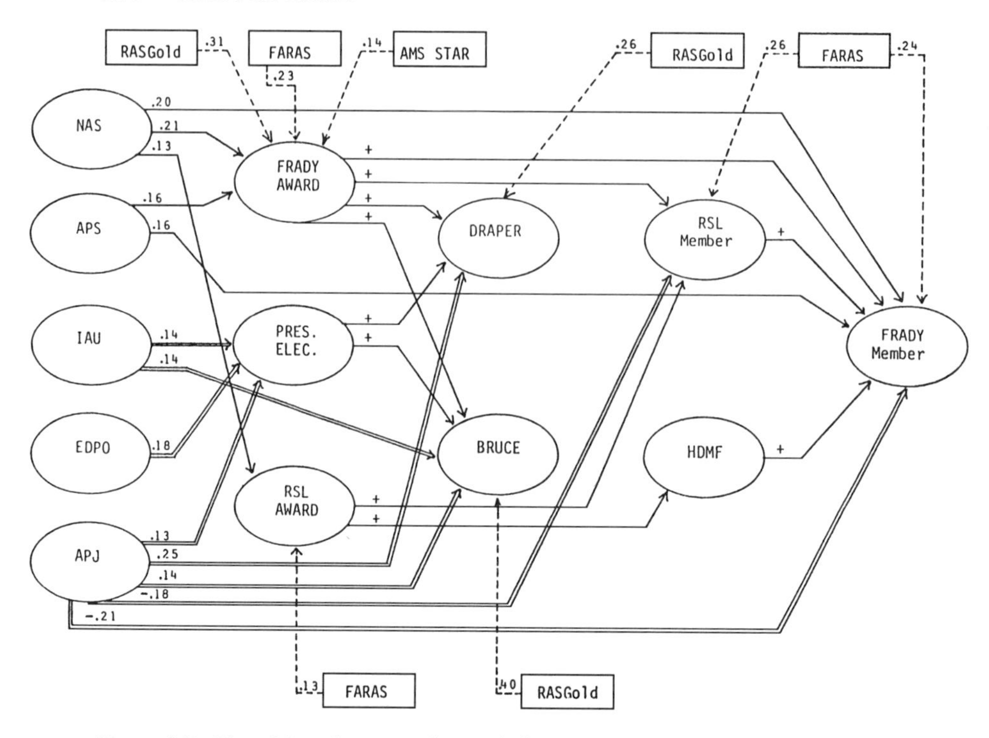

Figure 8.7. Transitions between elite and ultra-elite stages with paths shown from pre-elite to ultra-elite transitions.

ferentiation between statuses in the elite stage when viewed as part of the overall process of mobility. Membership statuses are more clearly aligned with the preferred patterns of the other stages, while gatekeeper positions are more closely aligned with the alternate patterns. The broken-line paths in this diagram represent effects of the pre-elite statuses that continue to directly affect mobility in the ultra-elite stage. *Only preferred statuses continue to have an affect.* This underscores the primacy of the preferred patterns and the terminal nature of the alternate patterns. Paths from the earlier European awards to the two major American awards support the notion of a triggering effect by European awards on these two statuses.

The evidence discussed in connection with the early and later phases of the careers of elite astronomers suggests that the stratification of careers exists in identifiable stages consisting of discrete statuses hierarchically arranged. Further, the evidence indicates two patterns of mobility operating at the later career stages. The preferred pattern provides a considerable advantage for future career enhancement while the alternate pattern provides no such advantage, and in fact, often appears to have a negative influence on further career development. The stages of a career are interrelated. Achieving the preferred pattern of mobility in the pre-elite stage is positively related to an advantageous resolution of the elite stage (by obtaining a membership status) and ultimately to attainment of the preferred pattern in the ultra-elite stage.

These findings support the idea that merit and patronage are complementary and together have important effects on mobility. As expected, productivity, institutional potential, and NAS contacts are important correlates of mobility at every stage of the career. Movement into a stage is dependent in part on productivity, institutional potential, and NAS contacts, regardless of patterning, but it is the combination of both patronage and merit that appears to be related most clearly to achieving the preferred pattern of mobility in the pre-elite and elite stages. In the ultra-elite stage, however, productivity emerges as a primary determinant in predicting which astronomers will achieve the honor of election as corresponding members of the French Academy of Sciences.

Chapter 5 examined the astronomical career as a social process and emphasized its nonlinear nature. Chapter 6 looked at the effects of patrons, mentors, and sponsors on the career. This chapter has examined careers from a different perspective. Here the career is conceived of as movement through the hierarchically ordered reward system of a modern scientific community. The pattern of movement is sequential and, remarkably enough, symmetrical. Movement through the reward system is, however, far too contingent to be understood as linear. The career is a complex unfolding of events that can be grouped into several distinct but related phases measured along the time line of increasing professional age. Patronage and merit function as interlocking norms (Mitroff 1974a, 1974b) guiding movement through the reward system. The action of one cannot be understood without reference to the action of the other.

It should come as no surprise that patronage dominates the careers of these elite astronomers until about professional age fifteen. As documented in chapter 6, mentorship, sponsorship, and the patronage of senior scientists are critical to the early years of the career. As members of successive cohorts achieved higher levels of educational attainment, the web of patronage stretched back into the pre-career training phase as well. When astronomers enter the pre-elite stage of the career at about professional age fifteen the balance between patronage and merit begins to shift. By this point, elite astronomers have achieved a level of maturity and independence, loosened ties to mentors and patrons and begun moving in a world in which performance becomes the metric by which they are evaluated.

If the stratification system in a modern scientific community is treated as a series of discrete statuses hierarchically arranged, new and often unsuspected aspects of the reward system are revealed. The symmetry of career development appears with the discovery that there are a limited number of well-defined routes by which elite astronomers achieve various levels within the stratification system as they move through each career stage. Because there are preferred and alternative entry statuses at each career stage, the process of mobility appears to involve a symmetrical two-track system. Differentiation begins at approximately professional age fifteen. The consequences of this original division unfold as the population enters the elite stage.

The model indicates that the process by which individuals achieve various levels of mobility is neither cumulative nor unidirectional. Indeed, acquisition of a terminal status suggests that the receipt of certain honors and awards may block further movement through the stratification system. The data indicate that early honors will define an individual's position on one track of a two-track system. The one track may well carry an individual successfully through each career stage because the original preferred status (foreign associate of the Royal Astronomical Society) is closely linked with preferred statuses at the elite and ultra-elite stage as well. The alternate entry status (election to the American Academy of Arts and Sciences) for the pre-elite stage *does not* necessarily lead to the next stage. In sum, a social historical approach to the study of mobility in the American astronomical community suggests that the acquisition of honors is a much more complex process than previously believed (Merton 1968; Cole and Cole 1973).

Evidence discussed in this chapter weighs heavily against the view that science is a meritocracy. On the evidence presented here, it would be difficult to assert that participants acted as if this were the case. To suggest that the reward system of American astronomy was something other than a meritocracy does not imply criticism or moral condemnation of scientists or the scientific enterprise. It does raise questions about the ideology of modern science; an ideology designed to provide science with a position of special privilege that places it above other human activities and guarantees science a unique status in a democratic polity. By the same token, this discussion raises questions concerning the ways in which historians and sociologists have either accepted this ideology without question or (just as reflexively) attacked science and scientists without mercy. Why must some historians give science a privileged position in society? Why must they pretend that scientists share little of our common humanity? And why must a whole school of sociology attack scientists because they show signs of sharing the faults and foibles of humanity? Answers to these questions entail a social history of the history of science and a historical sociology of the sociology of science (Kragh 1987: chaps. 1, 10, 17).

9 ✦

SCIENCE AND GENDER: WOMEN IN THE AMERICAN ASTRONOMICAL COMMUNITY

An unanticipated consequence of the new feminism has been the recovery of an important component of the history of science. Historians have found (Rossiter 1982) that women played an important, if often unacknowledged, role in the development of American science. In the pre–Civil War period, women were involved in such diverse activities as writing popular texts and collecting biological specimens (Kohlstedt 1978a, 1978b). After mid-century several developments opened the way for women to participate in astronomy (Kidwell 1990). First at Mount Holyoke and then Vassar, the new women's colleges offered instruction in both mathematical and observational astronomy. State universities such as California and private institutions like the University of Chicago provided opportunities for women to study astronomy at the undergraduate and graduate levels. Finally, the introduction of photography into astronomy, the development of large reflecting telescopes, and the reorganization of research observatories in keeping with the imperatives of efficient, large-scale production directed by scientist-entrepreneurs, stimulated the expansion of the labor force. The new factory observatories needed unskilled (female high-school graduates) and semiskilled (women with undergraduate degrees in astronomy, mathematics, or physics) workers to analyze and reduce the ever-increasing volume of observational data. Solar and stellar spectroscopy were especially labor intensive. So was photographic astrometry.

Between 1859 and 1940, 426 women were active in the American astronomical community. The vast majority of these women were at work for only a short time. As table 5.1 indicated, 30 percent of the women involved in American astronomy had careers that lasted less than one year and 50 percent were gone from astronomy after careers that spanned no more than five years.

Table 9.1 tabulates the distribution of women by cohort. For the

Table 9.1 Growth of the American Astronomical Community by Cohort and Gender, 1859–1940 (N=1205; Women=426)

Cohort	Male	Female	Percentage Female
1859 or Before	65	1	2%
1860–1899	250	56	18%
1900–1940	460	344	43%
Total	**775**	**401**	**34%**
Data missing on cohorts for 29 cases.			

Table 9.2 Distribution of Women by Research Area, 1859–1940 (N=426)

Cohort	Old Astronomy	New Astronomy	Mixed Astronomy
1859 or Before	1	0	0
1860–1899	21	13	5
1900–1940	142	134	15
Total	**164**	**147**	**20**
Data missing on astronomy type for women for 95 cases.			

cohort that entered astronomy in 1859 or before, Maria Mitchell was the lone woman. Cohort two, made up of those who entered between 1860 and 1899, included fifty-six women (18.3 percent). More than two-fifths (42.8 percent) of cohort three were women. Between 1859 and the coming of World War II, almost a third (34.1 percent) of the American astronomical community was female.

Table 9.2 shows the distribution of women by research area (old astronomy, new astronomy, or mixed). The figures for women indicate a rough parity between the old and new astronomy. For men, the balance tips in the third cohort in favor of astrophysics.

Table 9.3 presents data on levels of educational attainment for women. The largest group consists of those with no more than a secondary education. The numbers drop from that point. Levels of educational attainment differ between women and men. For example, at the baccalaureate and master's degrees, the difference is about 2:1 while for the Ph.D. it is 6:1.

Of all the differences between men and women, perhaps the most dramatic were those discussed in chapters 7 and 8. Women were, for the most

Table 9.3 Educational Levels for Women in the American Astronomical Community

Highest Level of Educational Attainment	Number
High School Only	266
Baccalaureate	62
Master's	53
Ph.D.	45
Total	**426**

NOTE: When compared with Table 9.12 differences are caused by women who never held a job after taking highest degree.

part, outside the reward system and virtually excluded from access to power in the American astronomical community. The social system of American astronomy was organized around gender stratification and segregation: women's work, institutions for women, and specific female career patterns. Chapter 9 explores aspects of gender stratification and segregation in American astronomy. However, first it is necessary to consider certain aspects of the larger American culture and the ways in which culture directly relates to the development of modern science.

SEPARATE SPHERES AND THE RISE OF WOMEN'S WORK IN AMERICAN ASTRONOMY

As historian Rosalind Rosenberg (1982:207) suggests in her discussion of the roots of modern feminism, "Throughout the Progressive era [ca.1890–1915] the Victorian concept of sexual polarity maintained its hold on the popular imagination." Sexual polarity suggested that women and men each had special talents (Welther 1966, Cott 1977). These ranged from cognition and emotion through motor skills. Women were, in this view, more delicate, moral, and religious, aesthetically sensitive, patient, and caring than men. Men were seen as more aggressive, politically skilled, better fitted to grasp and/or create new ideas, and less bound by custom and tradition. Thus the classic capitalist division of labor was augmented by a division based on gender. Within Victorian culture (and well into the twentieth century), the social world was divided into separate spheres. Women and men were seen to inhabit different moral and cognitive domains.[1]

Vassar's Maria Mitchell shared many of these views (Kendall 1896:

1. For a discussion of separate spheres in the context of the family of Naval Observatory astronomer Asaph Hall, see the biography of his wife, Angeline, by her son, Angelo Hall (1908: 82, 100–103).

237–38). "The training of a girl fits her for delicate work." Mitchell argued that "The eye that directs a needle in the delicate meshes of embroidery will equally well bisect a star with the spider web of the micrometer." Skill and care in delicate work such as embroidery and other forms of sewing trained the hand and eye of women in ways that men could never achieve. Further, these repetitive tasks prepare women to undertake "routine observations . . . [which] dull as they are, are less dull than the endless repetition of the same pattern in crochet-work." Mitchell pointed out that women were prepared for exactly the kind of precision work that was required in making observations with the transit circle or of the separation and position angle of double stars. The founder of the Vassar astronomy program concluded (Wright 1949: 238): "The girl who can stitch from morning to night would find two or three hours in the observatory a relief."

Williamina Paton Fleming, curator of astronomical photographs at the Harvard College Observatory, offered similar views in an address (1893: 688–89) to the Congress of Astronomy and Astro-Physics at the 1893 Chicago World's Fair. "While we cannot maintain that in everything woman is man's equal, yet in many things her patience, perseverance and method make her his superior." Fleming concluded, "Let us hope that in astronomy, which now affords a large field for woman's work and skill, she may, as has been the case in several other sciences, at least prove herself his equal."

Toward the end of the nineteenth century, a form of gender segregation developed in several branches of American science, and a separate female labor market came into existence (Rossiter 1982: 51–72). In astronomy and the biological sciences, employment categories known collectively as *women's work* emerged. In astronomy, the rise of factory observatories and the desire of astronomer-entrepreneurs to control wages and maximize production stimulated these developments. The idea of women's work in science reflected the separate spheres doctrine.

Woman's work in astronomy entailed a clear-cut division of labor. Men observed at the telescope. Women measured, classified, and reduced the data men collected. Women's work generally included the measurement of photographic plates or the reduction of material collected by observers using the transit circle. The data were then presented to male astronomers to interpret and prepare for publication. In certain research activities, the compilation of the Henry Draper catalogue at the Harvard Observatory, for example, women (Hoffleit 1991) had a great deal of control over the project.[2] Only gradually were women allowed to use tele-

2. Harvard Observatory director Edward C. Pickering reflected these ideas concerning women's work in astronomy when he established a fellowship for women to work at HCO.

scopes. This trend began in graduate programs at places like the Lick and slowly spread. Yerkes permitted women to make observations as a result of the shortage of male graduate students during World War I. At some institutions women were excluded from the observing until after World War II.

The notion of separate spheres and the idea of women's work in astronomy remained a potent force through the interwar years. In 1936 a freelance writer contacted Frank Schlesinger. She was working on an article for *McCall's Magazine* on women in astronomy and wanted the Yale director's opinion concerning "women astronomers you have worked with . . . [and whether they show] any marked aptitudes for certain phases of the work" and whether there are "other phases in which they have not been so successful?"[3] In his reply, Schlesinger discussed the area in which he felt women excelled: the reduction of astrometric data and the preparation of catalogues of stellar positions. "Such work . . . [is] better carried out by women than by men. This work calls for patience and great care to avoid errors." In conclusion, Schlesinger asserted that "women are not as creative as men of equal training."[4]

Mount Wilson astronomer Robert S. Richardson (1944: 6) also used the idea of separate spheres to justify sex segregation in astronomy. Characterizing observing runs, especially in cold weather, as tedious ordeals that impose a "severe strain on the observer's powers of endurance," Richardson argued that "few women have attempted active work at the telescope." Rather they are "engaged in the reduction and classification of observations taken by their more hardy masculine associates." Richardson concluded by referring to the special skills of women. "Women generally excel men in work demanding close attention to minute detail, and nowhere is this better shown than in astronomy."[5]

Maria Mitchell was a product of early Victorian culture and could never completely liberate herself from its thrall. She did, however, have

---

See Pickering to Robert H. Baker, 12 April 1917, p. 1. WHMC, Lawes Observatory Papers. Astronomers working in closely related research areas also praised women's special skills. As William A. Rogers wrote to Wellesley's Sarah F. Whiting, "I wish I could have one or two of your level-headed girls to take up some problems in this connection jointly with me." Rogers to Whiting, 12 October 1887, p. 1. WCA, Whiting Papers. Rogers' work involved various precision measurements.

3. Josephine Nelson to Frank Schlesinger, 13 December 1936. YUA, Astronomy Department Papers.

4. Schlesinger to Nelson, 19 December 1936. YUA, Astronomy Department Papers.

5. Vera Rubin of the Carnegie Institution of Washington broke the gender barrier at Mount Palomar in the mid-1960s. The final line of argument to keep women from using the Palomar telescopes involved lack of sleeping facilities and, of course, bathrooms! Vera Rubin was also the first female astronomer elected to the National Academy.

some understanding of the price women paid because of sex stereotyping and differential socialization that flowed from the idea of separate spheres. As Mitchell saw it, women's work in science involved a form of intellectual bondage (Wright 1949: 91). Women must escape from their reverence for tradition and authority (Kendall 1896: 187); and, at a practical level, they must be emancipated from "stitch, stitch, stitch"; only then would women "have time for studies which would engross as the needle never can." Women's work, a norm that found its way into the astronomical community from the larger culture, cut women off from many creative activities in science.

MALE ASTRONOMERS' PERCEPTIONS OF WOMEN AS SCIENTISTS

Evaluations of the intellectual and technical capacities of women by male astronomers involved certain fundamental themes. These evaluations reflected the separate spheres doctrine and justified a gender-based division of labor. Further, the evaluation of females involved a cluster of issues relating to domesticity. Marriage, home, and family were never far from the thoughts of male astronomers as they evaluated women. In recommending Edith E. Cummings (Taylor) (b. 1894) for a Lick graduate fellowship, Missouri astronomer Robert H. Baker concluded, "While unfortunately she is not a man, I believe she is handicapped by her sex less than any aspirant I have known. In our . . . photometric work she had done a man's share, securing plates as well as measuring them. Her computing is rapid and accurate. She had unlimited enthusiasm and the necessary health and strength to accompany it. She intends to make astronomy her life work."[6]

Male astronomers never lost sight of the gender-based division of labor. Heber D. Curtis of Allegheny sympathized with William Wallace Campbell, concerning difficulties the Lick director was encountering in his search for an astronomer to direct the Lick station in Chile. "Too bad A. E. Glancy has not Alfred instead of Anna for a first name."[7] Glancy was one of the best astronomy Ph.D.'s produced by the University of California, but because she was a woman would never be considered for the job of supervising the Lick southern station. Anna E. Glancy (1883–1975) was described as "brilliant, industrious and accurate" by Armin O. Leus-

6. Robert H. Baker to William Wallace Campbell, 3 April 1917, p. 1. Lawes Observatory Papers, WHMC. For candid and very positive evaluations of the abilities of two female doctoral candidates, see Armin O. Leuschner to William Wallace Campbell, 20 November 1911, pp. 3–4. Both eventually left astronomy: Waterman married and Levy moved to Stanford, seeking career mobility when it became clear that at Berkeley she would never attain faculty status.

7. Heber D. Curtis to William Wallace Campbell, 26 January 1922. Curtis Papers, SALO.

chner, chair of the department at Berkeley. However, in the next breath, Leuschner concluded, "She would be an excellent superintendent of assistants engaged in photographic work, including observation, measurement and reduction of plates and derivation of results."[8]

Characteristic of male evaluations of female graduate students is a report from Leuschner concerning a doctoral candidate, Lois Slocum. The committee gave Slocum high marks for her ability to handle telescopes, skill at astronomical photography, and competence in measuring plates, but had reservations. The male astronomers believed Slocum, "while accurate and persevering in practical work," had "some difficulty in obtaining a deeper insight into a given problem or in grasping the meaning or importance of each step of an investigation." Their conclusion was straightforward. "We . . . should hesitate to recommend her for independent research work" but "she would be quite good as a research assistant."[9]

At the end of the 1880s, Wellesley's Sarah Frances Whiting (1847–1927) toured Europe, visiting laboratories and observatories where she discussed science with leading physicists and astronomers. Her reception was generally courteous, but Whiting felt that often male scientists were embarrassed. "After talking with a woman who proved to be intelligent—it would come over them that it was unusual and might be revolutionary."[10] Sometimes, male scientists were able to articulate their fears. Whiting reported one such experience. She spent an afternoon talking with a distinguished physicist and touring his laboratory. Then they retired to his library. Apparently the great man was troubled. He ushered Whiting to a chair and then put more fuel on the fire. Finally he turned to his visitor and said, "I want to ask a question before we go down to *dinner*, if all the ladies should know so much about spectroscopes and kathode rays, who would attend to the buttons and the breakfasts?" Whiting, speaking in the third person, reported "He was assured that he was considering things mutually exclusive which were not."[11]

Married women had no place in science. Both graduate admissions committees and observatory directors preferred unmarried women. Given antinepotism rules, as well as tradition and custom in American academia,

8. Armin O. Leuschner to Frank Schlesinger, 26 February 1920. Astronomy Department Papers, BLUCB. Glancy had been involved in war work with the American Optical Company where she was making $175 per month. Whether any American observatory could or would have paid her a comparable wage is open to question.

9. Armin O. Leuschner to John C. Duncan, 17 February 1930, p. 1. Astronomy Department Papers, BLUCB.

10. Sarah Frances Whiting, "Notes for Speeches and Addresses," undated, p 5. Whiting Papers, WCA.

11. Ibid.

women were expected upon marriage to resign their posts and devote full time to domestic life.[12]

John A. Miller, professor of astronomy and director of the Swarthmore observatory often sent his best students to Berkeley for graduate training. It was, however, difficult to secure fellowships for women. As Miller confided to William Wallace Campbell, "I share your feeling regarding the investment in fellowships for women." He went on, however, "I know Miss Powell quite intimately however, and though I should hesitate to guarantee that she is not going to get married some time, I know that is not in her mind for the immediate future."[13] Within a year, Miss Powell succumbed to the lures of matrimony. Miller was very apologetic. "I am certain there was nothing further from her intentions than to abandon . . . [work on a Ph.D.] when she left here and from what I can learn it must have been a sort of whirl-wind courtship." Miller concluded, "I think you have been very generous in your attitude toward the whole affair."[14]

In discussing women in astronomy, Mount Wilson astronomer Robert S. Richardson commented on marriage. Richardson's remarks illustrate another aspect of the male perspective. Richardson (1944: 8) pointed out that there was a rough parity between the sexes in astronomy graduate programs, but that many of these female Ph.D.'s did not have long-term careers as research scientists. "The reason for this mysterious discrepancy is not hard to find. *Most of them become astronomers' wives instead of astronomers!*" The data are admittedly limited (in most cases this kind of information is not available in the sources on which the collective biography was based), but it appears that of astronomers (women and men) known to have married other scientists (N=38), the spouse was most often (N=31) an astronomer. Twenty-one of these spouses (55.3 percent) gave up their careers after marriage. Seven others married scientists who were not astronomers (physics, mathematics, engineering).

Male astronomers were concerned about the mental and emotional health of women. Clearly, they viewed women as more susceptible to emotional and physical illness than men. In the fall of 1921, William Wallace Campbell refused to permit Edith Cummings to take the oral examination for her doctoral dissertation. He believed Cummings to be "very close to

12. On antinepotism rules, see Rossiter (1982: 195–97). A discussion of the development and enforcement of antinepotism rules would make a valuable addition to our understanding of the history of higher education and the history of women.

13. John A. Miller to William Wallace Campbell, 26 January 1922, p. 1. Miller Papers, SALO. Miller concluded with an invitation for Campbell to present a paper at the April meeting of the American Philosophical Society in Philadelphia. Perhaps this put the Lick director in a better frame of mind concerning support for Powell.

14. Miller to Campbell, 8 February 1923. Miller Papers, SALO.

the point of a nervous collapse." Director Campbell traced the problem to "personal troubles" that began the previous March. "Notwithstanding her excellent thesis and her splendid general preparation" Campbell packed Cummings off to her home in Nebraska with orders to rest and recuperate. Her examination would be rescheduled for the following April.[15] In fact, Cummings was suffering from severe eye strain, but must have also had matrimony on her mind for within a few weeks she married.

In spite of her remarkable intellectual abilities, Cummings caused her professors concern. In recommending her to Robert H. Baker for the master's program at Missouri, Frank Schlesinger at Allegheny wrote of his computer that she was "mighty enthusiastic" and "has the makings of an astronomer in her." He hoped, however, that she would "soon learn to take a less emotional stand with regard to astronomical problems."[16] Later, in recommending her to Campbell for the California doctoral program, Schlesinger pointed to "her one serious drawback. . . . The emotional view she takes of astronomy and of life in general." Apparently, Cummings was so enthusiastic that she could not distinguish the wheat from the chaff; she could not divide the field into "questions of mere interest" and those that were desirable to solve.[17]

Sometimes women seemed so nervous as to be tongue-tied. Berkeley faculty member Russell Tracy Crawford described one such student. She could do a superior job on written examinations but "she can't answer a simple question orally." Crawford concluded, "*She is a puzzle to me.*"[18] Many years later, Harlow Shapley (1969: 113) recalled that, at Harvard, female graduate students did not make many contributions in seminars or discussion groups. Apparently, male astronomers never imagined their rough-and-tumble style of oral discussion could intimidate women.

To any one familiar with the personal correspondence of the leading male astronomers, there appears to be a double standard concerning mental and physical health. Campbell admitted to nerves and a bad back, Schlesinger could not sleep, Edward Emerson Barnard was plagued by ill health through most of his career, and the physical and mental problems of George Ellery Hale became almost as notorious as those of Charles

15. William Wallace Campbell to Armin O. Leuschner, 28 November 1921. Leuschner Papers, SALO.

16. Frank Schlesinger to Robert H. Baker, 6 December 1917, p. 3. Lawes Observatory Papers, WHMC.

17. Frank Schlesinger to William Wallace Campbell, 25 March 1918, pp. 1–2. Schlesinger Papers, SALO.

18. Tracy [Russell Tracy Crawford] to A.O. [Armin O. Leuschner], 9 March 1909, p. 4. Astronomy Department Papers, BLUCB. Emphasis added.

Darwin. Yet males never made an issue of their own mental or physical health. When it came to evaluating women, however, it was a very different story. They judged women by separate standards.

There were, to be sure, exceptions. Then as now there were superwomen who, like females in contemporary commercials, were able to pursue a successful career by day and be an efficient homemaker, mother, and wife as well as have time for sports and hobbies. Williamina Paton Fleming was, after her death, portrayed as a kind of superwoman. Oxford astronomer Herbert Hall Turner (1912a: 263) recalled, "As an astronomer Mrs. Fleming was somewhat exceptional in being a woman"; she not only carried heavy research and administrative responsibilities at the Harvard Observatory but undertook "on her return home those household cares of which a man usually expects to be relieved." As a single parent, Fleming saw a son through MIT on what was, at best, a meager salary, but as Turner wrote, "She was fully equal to the double task" of scientist and homemaker, "as those who have had the good fortune to be her guests can testify." The Oxford astronomer also noted that Fleming "yielded to none in her enjoyment of a football match, especially a match between Harvard and Yale."

Margaret Rossiter (1982: 57) discovered a diary kept by Fleming. While it describes her work in the spring of 1900, the document also records Fleming's reactions to the problems she faced after twenty years of service at Harvard. Her salary was apparently fixed, and director Edward C. Pickering, did not care to discuss raises. She had to spend much of her time supervising the work of eleven female computers as well as editing all manuscripts by male astronomers before the manuscripts were submitted for publication. Male assistants made two-thirds again Fleming's $1,500 per year salary. Fleming asked in her diary if director Pickering ever thought "I have a home to keep and a family to take care of as well as the men? But I suppose a woman has no claim to such comforts. And this is considered an enlightened age!" She was expected to work what Pickering defined as a seven-hour day: from 9 A.M. to 6 P.M. The pressure on Fleming was great and sometimes she felt "almost on the verge of breaking down." Why was the work of the observatory organized in such a way as to "throw so much of it on me, and [yet] pay me in such small proportion to the others, who come and go [referring to the turnover rate among assistants], and take things easy?" Yet Fleming kept up the pace until she died in 1911.

MARIA MITCHELL: EXEMPLAR

Maria Mitchell (1818–89) was the first woman astronomer in America. She became a beloved role model to generations of Vassar students and, in a larger sense, exemplifies many of the problems that women astronomers would face, not only in her own day, but in later generations as well. Further, Mitchell is unique in that she attained the status of an American heroine, the first female scientist to do so.

Mitchell was born and raised on Nantucket. Her father taught Maria the elements of observational astronomy. Her formal education was limited to private secondary schools on the island, but she read celestial mechanics and was well versed in the German and French astronomical literature. Her discovery of a comet in 1847 led to a great deal of publicity and subsequently to a gold medal from the king of Denmark as well as election to the American Academy of Arts and Sciences and the American Philosophical Society. Mitchell spent a summer observing and computing with Alexander Dalles Bache and a Coast Survey party in Maine and in 1849 joined the staff of the Nautical Almanac Office. She was given the task of computing tables for the planet Venus. Mitchell became a member of the original faculty at Vassar College in 1865 and retired a year before her death.

In spite of the values derived from her Quaker upbringing on Nantucket, where wives and daughters of whalers assumed a great deal of independence and responsibility in the absence of husbands and fathers, Mitchell was inescapably the product of early Victorian culture. She grew up in a world marked by gender polarity and, in the first flush of the romantic movement, a world that placed women on a pedestal. Even though she worked closely with her father and such leaders of the American scientific community as Alexander Dalles Bache, director of the Coast Survey, Mitchell was still deferential to males and understood that to succeed she had to be good at sewing on buttons and making breakfasts as well as solving differential equations.

In accepting appointment to the staff of the Nautical Almanac Office, Mitchell assumed a modest tone. While she and her father counted NAO director Captain Charles Henry Davis as a friend who knew her ability as an astronomer, Mitchell was diffident. "I am somewhat afraid that I may have undertaken a task for which I am not competent. When you send me my instructions, will you give me any advice or information which you can as to the best means of fitting myself for the work? I should not shrink from any amount of labour."[19] Apparently, Mitchell sought to keep a low profile and not make demands on her employer, but in 1856, having worn out her logarithm tables, she ask whether the NAO would provide her with a new book. Mitchell had also been buying her own paper and estimated that in the previous seven years the outlay would have paid for a round-trip ticket to Washington, D.C. She requested the Navy to supply her with paper on which to carry out computations.[20]

19. Maria Mitchell to Charles Henry Davis, 12 August 1849. Davis Papers, FML.

20. Mitchell to Davis, 24 October 1856. Davis Papers, FML.

An interesting description of Mitchell is contained in a letter to Matthew Vassar from an associate dispatched to New England by the wealthy Poughkeepsie brewer. Apparently there was some concern among members of the college's executive committee that Mitchell had actively sought a professorship at Vassar. Rufus Babcock's visit clarified the matter; she had only inquired on behalf of her brother-in-law. Assertive female behavior in the form of applying directly for a Vassar professorship probably would have led the all-male committee to reject Mitchell. As Babcock reported, "I had a long, and very interesting conference with her and her venerable father, the result of which I hope may be to our advantage. [That] she is by far the most accomplished astronomer of her sex in the world I have no doubt. And but few of our manly sort are any where near her equal in her loved and chosen pursuits."[21] Nor was this all. Mitchell possessed "a breadth of culture I was by no means prepared to expect. She has travelled a year in Europe with the best facilities of access to all the learned; and yet with all this ripeness, she is as simple minded, and humble as a child."[22]

Much of the day was spent discussing Matthew Vassar's plans for the college and trying to interest Mitchell in the project, but all the while, Babcock was studying the Quaker astronomer. He reported that "Miss Maria is not such a poor miserable 'blue-stocking' as to know nothing else but astronomy." The reference was clearly *not* to the high culture or knowledge of the sciences beyond astronomy and mathematics, rather it had to do with the eternal issue of buttons and breakfasts. "The day I spent with them their domestic [cook] was absent and Maria prepared dinner and presided in all the housewifery of their cozy establishment without parade, and without any apparent deficiencies."[23]

After the meal, Mitchell, her father, and their visitor from Poughkeepsie must have done some stargazing. "In her astronomical observatory, fitted up at the back end of her garden, she is still more at home, handling her well adjusted telescope with masterful ease, accuracy and success."[24] So impressed was Matthew Vassar's agent that he immediately obtained three recommendations "of her fitness for the place [professor of astronomy at Vassar] . . . as good and high as New England can furnish."[25]

21. Rufus Babcock to Matthew Vassar, 8 September 1862, pp. 1–2. Mitchell Papers, VCA. I have silently corrected the punctuation in this letter. The writer was given to the use of dashes rather than periods or commas.

22. Babcock to Vassar, 8 September 1862, p. 2. Mitchell Papers, VCA.

23. Ibid.

24. Ibid.

25. Ibid., p. 2, and 10 September 1862.

TEACHING VERSUS RESEARCH

As a young woman, Mitchell taught for a time in a private school and then briefly kept her own school on Nantucket, but she was soon lured away by the Athenaeum, over which she would preside as librarian, a job that included ample free time to read celestial mechanics and astronomy. Evenings were devoted to assisting her father make observations in order to adjust ships' chronometers or to her own researches. When she was called to Vassar at the age of forty-seven, Mitchell's classroom experience was distinctly limited in comparison with her years of research and private study.

In the Maria Mitchell papers in the Vassar College archives is a small nondescript notebook, perhaps three by five inches. There is only one date (1 August 1868) written in the little book and its pages are not numbered. The notebook includes descriptions of the Vassar observatory instruments and notes on their use as well as jottings on the history of astronomy. The book also contains some notes by Mitchell on the problem of teaching versus research. My inference is that these date from her early Vassar career. These notes clearly reveal the choices she faced in a teaching institution.[26]

Mitchell outlined two plans. The first related to her role as a scientist in a teaching institution. "Observations upon double stars—colors [of the stars] & spectroscope experiments—also sweep for comets & examination of planets & of large [that is, bright] stars for unknown companions—Also variable stars." The Vassar observatory director concluded that "The above would be popular & pleasing & would make a show." In other words, it would please the undergraduate students and would provide hands-on instruction in astronomical observation.

But there was a second plan, more ambitions than the first. This second plan was divided into to two sections. The first dealt with mathematical astronomy. "To settle down upon some specific point and *plod.* It might be one of the smaller planetoids [asteroids]." Here Mitchell was considering work on the definitive orbit of an asteroid. In the third quarter of the nineteenth century, discovery and observation of minor planets ran far ahead of rigorous mathematical treatment of their orbits. Mitchell continued, "This would take time and would be tedious. [But] it would be solid and endure for 100 years—." The second section of plan two was related to observational astronomy. "If I make a discovery it would be that of a small planet [asteroid] & would be very valuable."

The next portion of the entry compared the two plans. The first

26. Obviously there is no way to provide a proper reference for this little notebook since it is untitled, virtually undated, and without pagination. At the time I used the Mitchell Papers (1986) the notebook was in box 2, folder 3 of the collection.

"would be likely to interest the girls [Vassar students] & would probably result in *eye* discoveries [that is, discoveries made in the course of visual observing programs]—." The second plan, however, "would be more mathematical & would *in time* gain [Mitchell] a real solid reputation." She concluded that "in either case the inst[ruments] would be brought into adjustments & the work must be done scientifically." In the end, Mitchell opted for plan one. It was more popular and pleasing for the undergraduates. These activities became an important element of Mitchell's image. In addition, Mitchell did develop some long-term research programs.

An examination of Mitchell's observing books provides information on her research activities. She concentrated on solar system objects from sunspots to planetary markings. Her most ambitious program involved searching for new satellites of Saturn. This program employed the chronograph rather than the micrometer to measure the positions of suspected planetary satellites. This offset method entailed complex numerical reductions, and there is no evidence (work sheets, tables, etc.) that this was done on a systematic basis; nor are there any rough finder charts in the notebooks. Either Mitchell plotted the position of Saturn's moons on a chart such as the *Bonner Durchmusterung* and carried it to the telescope, or she did not feel this step to be necessary.[27]

Mitchell sketched planets at the telescope and then, the next day, in her sitting room, sought to color the sketches. Working from rough notes and memory was not satisfactory, as Mitchell often indicated in her notebooks. With her attention focused on the search for planetary satellites and the colors of planetary markings, other aspects of planetary phenomena were neglected. Mitchell may have made an early observation of Jupiter's great red spot in the winter of 1870.[28] She speculated on the possibility of a ring system for Jupiter as a result of observations made in 1874.[29] Numerous observations of white spots on Saturn provided an opportunity to improve the value of its rotation, but her attention was elsewhere.[30]

27. For an example of the search for new Saturnian satellites see "12-inch Observing Book, 1876–79," pp. 47–48, for the night of 7 November 1877. Mitchell Papers, VCA. Earlier entries in this observing book suggest that Mitchell did not calculate and plot the positions of Saturn's known moons in advance of an observing session. See "12-inch Observing Book, 1876–79," pp. 6–7. Mitchell Papers, VCA.

28. "Observations with the 12-inch Equatorial by Prof. Mitchell, 1866 April 13–1877 June 17," p. 95, for the observing session of 19 January 1870, when she wrote, "This peculiar oval figure . . . is new to me; it is very distinct . . . [But] I did not succeed in making others see it." Mitchell Papers, VAC.

29. "Observations with the 12-inch Equatorial by Prof. Mitchell, 1866–1877," p. 196, after observing satellites on 2 June 1874. Mitchell Papers, VCA.

30. "12-inch Observing Book, 1876–79," for the session of 15 October 1878, but observations of white spots extend from September 1877 through December 1878. Mitchell

As Mitchell indicated on the front flyleaf of the 1866–1877 observing book, the Vassar telescope was far from perfect and, apparently, had developed serious mechanical problems with age. The clock did not track celestial objects with accuracy and there was rust on the polar axis. These comments suggest that the 12.5-inch was difficult to handle and keeping it fixed on a celestial object could not have been easy. Of its optical condition, Mitchell wrote, "The equatorial telescope does not define well. It shows a good deal of color. . . . The micrometer eye-pieces are imperfect."[31] These comments make it clear why Mitchell could not measure the rotation rate of Saturn using the white spots. She could not hold the planet steady and her micrometer was too imperfect for the delicate task. Further, it suggests why she adopted the method of using the chronograph to determine the distance of suspected satellites from the limb of a planet.

By the 1880s, Mitchell's observing books had become a hodgepodge. Observations with the meridian transit, notes on the scientific literature, recipes for gingersnaps, and jottings for her dome parties are all interspersed.[32] The constraints and imperatives she found at Vassar had taken their toll. Mitchell was no longer the active young observer of the 1840s, discovering comets and camping out with the Coast Survey for a summer in Maine. The compromises demanded by the institutional realities she found at Vassar shaped Mitchell's scientific career after 1865.

Equal Pay for Women

It would be a mistake to believe that Vassar was, from the beginning, a feminist oasis. Conceived of by men, designed by men, and, for the most part, administered by men, Maria Mitchell's Vassar was no feminist institution. As she recalled (Kendall 1896: 191) after the death of the first president, J. L. Raymond, "He was not broad in his ideas of women, and was made to broaden the education of women by the women around him."

Apparently on the issue of equal pay, the policies of the Vassar administration were never broadened, in spite of efforts made by women faculty members. Mitchell's difficulties over equal pay began in the fall of 1862, when Rufus Babcock came for a visit. Babcock wrote to Matthew Vassar that Mitchell probably would continue to work for the NAO if she joined the faculty, implying that this consideration would influence her college

---

Papers, VCA. On the determination of Saturn's rotation rate from the observation of white spots, see Alexander (1962: chap. 15).

31. These comments are found in the front flyleaf of the 1866–77 observing book. Mitchell Papers, VCA.

32. "Rough Observing Notes, 1884–85," Mitchell, VCA. According to the *Royal Society Catalogue of Scientific Literature,* Mitchell's lifetime publication count is only eight papers. Four were published before she joined the Vassar faculty. Two of these dealt with observations of the comet she discovered, one with variable stars and one with double stars. The Vassar papers were all on observations of Jupiter and its four bright moons.

salary.[33] In 1865, Professor Charles S. Farrar, who was to teach mathematics, natural philosophy, and chemistry when Vassar opened its doors, wrote to President Raymond, "Would not a moderate salary from the College be both honorable & satisfactory to Miss M., since she will, of course, continue to serve the government and receive a government salary?"[34] There is no evidence that Farrar or President Raymond discussed the question of salary with Mitchell to see if she believed this arrangement "both honorable & satisfactory." Given the values of the day, it is improbable that these men would have ever thought it necessary. They simply assumed that they knew what was best for Maria Mitchell.

Thus the trustees fixed the salary of male professors at $2,500 with the provision that they were to pay the college for housing. The president earned $4,000 plus unspecified prerequisites. Mitchell's salary was fixed at $800 plus room and board for herself and her father. It is not clear how Mitchell learned of the inequities, but in June 1868 she and Dr. Alida C. Avery, professor of physiology and hygiene and college physician petitioned the executive committee of the board concerning "the injustice of the general practice of fixing compensation for women at a lower rate than that of men for services admitted to be of equal value." Avery and Mitchell expressed "the hope that the time may come when Vassar College will adopt a juster rule."[35] In 1869 Mitchell and Avery found their salaries increased to $1,400 plus room and board. Hanna Lyman, principal of the preparatory department earned $400 more than the two professors. This was doubtless justified because of her position as an administrator.

In 1870, after five years "of what we believe to be faithful working for the good of the college," Mitchell and Avery found out that "our pay is still far below that which has been offered at *entrance* to the other professors [males], even when they have been wholly inexperienced."[36] The trustees suggested that if Mitchell and Avery would only examine the data, they would see that in addition to their cash salaries ($1,400 per year) they also received room and board valued at least $450, making, on a conservative estimate, their actual compensation at least $1,850. Professors earning $2,500 a year had to pay the college between four and five hundred a year in rent for college housing. Thus their salaries were, in fact, reduced to about $2,000. At this point the trustees concluded "that there

33. Rufus Babcock to Matthew Vassar, 8 September 1862, p. 2. Mitchell Papers, VCA.

34. Charles S. Farrar to President J. H. Raymond, 13 February 1865. Mitchell Papers, VCA. Mitchell earned approximately $500 a year working as a computer with the NAO.

35. Selections from the Trustees Minutes Relating to Maria Mitchell, 22 June 1868, Meeting of the Committee on Faculty and Studies. Typescript of the minutes are in the Mitchell Papers, VCA. Salary information is from the same document.

36. Maria Mitchell and Alida C. Avery to President J. H. Raymond, 16 May 1870. Mitchell Papers, VCA.

is no substantial difference between the emoluments now enjoyed by Professors Mitchell and Avery and those allowed to other Professors in the College."[37]

In 1871, Mitchell and Avery, together with seven other female faculty and staff, petitioned the trustees for increases in salary or changes in the way salaries were paid. As Avery and Mitchell put the matter, if the trustees considered the salaries of all professors to be the same "we now respectfully ask that our salaries may not only be the same in amount, as those of other professors but may be paid in the same way, leaving questions of domestic arrangements [room and board] for separate & independent consideration." In short, the senior female professors wanted the same dollar figure as males occupying equivalent positions. They would pay for their own lodging and food. After some maneuvering, the trustees turned the issue over to a special committee which was to report at the next annual meeting.[38]

From this point, the story becomes difficult to follow. Mitchell and Avery wanted the trustees to provide an itemized breakdown of the cost of board rather than a flat figure of $16 per week. This the board would not do. Apparently tempers flared. As Mitchell wrote in December, "can it be possible that the Executive Committee [of the trustees] wishes to make us [Mitchell and Avery] uncomfortable?"[39] By mid-1872 Mitchell was ready to resign, but the president persuaded her to remain.

Mitchell early developed negative feelings toward administrators. As she wrote in the little notebook that contained her discussion of the two plans she might follow as a Vassar faculty member, "Devote myself more & more to my own department—try to be loyal to the stars, but keep away from the administration as much as possible—." After her bruising experiences in the late 1860s and early 1870s, Mitchell must have renewed her resolved "to be loyal to the stars." The experiences of Mitchell and her colleagues leaves us with a sense of déjà vu. These same questions are being discussed by committees on the status of women in major scientific societies in the 1990s.[40]

An American Heroine

In the early nineteenth century, writers and intellectuals sought to create heroes to serve as inspirational role models for the new American nation.

37. Selections from the Trustees Minutes, 3 June 1870, p. 2. Mitchell Papers, VCA.

38. Selections from the Trustees Minutes, 19 May 1871, p. 2. Mitchell Papers, VCA.

39. Maria Mitchell to Benson J. Lossing, 16 December 1871, and Mitchell to Lossing, 3 July 1872. Mitchell Papers, VCA.

40. A flyer circulated at the 172nd meeting of the American Astronomical Society (June 1988) by the Committee on the Status of Women in Astronomy raises many of the same questions that concerned Mitchell. See "Issues and Questions: An Open Meeting of the CSWA."

Parson Weems began this process with mythic tales of George Washington, while Andrew Jackson became the most popular hero in the early Victorian era. The Romantic movement, with its emphasis on great men, stimulated these tendencies in American culture. Before Maria Mitchell, however, there were no heroines or scientists in the national pantheon.[41] Only after World War I would America take another woman scientist (Marie Curie) to its heart (Rossiter 1982: 123–28). Mitchell's legacy is best understood if we consider her the first American heroine of science.

Maria Mitchell's rise to the status of an American heroine began on the night of 1 October 1847 when she discovered a comet. The discovery of a comet by an American caused a sensation. The new nation, determined to prove it was the equal of Europe, eagerly sought evidence of national intellectual prowess. Mitchell's discovery fed this cultural nationalism, demonstrating that America could compete with Europe in science. There was, however, a complicating factor. The great discovery had been made not by a man but by a woman. Americans would have to accommodate to this curious circumstance. Women were supposed to be minding buttons and breakfasts, not scanning the skies from a well-equipped observatory in the back garden.

One of the most interesting ways Americans legitimated this remarkable event was to link Mitchell with famous European women astronomers. As a story in the *New York Mirror* reported (Wright 1949: 81), "Down on the sea-girt Island of Nantucket lives one of the most remarkable women of our day, Miss [Maria] Mitchell, who might be the companion of Mrs. [Mary] Somerville and Miss [Caroline] Herschel, or any astronomers of the other sex." The English-born Somerville (Patterson 1983) translated and edited Laplace's great treatise on celestial mechanics. Caroline Herschel (Lubbock 1933) was sister to William Herschel and served as his assistant as well as becoming a successful observational astronomer in her own right.

Mitchell's fame grew in the 1850s and when, at the end of the decade, she returned from a European tour, a group of New England women presented her with a 5-inch Alvan Clark refracting telescope. This event is analogous to the much larger and more expensive drive by American women after World War I to purchase a gram of radium to present to Marie Curie on her visit to the United States.

The astronomer's status as a public figure is nowhere better illustrated than in the story (Wright 1949:70) of Mitchell and the butcher boy on the cars. On a railway journey, the boy selling newspapers paused to inspect the Quaker lady with deep set eyes. "Be you Mrs. Stowe?" he asked.

41. On the tendency to hero worship in American culture, see Wecter (1941). For an example of an American heroine (Jane Addams), see Davis (1973).

Mitchell replied no, she was not the writer, Harriet Beecher Stowe. Later the boy stopped again. "Be you Mrs. Stanton?" Again, no. Mitchell was not the leader of the women's rights movement. Finally, in frustration, the boy asked "Who be you then?" Mitchell told him her name. "I know'd you be somebody" the boy responded triumphantly.

Maria Mitchell as myth and symbol was interpreted to fit changing cultural needs. Post–Civil War American women moved, for the most part, away from reform and toward domesticity. Thus the hard edges of the image (Mitchell as competent scientist and independent woman who fought for equal pay) had to be softened. Maria Mitchell became the nurturer and mother figure to generations of Vassar students and, by extension, all American college women.

Mary King Babbitt (Vassar, 1882) provided the fullest exposition of Mitchell as myth and symbol for late Victorian women. At the beginning of her thoughtful analysis, Babbitt (1912:5) emphasized a startling point. "Far more interesting than anything Mitchell ever did [in science] was Mitchell herself." And again, toward the end of the discussion, Babbitt (1912:31) quotes a Vassar student. "It is not more astronomy that we want; it is more Maria Mitchell." Mitchell the scientist became less important than had been the case before the Civil War. Mitchell was presented as an inspired if unorthodox teacher, whose goal was to instill mental discipline and the habit of careful observation in students.

A central theme in Babbitt's discussion is domesticity. Mitchell's apartment in the observatory was characterized as simple and free from "fuss" (clutter, knick-knacks, etc). Babbitt (1912:22) recalled that there were "just two books on the parlor table; one was a very abstruse astronomical work, I do not know what; the other, as I well remember, was 'Twenty-Four Easy Patterns in Crochet.' " Babbitt sketches the scene in Mitchell's parlor, the professor and a few students in quiet conversation. "It was over crochet, and sometimes over tatting, that the intellect of our professor of astronomy was wont to unbend itself."

Babbitt portrayed Mitchell as a gifted female scientist who abandoned a research career to help other women. She had been convinced by the importance of the Vassar experiment to

> Forego thy dreams of lettered ease,
> Put thou the scholar's promise by,
> The rights of men are more than these.

Of course, as Babbitt made clear, it was the rights of women that were the issue for Mitchell. Babbitt's construction of Maria Mitchell's image begins by divorcing her from science. She is pictured in the parlor crocheting, not in the dome working with the 12.5-inch refractor or timing transits of stars with the meridian circle. We see Mitchell, as a teacher, leading women into more precise ways of thought, not instructing them in celestial mechanics or observational astrometry. Perhaps the most important element in the myth involves Mitchell sacrificing her scientific career for the good of other women.

It was in the annual dome parties that Maria Mitchell's image as mother figure and nurturer found its fullest development. Early in her Vassar career, Mitchell instituted the practice of inviting the undergraduate astronomy students, as well as alumnae who had taken her classes, to a very special breakfast at the observatory. In time, this event must have become one of the most important features of commencement week.

Tables and chairs were placed on the observatory porch and in the dome and transit rooms. Strawberries, coffee, and rolls made up the bill of fare. After the meal the college glee club might sing, but the center piece of the morning was the arrival of Mitchell with a large basket under her arm. In it were poems, written by the professor, describing each of her current students. The basket was passed around and verses drawn at random. Then one of the students was called upon to read the poem she had drawn. The person about whom that verse was written then read next, until all had taken part. Pencils and paper were then passed out and the women were invited to respond with their own extempore verses. Over the years, undergraduate women like Ellen Swallow Richards, chemist and founder of the home economics movement, psychologist and philosopher Christine Ladd-Franklin, and astronomers Antonia Maury and Mary Whitney sat at the tables under the dome and enjoyed the nurturing support and attention of Maria Mitchell.

Mitchell made no pretence at poetic skill, but her verses did, apparently, capture something basic about many of her students. This suggests the depth and complexity of Mitchell's relations with undergraduates as well as her skill in understanding character and personality. It was also enormously flattering for those about whom the professor took time to compose rhymes. Many Vassar women treasured the memory of Maria Mitchell's dome parties to the end of their lives.[42] In very private ways, these memories must have been of great importance to Vassar graduates who did not follow Mitchell's path of independence and scholarship but rather married and raised families. The softer image of Mitchell helped to legitimate their choice of marriage over a career and, at the same time, provided a private world of memory in which they could imagine roads not taken.

Babbitt (1912: 18) attempted to capture the importance of Mitchell for students in the 1880s. Mitchell offered women an escape from stereotypes. Women, who had been socialized "to feel that inaccuracy is amus-

42. I have relied on Babbitt (1912: 22–25) in this discussion of dome parties. In the Mitchell papers is a copy of Maury's "Dome Verses" written, perhaps, on one of these occasions.

ing, and that to be superficial is to be womanly" found in Mitchell's classroom a different role model. "Concentration, exactness, thoroughness" were inculcated and exemplified day after day. But there was something else. Mitchell's teaching put women in touch with larger views, with the vast and limitless subject matter of astronomy. "To women, then as now, too often wearied and narrowed by 'the trivial round, the common task,' she gave the inspiring vision of infinite spaces and cosmic law. What more could we ask?" As college women faced the choice between career and marriage, this image of Maria Mitchell, with its emphasis on nurturing and sacrifice, helped reconcile them to being educated women in what remained a male-dominated culture.

There is one other point. Out of the dome parties came a poem about Maria Mitchell that found its way into the *Vassar* College Song Book. Sung to the tune of "The Battle Hymn of the Republic," these stanzas form a remarkable tribute. In six stanzas, the song recapitulates the changing image of an American heroine. Only the first two verses deal with Mitchell as a scientist and professor of astronomy.

> We are singing for the glory of Maria Mitchell's name,
> She lives at Vassar College, and you all do know the same.
> She once did spy a comet, and she thus was known to fame,
> Good woman that she was.

The chorus is similar to that of the "Battle Hymn of the Republic" except that the last line of each repetition is a variation of "Good woman that she was.

In the second stanza, we learn of Mitchell the teacher.

> She leads us thro' the mazes of hard Astrono*my*,
> She teaches us Nutation and the laws of Kepler three,
> Th' inclination of their orbits and their eccentrici*ty*
> Good woman that she be.

Stanza three deals with Mitchell as a feminist and refers to her activities in connection with the Association for the Advancement of Women, of which she was a founding member.

> In the cause of woman's suff(e)rage she shineth as a star,
> And as President of Congress she is know from near and far,
> For her executive 'bility and for her silver ha'r
> Good woman that she are.

The final three stanzas deal with Mitchell's personal qualities beginning with her strength, gentleness, and calm, while the fifth extols her virtue. The closing stanza is an unabashed hymn to Mitchell's domestic qualities, especially her skill at nurturing undergraduates.

> Sing her praises, sing her praises, good woman that she is,
> For to give us joy and welcome her chiefest pleasure 'tis;
> Let her name be sung forever, till through space her praises whiz,
> Good woman that she is.

It is perhaps characteristic of Mitchell that she once remarked, "I have always liked that song because it is true. I am not a wise woman, but I am a good one."[43]

The honors and power that came to elite male astronomers was of a different order than that enjoyed by Maria Mitchell. A large painting of George Ellery Hale looks down from the wall of the ornate Member's Room at the National Academy of Sciences, but neither Hale nor any other male astronomer ever earned the status of an American cultural hero. Students never composed songs abut them. That was reserved for Maria Mitchell alone.

THE EDUCATION OF WOMEN ASTRONOMERS

A little more than a third of the women who worked in the American astronomical community attained at least a baccalaureate degree (see table 9.4, below). The vast majority of these were earned after 1900. For the other 60 percent, it is probable that formal education ended with common school or a high school certificate. The following discussion focuses on women who earned undergraduate or graduate degrees before taking a first full-time position in astronomy. Their experiences were far different from those of men discussed in chapter 4. The undergraduate years, most often at one of the women's colleges, were far more important in the intellectual and psychological development of most of these women than experiences they had in graduate school. It was as undergraduates that they were awakened to science and discovered their intellectual potential. Working closely with female mentors, these young women formed their basic scientific tastes and styles. In comparison to male astronomers, many of these women found the undergraduate experience far richer and more rewarding. Undergraduate mentors would remain more important than male dissertation supervisors and the undergraduate years would take a central place in the memories of these women.

The Women's Colleges: A Self-Conscious Culture

The women's colleges were self-conscious about their purpose and mission in ways (Horowitz 1987; Gordon 1990) that most other private, or in-

43. Mitchell, quoted in Babbitt (1912: 26). The song is found on page 94 of the 1881 edition of the *Vassar College Song Book*.

deed, public institutions were not. Pre–Civil War colleges like Oberlin, rooted in the evangelical-reform tradition, had a clear sense of mission, and the post–Civil War land-grant university movement was guided by a well-defined sense of purpose, but neither of these traditions encompassed the forces that shaped the women's colleges. Reform and public service were part of the ethos of the women's colleges, but there was something else. They saw themselves and were seen by society as experiments. On their success or failure rested the answer to a question much debated in Victorian America: can women be educated to the same level as men without disastrous physical and psychological consequences? In a frequently quoted passage (Wright 1949: 133), a Massachusetts physician insisted that "You can not feed a woman's brain without starving her body. . . . Open the doors of your colleges to women," wrote Dr. Edward Clarke, "and you will accomplish the ruin of the commonwealth." The Reverend John Todd joined the negative chorus. Women did not have the intellectual or physical stamina that would permit them to become "Newtons, Laplaces and Bowditches in mathematics and astronomy." Higher education was not part of God's plan for women. The Almighty seemed more interested in buttons and breakfasts. To be sure, there were supportive voices. Dr. Alexander Wilder argued that "There is no science which a man can learn that is impossible or improper for a woman." Wilder cited Maria Mitchell as a case in point.

Annie Jump Cannon (Wellesley, 1884), who spent more than forty years at the Harvard College Observatory, recalled in a speech at her fiftieth class reunion, the sense of excitement and challenge that surrounded her undergraduate years. On her first day at Wellesley, President Durant insisted that "You must take calculus," and "calculus seemed worth striving for."[44] Cannon and her classmates "realized that we were tools in a great experiment, believed in by few, doubted by even the wisest educators." The object of the experiment was to see "whether woman's health could endure years of study." Cannon remembered that "I heard President Eliot [of Harvard] declare before the Wellesley College Alumnae that his former doubts on that subject were dispelled by one glance at the women present."[45]

There was, however, another aspect to the mission of the women's colleges, an aspect that developed at the department level. Astronomy departments soon came to believe that a major reason for their existence was to supply factory observatories with women who had baccalaureates or master's degrees, to serve as computers; or to send a few of their graduates

44. Annie Jump Cannon, "50th Reunion, Wellesley College—June 1934," p. 2. Alumnae Biographical Files, Class of 1884: Annie Jump Cannon, WCA.

45. Ibid., p. 3.

on to take the doctorate in astronomy and, quite frequently, see them return to teach at their alma maters.

Cannon touched on this aspect of mission in her address. In the closing decades of the nineteenth century, astronomy "was undergoing the greatest transformation since the invention of the telescope."[46] Cannon pointed to the introduction of photography and the growth of astronomical spectroscopy as essential elements in the transformation process. She referred to Edward C. Pickering, HCO director and pioneer in the development of the factory observatory, as "a firm believer in woman's ability to do astronomical work, especially since photography transformed it into a daylight profession."[47] There was a demand for the product of these departments, and astronomers at the women's colleges viewed meeting this demand as an important part of their mission.

By the mid-1880s, Maria Mitchell was well aware of the important role her department played in supplying assistants to research observatories. She suggested, for example, that the astrophysical work of the late Henry Draper be continued by his widow with the aid of "a corps of lady assistants under your own [Mrs Draper's] direction," and that Mitchell "would be only too delighted to assist you in making a selection from the many bright young ladies that she has had under her instruction in astronomy."[48] In the end, the work was endowed by Anna Palmer Draper as the Henry Draper Memorial at the HCO and became an important employer of women, including Fleming and Cannon.

Half a century later, Professor Maud W. Makemson, chair of the Vassar Astronomy Department, demonstrated continuing concern for the mission of her department. A Vassar graduate reported that Walter S. Adams, Mount Wilson director and longtime employer of Vassar graduates, was critical of the program. Apparently, Adams believed that physics should play a greater role in undergraduate education. Makemson wrote to Adams, asking for clarification. Defending the undergraduate astronomy program at Vassar, Makemson stressed "general observatory experience" including the use of telescopes, the transit, spectroscope and spectroheliosope, and astronomical photography as "the best possible background for the student," a background that "enables her to understand observatory methods." Makemson, a California Ph.D. in celestial mechanics, continued, "For more specific training, the measurement and reduction of photographic plates and spectrograms and the computation of an orbit to enable her to acquire the technique of computing are of the

46. Ibid., p. 7.

47. Ibid., p. 8.

48. Daniel Draper to Anna Palmer Draper, 12 May 1885, pp. 1–2. John W. Draper Family Papers, LC.

greatest possible value." Makemson then confronted the main issue. "I can not imagine anything a student gains by majoring in physics that could offset these advantages."[49] She indicated that astronomy majors were required to take a two semester introductory course in physics as well as advanced work in light.

The response from Adams is instructive. He first paid homage to the tradition represented by astronomy departments in the women's colleges. "I think that anyone who comes to a large observatory should have just the kind of astronomical experience which you mention, and nothing can quite take its place." However, Adams entered a caveat. "In these days of intensive spectroscopy the more that one can learn about optics and the theory of spectra, the better it is . . . and the more knowledge a student can obtain in physics, especially in the theory of radiation, the more successful she is likely to be."[50] In spite of his compliments, Adams was raising important questions about the undergraduate preparation of these women. By the 1930s the offerings of the astronomy departments in the women's colleges may have appeared somewhat behind the times.

Alice H. Farnsworth (1893–1960), longtime member of the Mount Holyoke faculty, was asked in 1940 to report on developments in her department and job opportunities for those with degrees in astronomy. Farnsworth answered with a discussion of the careers of the last twenty astronomy baccalaureates. Half the group "have held positions in observatories or done grad[uate] work or both," Farnsworth reported to the Dean. Her second category involved the domesticity issue discussed earlier. Five "have married an astronomer or physicist as a direct consequence of [the] environment [that is, where they were employed]." Three graduates went on to teach in colleges and two in secondary schools. Two others used their skills working for science-based industries like Eastman Kodak. For six of the group, there was no information.[51]

Farnsworth went on to describe employment opportunities for astronomy graduates. She divided their options into three categories. The first included positions as computers or assistants in observatories. Farnsworth characterized the work as "interesting and [the] intellectual atmosphere stimulating" but cautioned that the pay was "20 to 25 dollars a week, with little hope of more later." These positions required a baccalaureate degree. There were a few college teaching positions open for women with master's degrees. Astronomy doctorates might find employment as research astronomers. The pay was described as moderate, but the work

49. Maud Makemson to Walter S. Adams, 24 April 1936. Adams Papers, MTWA.

50. Walter S. Adams to Maud Makemson, 28 April 1936. Adams Papers, MTWA.

51. Alice H. Farnsworth to Dean Harriet M Allyn, 15 March 1940. See also Allyn to Farnsworth, 6 March 1940. Farnsworth Papers, MTHCA.

Table 9.4 Leading Producers of Baccalaureate Degrees for Women in the American Astronomical Community, 1860–1940

Second Cohort, 1860–1899 (N = 34) Five institutions produced 68% of all undergraduate degrees for women.		**Third Cohort, 1900–1940 (N = 122)** Nine institutions produced 73% of all undergraduate degrees for women.	
Vassar	11 (32.4%)	Vassar	21 (17.2%)
Carleton Mount Holyoke Smith Swarthmore	3 @ (8.8%) for a total of 12 (35.3%)	Radcliffe	14 (11.5%)
The rest are scattered between eight institutions.		University of California	12 (9.8%)
		Smith	11 (9%)
		Mount Holyoke Swarthmore	9 @ (7.4%) for a total of 18 (14.8%)
		Northwestern	5 (4.1%)
		Wellesley Cornell	4 @ (3.3%) for a total of 8 (6.7%)
		The rest are scattered between twenty-seven institutions.	

fascinating. There were, however, "very few openings for women." Those interested "should have a very good mind, persistence and a Ph.D."[52] Farnsworth indicated that a position as a computer or assistant often provided the funds for graduate study.

Table 9.4 reports the leading producers of baccalaureate degrees for those women who entered the American astronomical community. For the second cohort (1860–1900) the number was relatively small. Three of the women's colleges (Vassar, Smith, and Mount Holyoke) were among the leading institutions. After 1900 these institutions continued to play an important part in the production of women who entered American astronomy, but they were joined by Radcliffe, Berkeley, Northwestern, and Cornell. Carleton and Swarthmore were coeducational colleges that made significant contributions to the pool of female astronomers.

At the center of the undergraduate astronomy program was the observatory. While other departments might share space in various college buildings, the observatory was freestanding and self-contained. Classrooms, a library, faculty offices and often the apartment of the director and quarters for one or more majors were all housed under a single roof. In addition there were the instruments: a transit circle and a large refracting telescope as well as smaller portable instruments to be taken out on the college lawn for use by students in the introductory course.

The separate and self-contained nature of astronomy at the women's colleges gave majors a special place among undergraduates. Seen through

52. Farnsworth to Allyn, 15 March 1940. Farnsworth Papers, MTHCA.

the lenses of Victorian romanticism, the separatist nature of astronomy tended to carry a certain element of glamour. Outsiders could only see the white dome and the ivy covered walls of the observatory. They had little idea of the difficulties associated with working out an orbit or making a series of micrometer measurements of the position of a comet on a January night in New England.

This romantic conception of an astronomy major is what Antonia Maury had in mind when she penned "Dome Verses."[53]

> A low-built tower and olden
> Dingy but dear to the sight,
> And they that dwell therein are wont
> To watch the stars at night.
>
> A wide and ample chamber,
> Where a clock on a marble pier
> With low beat follows the stars that track
> The slow-revolving year.

For Maury, astronomy students and faculty comprised a special family within the larger college community.

> And all that dwell in that chamber
> Or cross but its stone threshold,
> Whether for use or knowledge,
> All things in common hold.

This family was bound by the study of astronomy as a romantic discipline.

> And they that scan the heavens by night,
> Since truth's clear light they saw,
> No human meets and measures serve,
> But Nature's mightier law.

Knowledge of "Nature's mightier law" was gained at the telescope, where, "with searching glass I scanned / Those far, deep lanes of night." This-knowledge was also found "In woodlands wild or wind-blown dale / Far hid in the forest's girth" whence undergraduates roamed like the Sky Maiden.

Romantic philosophy of nature, as described by Maury, provided a bond that united all who studied at the observatory. These women knew nature in privileged ways; like gnostics, they had a special vision.

> For a nameless light and freedom
> Lifts the rude walls away
> And Nature blends with the spirit
> In love's dissolving day.

53. A. C. Maury, "Dome Verses." Mitchell Papers, VCA. This is a typescript and undated.

It is improbable that undergraduates, working at the Student's Observatory at Berkeley, under the supervision of Armin Leuschner and his graduate students, ever viewed their experiences in quite this way. Nor did the young Mary Byrd (1886: 264), speaking at the dedication of the Goodsell Observatory at Carleton College. For her, astronomical observation with the transit circle or filar micrometer was anything but romantic. It was hard work, demanding manual dexterity and physical stamina. The acquisition, reduction, analysis, and discussion of observational data were intellectually demanding and left little room for poetry.

But that is precisely the point. In other institutional contexts, the undergraduate study of astronomy was neither isolated nor romanticized. At both Berkeley and Carleton, the astronomy and mathematics departments were closely allied, and at California engineering students were required to take astronomy courses. Also, at both institutions, there were graduate students who served as a kind of practical, career-oriented and competitive leaven. In contrast to Berkeley or Carleton (during the latter's short but exciting period as a producer of graduate students in astronomy and mathematics), it should be recalled that astronomy students at the women's colleges were doubly isolated. The college was isolated from the larger society as it carried out its daring experiment in female education and the astronomy program was structurally and psychologically isolated from much of what went on within the college.

To say these things about astronomy at the women's colleges is not to suggest that either the faculty or students were deficient in mathematics, observing techniques, or a generally tough-minded, empirical approach to science. This discussion does, however, help us understand why these women looked back on their undergraduate years as unique and, on the whole, more important than graduate school. The sense of being a member of a select community, working together to understand nature, provided a bond that could not be replicated in the competitive, male dominated graduate school. Working side by side with their professors, astronomy majors at the women's colleges early developed a taste for scientific problems and a style for attacking them. In this sense, their professional skills were defined earlier than those of male astronomers, who generally had to wait for graduate school or the first job to provide mentors who would shape their approach to science.

Astronomy Education at the Women's Colleges

As table 9.4 indicates, Mount Holyoke, Vassar and Smith were major producers of women who entered the American astronomical community, supplying computers to the factory observatories and women who went on to take the doctorate in astronomy. Astronomy at Mount Holyoke takes pride of place. From its opening in 1837, founder Mary Lyon insisted that science should be a part of the education of women, and there was an astronomy unit in the course of study she devised. Soon Mount Holyoke students were using a 6-inch refractor to view the heavens. In 1881 the John Payson Williston Memorial Observatory was dedicated. It included an 8-inch Clark refractor and a 3-inch meridian transit. In 1902 a lecture room was added to the observatory. The 8-inch was modernized in 1929 and a Ross camera added in 1932.[54]

In 1866 Elizabeth Miller Bardwell (1831–99) was appointed to teach astronomy and mathematics and, in 1888, became professor of astronomy and director of the observatory. In 1869 solar physicist Charles A. Young of Dartmouth instituted a biennial course of lectures at Mount Holyoke, a practice he continued after moving to Princeton. Anne S. Young (1871–1961) took over from Bardwell and served until 1936. Alice H. Farnsworth, who earned a doctorate at Chicago, became professor and director of the observatory on Young's retirement. Bardwell and Farnsworth were both graduates of Mount Holyoke.

Beginning in 1887, the astronomy curriculum included an introductory course for juniors and a senior course on mathematical astronomy and the use of observatory instruments. This was much the same patten that obtained at Vassar under Maria Mitchell. A major reorganization of astronomy offerings took place at Mount Holyoke in 1893. A sequence of six courses was instituted, beginning with a two-semester introductory sequence (A1-A2). Astronomy 3 dealt with descriptive astronomy and A4 provided a historical overview. The capstone courses were A5 and A6, advanced mathematical astronomy and observatory training. Beginning in 1895, students were permitted to major in astronomy. In 1903 a course on celestial mechanics was added. Only after World War I was photographic astrometry added. Students used a Gaertner measuring machine to determine the positions of stars and asteroids. In 1924 work with the measuring engine was extended to include the measurement of spectrograms. The following year, A5-A6 were revised to include the study of astrophysics.

Astronomy at Mount Holyoke started as a one-woman department and slowly expanded. In 1901 a graduate assistant was added and in 1913 a full-time instructor. Some of these women went on to do graduate work in astronomy, others left for positions in one of the great factory observatories. Professor Anne S. Young (B.S., 1892, M.S., 1897, Carleton College; Ph.D., Columbia, 1906) described faculty research at Mount Holyoke. Her discussion began with an important qualification. "While the resources of the observatory are primarily devoted to the purposes of un-

54. Anne S. Young, "Annual Report" (1919, p. 23). See also Alice H. Farnsworth, "Astronomy at Mount Holyoke, 3 July 1950," p. 3. Farnsworth Papers, MTHCA.

dergraduate study, as much regular observing is done as is possible."[55] In fact, most research projects were selected in order to provide opportunities for student participation.

Research projects were of a routine nature (i.e., not driven by major theoretical considerations relating to either astrometry, celestial mechanics, or astrophysics) and within the capabilities of available instrumentation. The department provided time to the college. The 3-inch transit was used for accurate determination of time in order to correct the master clock in Mary Lyon Hall. In 1900 the department started collecting sunspot data as part of an international program. Observation of long-period variable stars was another departmental project. Micrometer measurements were made of the four bright jovian satellites. Reduction and analysis of data collected in these various projects were carried out elsewhere.

Young admitted that one important field of instruction and research was closed to students and staff at Mount Holyoke. There was no instrumentation suitable for astronomical photography and "no photographs are made here."[56] However, faculty and students used plates lent by the Yerkes Observatory to determine the positions of comparison stars for the great variable star atlas (*Atlas Stellarum Variabilium*) being prepared by the Vatican Observatory. They also measured the proper motions of long-period variables using Yerkes plates.

It is important to understand the commitment to teaching held by astronomers at Mount Holyoke and the other women's colleges. In the *Annual Report of the Department of Astronomy and the John Payson Williston Observatory for the Year 1934–35*, Professor Young noted that "we have continued our policy of giving a course, even though the registration is limited to a very few or in special cases one student. This does, of course, make a very heavy demand on our time and cuts out the possibility of a research program while college is in session."[57]

Samples of examinations in the Mount Holyoke archives, suggest the depth of technical knowledge and understanding of observatory procedures expected of students. Mary Hunt's 1936 honors examination was devoted to photographic photometry. Questions ranged from the physics of the photographic plate, observing techniques, and methods of reduction of photometric data, to the process of establishing magnitude sequences. After the written portion, the candidate faced an oral examina-

55. Young, "Annual Report" (1919, p. 22). Copy in the Astronomy Department Papers, MTHCA.

56. Young, "Annual Report" (1919, p. 23). See also Alice H. Farnsworth, "Astronomy at Mount Holyoke, 3 July 1950," p. 3. Farnsworth Papers, MTHCA.

57. Young, "Annual Report" (1934–35, p. 1). Copy in the Astronomy Department Papers, MTHCA.

tion. Hunt's honors project was on photographic photometry and she was cautioned to be ready to take the committee to the dome and demonstrate her use of the equipment.[58]

Alice H. Farnsworth, Young's successor at Mount Holyoke, graduated from high school in 1910 with highest honors. She went on to take an A.B. (1916) and M.A. (1917) at Mount Holyoke before moving to Chicago for the doctorate (1920). Farnsworth returned to her alma mater in 1920 and, with the exception of a research leave spent at Lick (1930–31), remained for the rest of her life. In the Mount Holyoke archives there is a notebook covering Farnsworth's high school and undergraduate years. The entries began with French grammar and moved on to algebra. Latin was added and geometry joined algebra. Soon we find the high school student working on lists of German verbs; advanced algebra was mingled with English history. At some point, perhaps late in high school or as a freshman, Farnsworth read a book by Garrett P. Serviss, one of the great popularizers of observational astronomy. Soon, her notebook indicates, she was reading widely on astronomy. The end papers were reserved for lists of "Books to Look Up." These included poetry about the sky, books on religion, and observing manuals.[59] Her reading was both wide and deep, including not only mathematics and astronomy, but the humanities as well. It is especially interesting to watch Farnsworth's interest in astronomy grow as she moved from popular books to specialized monographs.

Astronomy at Maria Mitchell's Vassar moved quickly from a single course for juniors to a sequence for juniors and seniors. The first year provided an introduction to astronomy and the second was devoted to mathematical astronomy and the use of observatory instruments. Mitchell clearly expected her students in the senior course to be well versed in mathematics. When Mary Whitney took over from her teacher, she introduced changes in the curriculum. Whitney, who earned an A.B. (1868) and M.A. (1872) at Vassar gained practical experience as a volunteer observer at the old Dearborn Observatory in Chicago, before returning to Vassar in 1881. She succeeded Mitchell in 1888. In 1890, A1 was offered for sophomores and proved popular. The junior course became an introduction to mathematical astronomy and the senior course dealt with observational astronomy. Soon a sequence of six courses made up the undergraduate major. Most notably, Whitney instituted a senior course on the solar spectrum, the first indication that the new astronomy was being taught to undergraduates at Vassar.

In 1894, Caroline E. Furness (1869–1936) joined the Vassar staff as

58. "Honors Examination in Astronomy. Mary Hunt. Written Portion. Friday 29 May 1936, 9 AM. 1½ hrs." Astronomy Department Papers, MTHCA.

59. Alice H. Farnsworth, Notebook, ca. 1901–13. Farnsworth Papers, MTHCA.

an assistant. Furness took her undergraduate degree with Whitney in 1891 and earned a Columbia Ph.D. in 1900. She spent research leaves as a volunteer assistant at Yerkes and also at the Kapteyn Astrographic Laboratory at Groningen in the Netherlands. Only in 1915 did Furness achieve the rank of full professor. In 1897, she added astrophysics to the Vassar offerings.

Whitney introduced the first graduate course in 1900. It dealt with astronomical spectroscopy and the measurement and reduction of stellar photographs. Here, Vassar led the way. It would be twenty years before Mount Holyoke offered such a course. Smith would not teach this subject until 1923, and Wellesley apparently never did. Courses that introduced majors and graduate students to methods for measuring astronomical photographs made these women especially desirable as computers in factory observatories and prepared them for graduate work at institutions like Lick or Yerkes.

In 1901 the offerings were revised and expanded. Two semesters of introductory astronomy were followed by two of observational work. Mathematical astronomy and the use of instruments the subjects of three courses. A readings course, which introduced students to the research literature, capped the undergraduate sequence. New graduate courses were also added. One dealt with the calculation of cometary orbits and the other with variable stars. Furness became a recognized authority on the study of variables. Later, an undergraduate course on variable stars was added and work in mathematical astronomy expanded to three semesters. Through the 1920s Furness continued to modify the curriculum. Readings in astrophysics appeared in 1922, and in 1925 a course was added that emphasized the study of solar and stellar spectra and the measurement of spectrograms.

The range of offerings, the fact that courses kept pace with developments in the field (including the demands of the factory observatories), and the quality of both faculty and students help explain why Vassar was the leader in the production of women who entered the American astronomical community.

Astronomy at Smith College began with the opening of the observatory in 1887. Equipped with an 11-inch refractor, the observatory and department were directed by Mary Byrd (1849–1934), who held an undergraduate degree from Michigan (1878) and a doctorate from Carleton College (1904). Byrd published extensively on what today would be called astronomy education and was especially concerned with developing laboratory methods for teaching the introductory course. Before coming to Smith in 1887, Byrd served as an assistant in astronomy and mathematics at Carleton and as a volunteer assistant at Harvard. Byrd left Smith because she did not think it morally correct for the college to take funds from Andrew Carnegie, even for such a worthy project as faculty pensions. She was succeeded by Harriet Bigelow (1870–1934), who earned an undergraduate degree in astronomy at Smith in 1893 and a Michigan doctorate in 1900. From 1896 to 1904 Bigelow served as an assistant and became an instructor in 1904. She was elevated to professor of astronomy and director of the Smith College Observatory in 1906.

The department at Smith was the largest among the women's colleges. Lois T. Slocom joined the faculty in 1932, replacing Priscilla Fairfield Bok (Berkeley Ph.D., 1920), who had served for a decade. N. Wyman Storer (b. 1900), a Berkeley doctorate, became the first man on the Smith faculty but remained for only a year. A series of instructors and demonstrators assisted in teaching astronomy at Smith. For example, Hazel M. Losh, was an instructor in 1925–26, before returning to Michigan where she spent the rest of her career. Mary M. Hopkins (1878–1922) took her undergraduate (1899) and master's (1911) in astronomy at Smith and a Columbia doctorate (1915). She served as instructor and assistant professor from 1911 until her suicide in 1922.

Astronomy at Smith tended to emphasize celestial mechanics and astrometry. Offerings changed slowly in comparison to Vassar. Only in 1924 did the catalogue list an introductory course in astrophysics while the year before, a course on the measurement photographic plates was introduced. To be sure, A7 and A8 were "mathematical courses involving theory, observations and reductions of high precision that would be recognized in any working observatory."[60] It appears that in comparison to Vassar, Smith's astronomy offerings remained relatively static.

Smith College astronomy was apparently the only program to maintain a support group in the form of an Astronomical Society composed of alumnae. The society met each year at commencement. Its annual newsletters provide an interesting window into the history of astronomy at Smith. In addition, there was an undergraduate astronomical society, *Telescopium*, open to majors and other interested students.[61]

At Wellesley, astronomy flourished under Sarah Frances Whiting. Students concentrated on spectroscopic astronomy and laboratory work in spectroscopy. The program also gave special emphasis to advanced observatory work in photometry and the study of variable stars as well as sunspots. By 1905, a master's degree in astronomy was available. Only in 1906 did Ellen Hayes (b. 1851) offer celestial mechanics at Wellesley. Hayes, an Oberlin graduate (1878) was a member of the Wellesley faculty

60. "Report to the Trustees, Astronomy Department, May 1909," p. 2. Astronomy Department Papers, SCA.

61. The annual newsletters of the Smith College Astronomical Society extend from 1900 to 1923. Astronomy Department Papers, SCA.

from 1904 until 1916. She was replaced by John C. Duncan, who is remembered as the author of one of the three most successful undergraduate introductory astronomy texts published between the wars. The Wellesley program remained small and did not produce as many students as the other women's colleges.

Radcliffe's role in producing women who took jobs in the American astronomical community is complex. The Willson Astronomical Laboratory was a part of Harvard's undergraduate astronomy program through World War I, but between the wars there were very few offerings for undergraduates (Hoffleit 1979: 18; Kidwell 1986). Of the Radcliffe women who entered the American astronomical community, all but one did so after 1900, and eight entered after 1920. It is doubtful if these were astronomy majors. Rather, they probably concentrated in mathematics and/or physics. All of the Radcliffe undergraduates who entered the American astronomical community went to the Harvard College Observatory for their first job. Radcliffe served as a source of computers for HCO in the later Pickering and early Shapley years. Distinguished alumnae include Doritt Hoffleit (b. 1907), class of 1928, who went on to earn a Harvard Ph.D., and Henrietta Swan Leavitt (1868–1921), who graduated in 1892 and spent her career at the Harvard College Observatory.

Table 9.5 indicates institutions that produced women with master's degrees in astronomy, who entered the American astronomical community. For the second cohort, only five women earned master's degrees. Vassar and George Washington University in Washington, D.C., led the way, followed by Johns Hopkins University. For the third cohort, the number of women earning the master's rose sharply. Seven institutions produced nearly two-thirds of the master's degrees. Radcliffe led the way, followed closely by the University of Chicago and the University of California. Smith College stood in fourth place while Carleton, Swarthmore, and Vassar tied for fifth. The rest were scattered across twenty institutions. That Smith, Carleton, Swarthmore, and Vassar should produce almost a quarter of the master's in astronomy is, at first glance, surprising. However, it indicates an interesting pattern. Women, most often with baccalaureate degrees from one of the women's colleges or a coeducational college like Swarthmore or Carleton, appeared to delay the move to a major university for graduate work until the Ph.D. level.

Between 1860 and 1900 only six women earned doctorates (table 9.6). Half of these came from the short-lived Carleton College experiment in graduate education. Yale produced two and the final doctorate came from the Sorbonne. For cohort three, the number of doctorates in astronomy increased more than sixfold. Five institutions produced 90 percent of the female Ph.D.'s with California leading the way (36 percent) followed by Michigan (20.5 percent) and Radcliffe (15.4 percent). Chi-

Table 9.5 Leading Producers of Master's Degrees for Women in the American Astronomical Community, 1860–1940

Second Cohort, 1860–1899 (N=5)		Third Cohort, 1900–1940 (N=73) Seven institutions produced 62.8% of the master's degrees.	
George Washington University Vassar	2 @ (40%) for a total of 4 (80%)	Radcliffe	12 (16.4%)
Johns Hopkins	1 (20%)	University of Chicago	9 (12.3%)
		University of California	8 (11%)
		Smith	5 (6.8%)
		Carleton Swarthmore Vassar	4 @ (5.5%) for a total of 12 (16.5%)
		The rest are scattered between twenty institutions.	

Table 9.6 Leading Producers of Ph.D. Degrees for Women in the American Astronomical Community, 1860–1940

Second Cohort, 1860–1899 (N=6)		Third Cohort, 1900–1940 (N=39) Five institutions produced 90% of Ph.D. degrees	
Carleton	3 (50%)	University of California	14 (35.9%)
Yale	2 (33.3%)	University of Michigan	8 (20.5%)
Sorbonne	1 (16.7%)	Radcliffe	6 (15.4%)
		University of Chicago	4 (10.3%)
		Columbia University	3 (7.7%)
		The rest are scattered between four institutions.	

cago (10.3 percent) and Columbia (7.7 percent) bring up the rear. This distribution reflects some of the patterns that were discussed in chapter 4. California was the leading producer of male doctorates who entered the American astronomical community. Chicago came second. Harvard and Michigan were newer programs, producing the bulk of their doctorates after World War I. The astronomy program at Columbia was in decline for the first half of the twentieth century.

"But What Are Their Chances?" The Careers of Women in the American Astronomical Community

In addressing her fiftieth Wellesley class reunion, Harvard astronomer Annie Jump Cannon recalled that, during senior year, the English poet and critic Matthew Arnold came to lecture. "When Mr. Arnold was being driven up through the college grounds, he exclaimed, 'Extraordinary. All for young ladies.' Then, adjusting his monocle for a closer look, Arnold asked 'But what are their chances?' " Cannon insisted that "the chances were really excellent in those days. Many doors were open to women . . . the roads were not crowded."[62] Cannon's views were colored by her own remarkable talent and good fortune. In fact, the chances for a satisfying career in science depended on the gender of the intending scientist.

Understanding the careers of women in American astronomy involves confronting several fundamental tensions specific to the science. The first can be characterized as the tension between the role of computer and that of observer. The computer is a day job, carried out in an office. The women and men (there were a few) who worked as computers measured photographic plates using devices that ranged from a simple hand lens to a sophisticated measuring engine. At institutions like the Dudley or Naval observatories, computers reduced observations made with meridian circles, applying various corrections. The adjusted observations were analyzed and then prepared for publication in the form of catalogues.

Observers, on the other hand, might spend time during the day with various numerical computations and measuring plates, *as well as* work with telescopes at night. Understood in this way, the role of the observer was active and that of the computer passive. From here, it was only a short step to explaining these differences as a manifestation of gender polarization. Men are active and strong, working at the telescope during the long night hours. Women are passive and delicate, not fit for night work and confined to an office developing the products of male observers.

Ironically, undergraduate education in astronomy, at least at the women's colleges, prepared students for both computing *and* observing. Mitchell, Byrd, Farnsworth, and Young all expected their graduates to know how to use telescopes for data acquisition as well as the best methods of data reduction and analysis. Women astronomy majors were socialized into the roles of both observer and computer. However, if they opted for a position as a computer at a factory observatory, their education proved, at least in part, dysfunctional. As computers they were excluded from the masculine world of observing, for which their college professors had so carefully prepared them.

62. Annie Jump Cannon, "50th Reunion, Wellesley College, June 1934." Alumnae Biographical Files, WCA.

The issue is even more complex. The special nature of astronomy education at the women's colleges endowed observing with a romantic aura. Antonia Maury's images in her dome verses make it clear that observing was more than simply learning to read a sidereal clock, calculating the hour angle of a celestial object, and setting the telescope. The manipulation of instruments led the women into a kind of romantic communion with nature. Although they became expert observers, they were socialized into this role in a far different psychological context than students taking astronomy as undergraduates in universities or coeducational colleges.

Thus the women were in a double bind. They had been trained as *both* computers and observers. And, to make matters more complex, they endowed the role of observer with certain romantic overtones. The prevailing male conception of women's work in astronomy cut them off from the telescope, even though they had received excellent training as observers. On taking jobs in factory observatories, graduates of the women's colleges found themselves using only half their education. They were, in a very real way, estranged from nature. This must have led to frustration and a sense of wasted talent and skill.

A second tension marked the careers of women in the American astronomical community. With few exceptions, avenues of mobility within the community were closed. Computers seldom rose to any higher status within an observatory. This stands in sharp contrast to some of the fundamental beliefs of American culture. Americans shared a deep faith in mobility as a reward for talent, training, and productivity. Yet, within astronomy, the conception of women's work created a ghetto from which few women ever escaped. This, together with the underutilization of women's skills, helps explain the level of frustration that must, from time to time, have overcome even the strongest of these women.

The consequences of these fundamental tensions are seen when we look at the allocation of women to various positions in the American astronomical community. Controlling for level of educational attainment, we find women with either a high school certificate, or at best an undergraduate degree, working in major research institutions while women with the doctorate are generally found in institutional locations where the potential for making serious research contributions was vanishingly small. In short, before 1940, the better-trained a woman, the greater the likelihood that she would spend most of her career in a collegiate teaching situation whose institutional potential was low. Even the most able and ambitious women were ultimately trapped by a system that allocated positions on the basis of gender.

The Caroline Furness papers in the Vassar College archives provide a window through which we can glimpse the ambition of a very able and articulate young woman. Writing to her family in the spring of her senior year, Furness stated her position forcefully. "I am ambitious, and if my health continues, there is nothing to stand in the way of my realizing my plans. In a year or two, I am going to Yale to study for the Ph.D. in mathematics and astronomy."[63] Furness was already in contact with a Vassar graduate working at the observatory in New Haven. Indeed, she believed that her Vassar contact would help her secure a position as computer in order to finance graduate study. Furness summed up the psychological value of ambition and education, "It is such an inspiration . . . to have something ahead of me."[64]

Perhaps one of the first indications of ambition came in 1889 when she asked her sister Mary for a specific Christmas present, a copy of William Chauvenet's classic handbook on practical and spherical astronomy. "I am really in earnest about my zeal for astronomy," and "I hope when I leave college I shall be fitted to assume active duties as assistant in some observatory."[65]

By senior year, Furness had set her sights higher. She no longer desired to be an assistant (computer) in an observatory, but rather for the Ph.D. and a teaching position. Apparently her widowed father and younger sister were troubled. They did not know what to make of her ambition and had little understanding of the world of professional astronomy. Further, they assumed that after graduation, Furness would return to Cincinnati to teach high school and care for her father. Clearly, this course of action did not appeal to the ambitious Vassar senior.

In the spring of 1891, Furness wrote her father a long and revealing letter. She began with a candid assessment of the limits the male-dominated astronomical community placed on female careers. She defined her goal as a teacher of astronomy, not a research astronomer. "The opportunities for becoming strictly an observer would never come to me. I aspire only to be a teacher."[66] Furness made it quite clear that she was talking about college teaching and that she needed the doctorate to achieve that goal.

Letters from her father and sister do not survive, but something of their attitude may be inferred from the lengths to which Furness went to explain herself. She did not want them to think her impractical, "studying

63. Caroline E. Furness to Mary and Papa, Sunday [Spring 1891], p. 2. Furness Papers, VCA. Shelda Eggers (1995) locates Furness in the context of women's history in a University of Missouri-Columbia senior honors thesis on the basis of a careful reading of the Furness papers at Vassar.

64. Furness to Mary and Papa, Sunday [Spring 1891], p. 4. Furness Papers, VCA.

65. Furness to Mary, 1 December 1889. Furness Papers, VCA.

66. Furness to Papa, 23 April 1891, p. 1. Furness Papers, VCA.

just for the sake of studying." She insisted that "I only want to prepare myself for the highest place—just as any young man might." Furness did not intend to "dilly dally along waiting for some young man to come along to marry me. . . . I want to prepare myself to live a useful & happy life without marriage, and then, if the right one comes along, well and good . . . but I shall not be obliged to take a man just for the sake of a home." Furness protested that she put as much value on domestic life as did her father and sister, but insisted "I do not wish to be a party to an unhappy marriage. Therefore, I want to be ready to live a useful & contented life without it."[67]

Furness brought the letter to a powerful rhetorical climax with the following argument. "Now, my dear Papa, do not think I am selfish in planning so for myself. I must make my own way in the world. I have ability and interest, and why should I not rise to as much distinction in my profession as any man. *If I were your son instead of your daughter, you would fully approve of my ambition.*"[68] The letter concluded with an analogy which the daughter hoped might soften her father's heart. "When I am a professor in some college, with a large salary, you shall live with me and be an advisor of all the girls. Just as Prof. [Maria] Mitchell's father lived here [at Vassar] with her."[69]

Twenty years later, one of Furness's students, Phoebe Waterman (A.B., 1904; M.A., 1906) was awarded a Vassar fellowship to pursue doctoral studies in astronomy at Berkeley. After the master's, Waterman spent a year as an assistant to Furness before moving to Mount Wilson (1909–11) as a computer. She earned the doctorate in 1913. Her scientific career came to an end when she married in 1914. Phoebe Waterman Haas remained active as an amateur throughout her long life, and after World War II reduced observations for the American Association of Variable Star Observers.

So well prepared was Waterman that the Berkeley faculty required that she take only three additional courses in mathematics and astronomy before beginning doctoral research on Mount Hamilton. Waterman's ambition differed from her teacher's. No less intense and probably as able, Waterman rejected teaching as a career goal. "But *oh*, I do want so much a position as astronomer, part of my work with the instruments and part with the reduction of my plates, as the men here [at Mount Wilson] have; I never did want the teaching." Clearly she had discussed this possibility with staff members at Mount Wilson, but the male astronomers were not supportive. "It is very bold & presuming of a woman to think of such a

67. Ibid., p. 2.
68. Ibid., pp. 3–4. Emphasis added.
69. Ibid., p. 4.

position I suppose!! They think so here."[70] Waterman was fired with ambition and looked forward to "a little share, just a little one, in the big work you are all busy in!"[71] We shall never know what went on in the mind of this talented woman when she gave up a career in astronomy. Perhaps it had something to do with the realization that she would never attain the status of observer. Marriage may have seemed a wiser choice than life in the ghetto of women's work.

The Careers of Women in Astronomy

Material drawn from the collective biography illuminates the careers of women in the American astronomical community. In developing these data, I have controlled for highest level of educational attainment as well as location of first job. A single job tended to be the norm for most women. Findings are suggestive; they indicate complex and often asymmetric patterns relating employment to highest level of educational attainment and point to issues concerning the tastes of employers and the ability of the women's colleges to supply semiskilled workers for the factory observatories.

Table 9.7 reports data on the employment of women whose highest level of education attainment was no more than a high school certificate.

For cohort two (1860–99) the Harvard College Observatory was the primary employer of women with this level of education. Eighty-five percent of the women in this class worked at Harvard. The rest were scattered between four institutions. During the last two decades of the nineteenth century, the HCO became the prototype of the factory observatory. Edward C. Pickering initiated large-scale programs in photometry and spectroscopy and utilized an unskilled labor force to reduce and analyze rapidly accumulating observational material.

Data for the third cohort reveal rapid growth in the size of this labor pool as well as changing patterns of employment. While Harvard remained an important employer of unskilled women, the Dudley and Mount Wilson observatories took a commanding lead. These private research institutions, funded by the Carnegie Institution of Washington, rank at the top of the institutional potential scale. Private research institutions had the most secure resource base and state-of-the-art instrumentation that enabled them to lead the way in astronomical research. They were organized as knowledge factories.

Three institutions accounted for about two-thirds of the jobs for un-

70. Phobe Waterman to Caroline Furness, 14 May [1911]. Furness Papers, VCA. It is virtually impossible to provide pagination on this letter given the way it was written and folded. By my count, the quotation is from pages 7–8.

71. Ibid. By my count, the quote is found on page 6.

Table 9.7 Employment Patterns for Women Whose Highest Level of Educational Attainment Was No More Than a High School Certificate

Second Cohort, 1860–1899 (N = 27)		**Third Cohort, 1900–1940 (N = 234)** Missing data = 20 cases	
Harvard College Observatory	23 (85.2%)	Dudley Observatory	81 (34.6%)
The rest are scattered between four institutions.		Mount Wilson	38 (16.2%)
		Harvard College Observatory	28 (12%)
		Three institutions employed 62.8% of women with no more than a high school education.	
		Smithsonian Astrophysical Observatory	10 (4.3%)
		Yale University Observatory	8 @ (3.4%) for a total of 16 (6.8%)
		Allegheny Observatory	
		U.S. Naval Observatory and the Nautical Almanac Office	7 (3%)
		The rest are scattered between fifteen institutions.	

skilled women workers after 1900. The Dudley Observatory took the lead with its labor-intensive astrometric research program that would eventually produce a definitive multivolume star catalogue and important studies of stellar proper motions. Mount Wilson, the great astrophysical research factory, followed the Dudley as an employer of unskilled women. Harvard comes next on the list followed by the Smithsonian Astrophysical Observatory, Yale, Allegheny, and the USNO/NAO. These data indicate that about 45 percent of women with this level of educational attainment were associated with institutions that engaged in astrometric work, while about a third worked in astrophysical research factories.

Table 9.8 presents information on the labor force at two of the great factory observatories, Harvard and Mount Wilson. Of primary interest are those women whose highest degree was a baccalaureate and whose first job was at one of these institutions. The table is organized to indicate the baccalaureate origins of semiskilled workers. Eleven of the women who worked at Mount Wilson from its founding through 1940 fall into this class. Vassar supplied three of these women and Smith two; the rest came from public or private coeducational institutions, primarily on the West Coast.

Table 9.8 Institutional Origins of Women Computers at the Mount Wilson and Harvard College Observatories, Where the Highest Level of Educational Attainment was a Baccalaureate Degree

Mount Wilson Observatory N=49. Of these women, 11 (22.4%) had baccalaureate degrees, but for 38 (77.6%) the highest level of educational attainment was a high school certificate.		**Harvard College Observatory** N=42. Of these women, 14 (33.3%) had baccalaureate degrees, but for 28 (66.6%) the highest level of educational attainment was a high school certificate.	
Vassar	3	Radcliffe	10
Smith University of California at Berkeley Pomona	2 @ for a total of 6.	Vassar	2
Drake University University of California at Los Angeles	1 @ for a total of 2.	Cornell Miami	1 @ for a total of 2.

If we examine the unskilled female labor force in these two institutions, we see that women whose highest level of educational attainment was no more than a high school certificate predominated. It would appear that by a margin of 4:1 unskilled women led semiskilled female workers at Mount Wilson. The same pattern obtains for HCO, where unskilled women outnumbered skilled by 2:1. Further, table 9.8 indicates Harvard's special relationship with Radcliffe. Vassar supplied only two of the fourteen while one each came from Cornell and Miami of Ohio.

Data reported in table 9.8 raise important questions. Do these patterns reflect employers' preferences for unskilled over semiskilled women workers or do they suggest that higher education could not meet the needs of the factory observatories? Available data do not provide clear-cut answers. Several factors may have been at work. The Dudley Observatory employed unskilled women almost exclusively. This represented a decision on the part of the observatory management.[72] Given the scale of operations, expenses had to be held down. Controlling labor costs was one way to achieve this goal. Harvard apparently never could secure enough female workers by relying exclusively on semiskilled college graduates. Mount Wilson must have been in the same position. From the perspective of both Cambridge and Pasadena, the problem may have seemed one of

72. Benjamin Boss to William Wallace Campbell, 18 April 1922. B. Boss Papers, SALO. See also "Department of Meridian Astrometry, Estimates for 1910." DOA. Lewis Boss indicates in this document that he is holding down costs by not promoting his female assistants even though "promotion for five of the six young women is long overdue."

Table 9.9 Institutional Potential (IP) of First Job for the Population of Women Whose Highest Level of Educational Attainment Was a Baccalaureate by the School Awarding the Degree

Vassar N=11	**Radcliffe** N=10	**Wellesley** N=2	**Smith** N=5	**Mount Holyoke** N=3	**Swarthmore** N=4
IP 5 = 3	IP 5 = 0	IP 5 = 0	IP 5 = 2	IP 5 = 0	IP 5 = 0
IP 4 = 5	IP 4 = 10 *(all at the Harvard College Observatory)*	IP 4 = 0	IP 4 = 1	IP 4 = 0	IP 4 = 2
IP 3 = 1	IP 3 = 0	IP 3 = 0	IP 3 = 0	IP 3 = 2	IP 3 = 0
IP 3 = 0	IP 2 = 0	IP 2 = 0	IP 2 = 0	IP 2 = 0	IP 2 = 0
IP 1 = 2 *(both at Vassar)*	IP 1 = 0	IP 1 = 2 *(both at Wellesley)*	IP 1 = 2 *(both at Smith)*	IP 1 = 1 *(at Smith)*	IP 1 = 2 *(one at Swarthmore and one at Smith)*

supply rather than demand. But there is a complicating factor. The tastes of women graduates, especially from the women's colleges, may have restricted their employment horizons. The West Coast was a long way from South Hadely or Poughkeepsie. Relocation may have proved an insurmountable barrier to many of these graduates.

Data discussed in this chapter suggest that, overall, the women's colleges never produced enough semiskilled women to meet the demands of the American astronomical community. Presumably, when the supply of semiskilled workers was exhausted, directors at the factory observatories dipped into the pool of women high school graduates. If this was, indeed, the case, it suggests that observatory directors made few distinctions between unskilled and semiskilled female workers. A high school graduate and a graduate from one of the women's colleges were treated as interchangeable by the CEOs of the great factory observatories.

Table 9.9 provides another perspective on employment patterns for women whose highest degree was a baccalaureate. A.B.'s from five women's colleges and Swarthmore are arranged according to the institutional potential of the first job. For Vassar and Radcliffe, the concentration is toward the top of the institutional potential scale. Smith graduates went to either extreme and Mount Holyoke baccalaureates tended to concentrate from the middle of the IP scale downward. Like Smith, Swarthmore also placed its graduates toward the extremes of the IP scale. Wellesley stands as an exception. The undergraduate astronomy program at Wellesley produced few students, and they either went directly to grad-

Table 9.10 The First Job for Women Whose Highest Level of Educational Attainment was a Master's by Employer and Institutional Potential of Employer

Second Cohort, 1860–1899 (N=4)		Third Cohort, 1900–1940 (N=49) Missing data = 3 cases	
Columbia University (IP = 4)	1	Harvard College Observatory (IP = 4)	11
U.S. Naval Observatory (IP = 2)	2	Mount Wilson Observatory (IP = 5)	3
Vassar College (IP = 1)	1	Yerkes Observatory (IP = 4)	3
		Lick Observatory (IP = 3)	3
		U.S. Naval Observatory (IP = 2)	3
		Wellesley College (IP = 1)	3
		The rest are scattered between sixteen institutions.	

uate school or did not enter the American astronomical community after graduation.

Table 9.9 provides important evidence concerning employment patterns of women. Controlling for highest degree and institutional potential of first job, these patterns appear remarkably asymmetrical. Higher levels of educational attainment do not provide access to positions with greater institutional potential. These patterns will become even more obvious as we examine the employment of M.A.'s and doctorates.

First jobs for women whose highest degree was an M.A. are indicated in table 9.10. In cohort two, only a few women earned a master's degree. Half of these were employed at the Naval Observatory. In the third cohort the situation changes. The numbers grow rapidly and their distribution becomes more interesting. Harvard was the largest single employer of M.A.'s and its special relationship with Radcliffe continues at this level. The group of five institutions (Mount Wilson, Yerkes, Lick, USNO, and Wellesley) that employed almost a third of the M.A.'s working in American astronomy after 1900 is interesting. In this group there is *one institution representing each of the five levels of the institutional potential scale.* This suggests that women with master's degrees found employment across the institutional spectrum. It stands in contrast with the asymmetric tendencies indicated in table 9.9 and in even sharper contrast with data presented in table 9.11.

Employment patterns for Ph.D.'s are reported in table 9.11. While data for cohort two indicate no particular pattern, the third cohort suggests a different story. The institution employing the most Ph.D.'s (Smith College) has an IP of 1; that is, it ranks lowest on the IP scale. Berke-

Table 9.11 First Job for Women Whose Highest Level of Educational Attainment was a Ph.D. by Employer and Institutional Potential of Employer

Second Cohort, 1860–1899 (N=6) Missing data=2 cases		**Third Cohort, 1900–1940 (N=37)** Missing data = 4 cases	
Yale University (IP = 4)	1	Smith College (IP = 1)	5
Columbia University (IP = 4)	1	University of California* (IP = 3)	4
Paris Observatory (IP = 2)	1	Vassar College (IP = 1)	4
Mount Holyoke College (IP = 1)	1	Lick Observatory* (IP = 3)	2
		Argentine National Observatory (IP = 2)	2
		Mount Holyoke College (IP = 1)	2
		The rest are scattered between fourteen institutions.	

*While the Lick Observatory and the Astronomy Department at Berkeley sometimes hired their own female Ph.D.'s, these women were never placed in tenure track positions. They were paid grant money and in most cases remained for only a few years.

ley and Vassar vie for second place with four female Ph.D.'s each. Like Smith, Vassar ranks at the bottom of the IP scale. The data on Berkeley and the Lick Observatory must be examined closely. Female Ph.D.'s at Berkeley or the Lick worked as computers or assistants. They were paid from grants or one-time appropriations rather than permanent budget lines and had no prospect of professional advancement similar to that of a male astronomer in a tenure-track position. The women working at the Argentine National Observatory at Cordoba remained for only a short time. Considering these caveats, the list takes on a new shape. In fact, professional positions for female Ph.D.'s were at the women's colleges where the institutional potential was lowest. The women with the most advanced training in astronomical research (Ph.D.'s) were shunted into institutional contexts in which they could not effectively pursue research. This is a remarkable asymmetry. Ideally, training and opportunity should be matched. It would seem, that for women, the doctorate provided disadvantages rather than advantages. The best-educated women were segregated in institutions which made advanced research virtually impossible.

In addition to gender segregation, including the development of specific career paths for women, there were other factors that constrained the careers of women in the American astronomical community. Three of the most important of these constraints are examined in this section. They include the pressure to choose teaching over research, the choice between filial duty and a career, and the professional isolation of women.

The women's colleges valued teaching over research. Neither institutional traditions nor policies encouraged or, indeed, permitted faculty to devote large blocks of time and energy to research. Instruction came first. The reward system of the American astronomical community, however, honored scientists who provided new knowledge, e.g., discoveries or large quantities of data (in the form of catalogues, for example) which could be used to develop new systems of classification and analysis. Routine sunspot counts and occultation timings were not the kind of research that attracted the attention of male colleagues. Thus women astronomers were often isolated from the mainstream of their science because of the institutional location in which they found themselves.

Sarah Frances Whiting noted that the new Wellesley observatory was "primarily a students laboratory in astronomy, but later we hope to make some contributions" through original research.[73] A generation later, Anne S. Young wrote of the Mount Holyoke astronomy department, "This has been another year in which the demands of class room instruction upon every member of the department have been so time-consuming that little could be done in observing work beyond that shared by students in courses."[74]

In the early 1890s, Smith astronomer Mary E. Byrd summarized her teaching responsibilities. She estimated that the observatory was open from twenty-five to forty-five hours per week for daylight work including observations of sunspots, the examination and testing of instruments, and planning for night work. With the exception of six weeks in mid-winter, the Smith College Observatory was open for instruction from sunset until at least ten P.M. As Byrd noted "Except in bitter mid-winter weather . . . there have not been more than three or four nights a term when I could handle a telescope for any purpose save to adjust it for students until after ten o'clock at night; and my work begins in the morning at eight or half after."[75] In order to work on personal research projects, Byrd had to observe late at night, in very cold weather, or during summer vacations. In spite of these limitations, she managed to keep up a modest program, measuring the position of comets, which she reported in the *Astronomical Journal.*

Another factor could sometimes constrain the careers of women in American astronomy. Children were seen as providing support, care, and

73. Sarah Frances Whiting to William Wallace Campbell, 26 November 1901. Whiting Papers, SALO.

74. Anne S. Young, "Annual Report" (1932, p. 1), Astronomy Department Papers, MTHCA.

75. Mary E. Byrd, "Smith College Observatory, 1892–93," 30 May 1893. Typescript copy, pp. 4–6, 7. See also Mary E. Byrd, "Annual Report of the Smith College Observatory, 1900–01." Typescript, esp. p. 5. Astronomy Department Papers, SCA.

comfort for their parents in old age, at times of ill health, or in widowhood. These relationships were rooted in preindustrial society, but they carried over into late- nineteenth-century middle-class urban America. Increasingly, however, these obligations were seen to effect daughters more than sons. It was frequently the case that unmarried daughters sacrificed their ambitions in the interests of their parents.

Sometimes, these cases are documented in detail. After graduation, Martha Renner (Vassar, 1910) secured a position at Mount Wilson. Within a few months, Renner informed Caroline Furness, that she would leave at the end of a year. Her mother was "so homesick and so dissatisfied here that I do not feel it would be right" to remain in Pasadena. Renner expressed disappointment: "I like the work and would like to stay."[76] Furness evidently thought highly of Renner and offered her a graduate assistantship. Renner declined. "I would like very much to try—if I had only myself to consider. But I think my mother would consider that worse than our staying out here, for though she would be home, I would have to be away from her the greater part of the year, and I simply could not leave her to live there by herself all that time." Renner must have realized that she was saying farewell to a career in astronomy. "I know it would be a splendid opportunity. . . . It would be good to get back to the transit instrument and the five-inch house again! I appreciate your asking me, but I do not feel that I can consider it."[77]

The imperative of a daughter's transcendent duty to her widowed, elderly, or infirm parents haunts the careers of women scientists. It posed one of the most difficult choices imaginable. How many Martha Renners were there? How many chose the path of filial duty over a career in science?

The theme of isolation for women astronomers is implicit in much of the material discussed in this chapter. It difficult to overestimate the degree to which most women, isolated in women's colleges, relegated to areas of women's work in science, segregated on career paths for women, stood outside of the specialist research communities or the larger generic community.

Frequently, the isolation of women resulted from the direct action of men. These actions were, in turn, predicated on the perceptions men had of women. For example, William Wallace Campbell wrote to acting HCO director Solon I. Bailey concerning the appointment of Annie Jump Cannon as a member of the American delegation to the first meeting of the International Astronomical Union. While Campbell was sure she would be welcomed by the other members of the delegation, "I do not feel like

76. Martha Renner to Caroline E. Furness, 15 November 1911, p. 1. Furness Papers, VCA.
77. Renner to Furness, 15 January 1912, p. 1. Furness Papers, VCA.

encouraging her as she would be the only woman on the delegation and probably the only woman to be a member of the meeting. . . . I fear she would not feel at home."[78] There is no indication that Campbell or Bailey ever discussed the matter with Cannon. They simply made a major career decision for her.

The experiences of Alice H. Farnsworth of the Mount Holyoke astronomy department illustrate another aspect of the isolation of many female astronomers. Farnsworth spent a research leave on Mount Hamilton and later prepared her results for publication as a *Lick Observatory Bulletin.* The Lick staff members were very rough on Farnsworth, subjecting the manuscript to harsh criticism and refusing to let her read proof before publication.[79] Clearly angry, Farnsworth concluded that part of the problem with the senior Lick staff resulted from her isolation and inability to consult with other research workers.[80] Lick astronomer Robert G. Aitken responded with enthusiasm. "Frequent consultation during the course of writing papers or articles is, in almost all cases, a decided advantage. . . . All the others here [at the Lick] feel the same way about it. I know we read each other's papers and talk about them a good deal during the course of their preparation."[81] The astronomy departments in the women's colleges, with perhaps two professors and a graduate assistant, could hardly provide the kind of intellectual environment Aitken referred to. Women were isolated from the opportunity for on-going discussions with individuals who represented various perspectives and different research experiences.

In the Smith College archives there is a notebook that records meetings of women astronomers working in New England and the northeast. This group was the women's counterpart to the Neighbors founded by Frank Schlesinger after World War I. Apparently organized in 1929, the women met at least though 1938.[82] Regrettably, the women's Neighbors left even fewer records than did their male counterparts. Their meetings were generally at Yale or Smith, with one excursion as far south as New York City. Attendance ranged from nine to a high of thirty-two. The schedule called for a meeting from 2 to 4:30 P.M. followed by tea and dinner at 6:30. The meeting continued in the evening. Sometimes there was an invited speaker. Available records do not list detailed activities or

78. William Wallace Campbell to Solon I. Bailey, 3 June 1919. Bailey Papers, SALO.

79. While examining the Shane Archives of the Lick Observatory, I never found examples of male astronomers subjected to similar editorial criticism.

80. Farnsworth to Aitken, 3 January 1934. Farnsworth Papers, SALO.

81. Aitken to Farnsworth, 8 January 1934. Farnsworth Papers, SALO.

82. Record Book of New England Women Astronomers Meetings. Astronomy Department Papers, SCA.

participants. It is impossible to know whether the women used the same informal organization as the men.

By the mid-1930s there was pressure to include women in the Neighbors. Harlow Shapley asked Schlesinger in the spring of 1936, "I wonder if you want to consider also the possibility of increasing the number of sexes represented." Shapley pressed for invitations for Annie Jump Cannon and Cecilia Payne-Gaposchkin. He suggested that since both were members of the American Philosophical Society, they might also be eligible to attend the Neighbors.[83] Schlesinger shot back: "Ah, the entering wedge!" The Yale astronomer admitted that he was not opposed "to the occasional attendance of women at the meetings of the Neighbors" but wondered if this would be an appropriate time to begin. An influx of women, some of whom would be wives rather than astronomers, "would spoil the character of the meeting from the point of view of the dry-as-dust people."[84] At all events, the Neighbors remained an exclusive male club at which much of the business of the American astronomical community was transacted.

The career of Maud W. Makemson (1891–1978) vividly highlights some of the ways in which women were isolated from the mainstream of the profession. Makemson entered the undergraduate astronomy program at Berkeley as a single parent of three children after several years of newspaper experience, work in advertising, and as an office manager in Arizona. By her junior year, she had to withdraw from school in order to earn money.[85] Makemson was awarded a bachelor's degree in astronomy in 1925. The master's followed two years later and in 1930 the Ph.D. in celestial mechanics under A. O. Leuschner.

Makemson's progress toward the doctorate was difficult. Money problems were pressing. Her invalid mother lived with her; the children lived with relatives. What kept her going was "the hope of getting a degree and a position where I could have my children and we could all have enough to eat and a chance to find some playtime together." For seven years she worked "night and day, summer vacations and all, toward that ideal."[86]

Aitken recommended Makemson for a postdoctoral position at Yale. He described her as "a woman of high character" but "her fight for an education under handicaps that would discourage most women has made

83. Harlow Shapley to Frank Schlesinger, 27 April 1936. Astronomy Department Papers, YUA.

84. Schlesinger to Shapley, 1 May 1936. Astronomy Department Papers, YUA.

85. Maud W. Makemson to Robert G. Aitken, 23 November 1923. Makemson Papers, SALO.

86. Makemson to Aitken, 3 December 1929, p. 1. Makemson Papers, SALO.

her rather aggressive in manner, but of late she had not been disagreeably so." Aitken characterized Makemson as "one of the best graduate students in astronomy . . . in recent years."[87]

Makemson's search for a job during the depths of the Great Depression was difficult. As one possibility after another failed to materialize, she supported herself as a computer and later taught in a junior college. At length she secured a position at Rollins College in Florida, and, in 1932, became an associate professor of astronomy at Vassar. In time, she succeeded to the headship of the department and became director of the observatory.

In 1939 Makemson took a sabbatical leave. She asked Walter S. Adams for permission to spend the year at Mount Wilson, working on an astrophysical problem. She planned to attend the Harvard summer school in astrophysics, were she would formulate a research problem.[88] However, Makemson's plans soon started to unravel. She did not attend the Harvard summer school. The gift of a new telescope and a project on the history of Polynesian astronomy kept her in Poughkeepsie. As a consequence, she asked Adams to suggest a problem in celestial mechanics that she might work on. Makemson knew little about astrophysics and had missed the opportunity to take a crash course. Adams was gracious in his response, suggesting that she work with either Ralph Wilson or Seth B. Nicholson. Nicholson proposed a revision of the orbits of the sixth and seventh satellites of Jupiter.[89]

Makemson lived in Pasadena with her daughter but was slow to contact Adams. "I am so deeply involved in my work on Polynesian astronomy and astronomical myths that I can think of nothing else."[90] While Makemson occasionally visited the Santa Barbara offices of the Mount Wilson Observatory, she never undertook active research in celestial me-

87. "Confidential Statement Concerning Maud Worcester Makemson, Sent to Yale University," 15 February 1930. Makemson Papers, SALO. In the days before rhetorical inflation and hyperbole became common in letters of recommendation, this was high praise for a woman. However, the department at Berkeley did not think highly enough of Makemson to award her a Morrison fellowship.

88. Maud W. Makemson to Walter S. Adams, 19 February 1939. Adams Papers, MTWA. In the same letter, Makemson recommended a student for a computer's position at Mount Wilson. She noted that the student had been forced to take off a year to care for her dying mother but could now seek employment. The young woman was engaged, but her fiancé lived in Southern California and the woman assured Makemson that if they married she would not leave Mount Wilson but "just keep on working." See also Maud W. Makemson to Alice Farnsworth, 11 May 1939. Farnsworth Papers, MTHCA.

89. Adams to Makemson, 18 July 1939. Adams Papers, MTWA. See also Makemson to Adams, 2 August 1939, Adams Papers, MTWA.

90. Makemson to Adams, 7 October 1939. Adams Papers, MTWA.

chanics, and the new year found her in Berkeley, "keeping house with my son Don, who is a graduate student."[91] Makemson had decided to spend full time on her study of Polynesian astronomy and needed the resources of the Berkeley libraries in order to complete the task.

Securing a First Job

Compared to the experience of males (chapters 5 and 6), the process by which women secured a first position in the American astronomical community is not well documented. Extant sources suggests that securing a first position involved two different tracks: one in the women's colleges or, in a few cases, a coeducational college, and the other as computers in the great factory observatories.

The recruitment of computers appears to have been relatively informal. Directors of factory observatories sent letters to chairs of astronomy departments at the women's colleges, coeducational colleges, and universities, asking for nominations. These letters described the job and the salary.[92] Sometimes, departmental chairs at the women's colleges were forced to reply that the number of majors was either too small to meet current demands or that there were none at all.[93] Generally, directors selected candidates on the basis of recommendations. Only rarely did they seek to interview prospective computers.[94] Retrenchment in the 1930s must have virtually curtailed expansion of the unskilled and semiskilled labor force at the factory observatories. The Mount Wilson director advised one young woman to seek employment as a teacher of astronomy rather than look forward to a career as a computer.[95]

Rarely did female Ph.D.'s seek positions at a factory observatory. Only in the 1920s was the first female Ph.D. appointed to the Mount Wilson computing staff. The application of Hazel Marie Losh (b. 1898), who earned a Michigan doctorate in astrophysics in 1924, is revealing. Losh summarized her professional qualifications and also indicated that at Michigan she had served as Professor Hussey's secretary.[96] Losh eventually returned to Ann Arbor, where she spent her career.

The issue of a candidate's marital status and/or intentions played a part in the hiring process. In 1928, Mount Wilson astronomer Seth B.

91. Makemson to Adams, 22 January 1940, p. 1. Adams Papers, MTWA.

92. See, for example, Robert G. Aitken to Anne S. Young, 10 April 1922. SALO, A. S. Young Papers.

93. See, for example, Harriet W. Bigelow to Robert G. Aitken, 14 May 1924. SALO, Bigelow Papers.

94. Walter S. Adams to Elizabeth Cornwall, 22 November 1935. Adams Papers, MTWA.

95. Walter S. Adams to Melba R. Love, 2 September 1939. Adams Papers, MTWA.

96. Hazel M. Losh to Walter S. Adams, 9 February 1924. Adams Papers, MTWA.

Nicholson lectured in the Berkeley summer session. He was on the look-out for young women to recruit for the computing corps. In describing a candidate, Nicholson indicated "As far as I can see she had no matrimonial intentions!"[97] Writing from Berkeley were she was working as a computer in the astronomy department, Katherine P. Kaster assured Walter S. Adams that if employed at Mount Wilson she would remain for some time. "I finally got a divorce last spring and so am more likely to continue on a job than I would be if I had a husband to consider."[98]

Filling positions at the women's colleges was more complex. Professors sometimes identified outstanding undergraduates and encouraged them to earn a Ph.D. in astronomy with the implicit promise of employment at their alma mater. Senior faculty in the women's colleges had several ways of facilitating the graduate education of favored candidates. The first was an assistantship. By teaching the freshman survey, young women could gain valuable experience and earn money to finance graduate school. There were also a few fellowships awarded to alumnae to support doctoral work. After residency requirements had been met and data collected for a dissertation, a woman might return as an instructor. On completion of the doctorate, she could look forward to moving up through the ranks to a professorship.

Not all job offers were acceptable. Alice Maxwell Lamb graduated from the University of Wisconsin in the early 1880s and remained at the Washburn Observatory as an assistant, becoming a protegee of director Edward S. Holden. In 1885 Lamb reported to her patron that she had been offered an instructorship in mathematics at Yankton College in South Dakota "with the inducement of a possible observatory to direct in the course of a year or two." The Yankton trustees apologized for the salary, (it was twice what Lamb earned at Madison!), but she refused the offer. "I had far rather work in a true observatory than direct a toy one."[99] Lamb spent thee years at the Argentine National Observatory in Cordoba where she married Milton Updegraff (b. 1861). When they returned to the United States in 1890, Mrs. Updegraff turned her attention to buttons and breakfasts.

In comparison with appointments for men seeking first positions, appointments for both of these tracks appear to have been very informal. This may have had something to do with the power of the directors of the great factory observatories as well as the early identification of candidates for positions at the women's colleges.

97. Seth B. Nicholson to Walter S. Adams, 17 July 1928. Adams Papers, MTWA.

98. Katherine P. Kaster to Walter S. Adams, 19 January 1938. Adams Papers, MTWA

99. Alice Maxwell Lamb to Edward S. Holden, 19 December 1885, pp. 3–4. Washburn Observatory Papers, SALO.

The Life of a Computer

The definition of women's work in astronomy began with the appointment of Maria Mitchell as a computer with the Nautical Almanac Office in 1849. By the 1880s, Mitchell's department at Vassar and the other astronomy departments in the women's colleges had become part of the system, supplying semiskilled workers for the new factory observatories. Mitchell had mixed feelings about these developments.

In the early 1870s, Mitchell wrote physicist Joseph Henry, secretary of the Smithsonian Institution, asking whether there might be any openings for trained female astronomers in the expeditions that were being organized to observe the transits of Venus. Henry turned the letter over to Mitchell's old friend, Admiral Charles Henry Davis, who had served as the first director of the NAO. Davis was president of the Transit of Venus Commission. The admiral went to some lengths to convince Mitchell that it was impossible to include women as members of the parties being sent to Asia and the Pacific to observe the transit of 1874. The teams were traveling as supercargo on naval vessels, and Davis could not imagine subjecting a "cultivated woman" to such hardships. Davis did suggest an alternative. "There is, however, another and, I may say, higher field of work in the subsequent calculations [that is, the complex and laborious reduction of observational data], eminently suited to your genius, the occupation of which will confer much higher distinction upon my honored friend and her fellow [women?] scientists, than the observations in the field."[100] The men and boys would have the fun of going on expeditions to the far corners of the globe to collect data, while the women remained at home with their log tables, computing.

Mitchell grew increasingly distressed with the emerging field of womae's work in astronomy. The Vassar astronomer (Wright 1949: 236) did not like the idea of women being relegated to sedentary desk jobs. "In the half-lighted and wholly unventilated offices women work patiently at the formulae, and pile up logarithmic figures; in the open air, under the blue sky or the star-lit canopy, boys and men make the measurements!" One of the great ironies of Mitchell's life is that in spite of objections, she was instrumental in the creation of women's work by supplying factory observatories with semiskilled workers in the form of Vassar astronomy graduates.

As the career data suggest, factory observatories tended to employ

100. Charles Henry Davis to Maria Mitchell, 17 March 1874. Transit of Venus Commission Letter Book, p. 108. Transit of Venus Commission Papers, U.S.Naval Observatory Records, NA. The calculations of which Davis spoke were so complex and extensive that the American data from the 1874 and 1882 transits were never fully reduced or analyzed. Lack of final results is also related to serious doubts about the reliability of the photographic observations of the transits (Lankford 1987c).

more unskilled than semiskilled workers. Both Harvard and the Dudley observatories provide examples of this practice. Benjamin Boss, who succeeded his father as director of the Dudley, wrote to William Wallace Campbell, "All of the girls we employ are merely high school graduates with no experience in astronomical work and no scientific training. These girls start with a salary of $720 a year [$60 per month]." Advancement depended on "efficiency, longevity of service and vacancies." Boss insisted that "Under this system the girls of promise soon acquire skill and training and are encouraged as far as possible through promotion. The nonproficient ones at least earn the stipend accorded them and have the privilege of resigning if not satisfied with it. Only two out of ten girls are paid a salary over $100 per month."[101] A decade earlier, Lewis Boss indicated that he was able to move forward rapidly with the reduction of large quantities of astrometric data because "I have a well organized corps of young women for the routine work. Everything is always prepared for them to do in large blocks from end to end. It is an extremely economical system."[102]

In effect, the elder Boss was describing the rationalization of work at the Dudley Observatory. Reduction of astrometric data was broken down into units and subunits so that the task could be mastered by unskilled workers. These tasks were repetitious and quickly learned. Calculations were done on standardized forms according to specified procedures. Duplicate computations provided a check against errors. At no point were the women expected to envision, let alone understand, the final product. These women could not look forward to promotion based on the acquisition of more specialized skills and technical knowledge.

The experiences of Dorothy Applegate (Whitman College, 1924) illustrates some of the differences and similarities between unskilled and semiskilled female workers. In October 1924, Robert G. Aitken offered Applegate the position of assistant at the Lick Observatory. She would earn $90 per month plus free room. Board was estimated at $1.10 a day, approximately one-third of the annual salary. Aitken apologized for the salary, "but in view of your lack of experience [skills] we think that it will be a proper compensation, keeping in mind the fact that you will also have the opportunity to get some acquaintance with practical astronomical work." Aitken described Applegate's duties as "largely routine computing," and admitted "the work may not be, in itself, very inspiring," but held out the promise that, in addition to the drudgery of computing, Applegate could make observations.[103]

101. Benjamin Boss to William Wallace Campbell, 18 April 1922. B. Boss Papers, SALO.

102. Lewis Boss to David Gill, 16 March 1911, p. 3. DOA.

103. Robert G. Aitken to Dorothy Applegate, 16 October 1924. Applegate Papers, SALO. See also Aitken to S. L. B. Penrose, President of Whitman College, 4 October 1924. Applegate Papers, SALO. In this letter Aitken inquires about Applegate's personality and character.

It is difficult to believe that many female assistants availed themselves of these opportunities. In order to gain the observatory experience Aitken spoke of, an assistant would have to develop informal relationships with staff astronomers and, perhaps, graduate students. One could not simply walk into the dome of the 36-inch Clark refractor and start making observations! The development of these connections depended on personality rather than contractual obligations. Applegate apparently was able to establish working relationships with staff astronomers and published a *Lick Observatory Bulletin* based on her observations of a spectroscopic binary. Whether less determined women could have achieved this goal is doubtful.

Compared to the pay of high school graduates, Applegate's salary at the Lick was only marginally higher and the work virtually identical. Apparently, possession of an undergraduate degree did not enlarge her duties or responsibilities, let alone provide independence in comparison to the tasks assigned computers with less education. Only the chance to learn about the work of the observatory (after the close of the business day), differentiated Applegate from computers with less formal education. Differences between unskilled and semiskilled women workers in the great factory observatories were often blurred. Pay was low and hours were long. The work was frequently routine and measuring instruments sometime posed physical problems for the women who used them. In no case were these female employees viewed as permanent. Comments by Harlow Shapley, fifth director of the Harvard College Observatory, indicate ways in which management viewed unskilled and semiskilled workers. When Shapley took over at Harvard in 1921, he instituted a number of new lines of research. Of special importance was the study of variable stars in both the Milky Way and other galaxies. These projects "required a tremendous amount of measuring. I invented the term 'girl-hour' for the time spent by the assistants. Some jobs even took several kilo-girl-hours. Luckily Harvard College was swarming with cheap assistants; that was how we got things done."[104] Shapley made no attempt to distinguish between "cheap assistants" who were high school graduates and those with undergraduate degrees from Radcliffe.

No one has yet developed a reliable time series indicating trends in the pay of women who worked as computers. The evidence is impressionistic, contained, for the most part, in letters rather than official reports of observatories; however, certain patterns appear. From the beginning of the century to World War I, the yearly salary of computers (whether graduates of the women's colleges or high schools) appeared to range between $655 per year for beginners through $800 for women with experience. During

104. Shapley (1969:94). Astronomers like Yale's Frank Schlesinger were openly critical of Shapley's policy of paying low wages to female computers.

the interwar years, the median appears to have climbed to between $1,200 and $1,500 per year. Of course, these figures have little meaning without an understanding of the cost of living at both the national and local levels.

Pay scales at the Dudley Observatory suggest the differentials between males and females in 1910. Richard H. Tucker, on leave from the Lick to direct the Dudley expedition to San Luis, Argentina, earned $3,600 per year. Tucker, with only an undergraduate degree, was one of the leaders of observational astrometry. A. J. Roy of the Dudley staff was second in command at San Luis. With an undergraduate and master's from Union College, the younger man (b. 1869) earned a similar salary. The male observers at San Luis earned between $1,200 and $1,800 per year. Most had baccalaureate degrees. The highest-paid woman computer at the Albany headquarters of the Dudley Observatory earned $875 per year and the lowest $655. Roughly speaking, the basic differential for male and female entry-level positions at the Dudley Observatory was about 100 percent; males tended to earn at least twice as much as females.[105] For purposes of comparison, from about 1900 to 1940 observatory directors earned between four and six thousand dollars a year.

One other category of evidence suggests gender-based differences in hiring and rewarding staff members at the factory observatories. Men who were appointed as computers or assistants had, from the beginning, a wider range of duties and responsibilities. For example, they were hired with the promise that they could also observe.[106] Sometimes men were appointed as computers with the clear understanding that this was an entry-level position which would provide mobility. In 1923 director Adams tried to lure William H. Christie from the Dominion Astrophysical Observatory in Canada. He had nothing other than a computer's position to offer, but indicated that the pay would begin at $2,000 per year "and of course I would count on transferring you to our regular staff whenever any opening developed."[107] As far as I know, no female computer made such a large salary in the 1920s or had ever been promised advancement to a regular staff position. More than a decade later, Adams was offering male computers starting salaries comparable to those of senior female members of the computing staff.[108]

The Vassar College archives contain letters to Caroline Furness from alumnae working as computers in the great factory observatories. These

105. "Department of Meridian Astrometry, Estimates for 1910." DOA. Lewis Boss indicates in this document that he is holding down costs by not promoting his female assistants.

106. See, for example, Walter S. Adams to W. M. Gilbert, 1 May 1928, where he describes the appointment of Robert S. Richardson as a computer who is also to take part in the solar observation program. Adams Papers, MTWA.

107. Walter S. Adams to William H. Christie, 15 November 1923. Adams Papers, MTWA.

108. Walter S. Adams to Boris Karpov, 27 January 1936. Adams Papers, MTWA.

letters, written in the early twentieth century, provide important insights into the day-to-day experiences of women in these institutions. For most Vassar graduates, working at Mount Wilson, the Lick or Yerkes was little more than an extension of college. Their world was one of hard work but also of picnics, hikes in the mountains, golf and tennis parties, swimming in Lake Geneva, and visits to the California coast. With few exceptions, these women did not see themselves as having permanent careers in science. The experience was an interlude between graduation and marriage. Frequently computers clubbed together, renting a house and taking turns cooking. At the domestic level, their experiences resembled dormitory life at college. Further, their letters suggest the position of computer provided advantages in looking for a spouse. Male astronomers recently out of graduate school were in plentiful supply and came into contact with intelligent, well-educated young women on a daily basis.

These women often described their work to Furness, but they did so in a way that suggests only casual involvement with science. When a woman looked on astronomy as a career, she was not happy with conditions at Mount Wilson. Phoebe Waterman, for example, found the work far too confining and moved to Berkeley to pursue a doctorate. Only a few of these women wanted more responsibility—above all, the opportunity to use telescopes. In short, most of these women's college graduates stood in a passive relation to science. Their work was defined and directed by men who set the goals and generally wrote up the results. Nowhere is this passivity more clearly illustrated than at Mount Wilson. Sometimes, in good weather, a group of women would spend the weekend in one of the cottages on top of Mount Wilson. Aside from hikes in the woods, their biggest treat was to sit quietly in a corner of the darkened dome, watching a male astronomer guide the 60- or 100-inch telescope as he photographed the spectra of some distant star or galaxy.[109] Women were passive spectators as males made observations and collected data.

On Mount Hamilton, the social situation was more complicated than at the Santa Barbara offices of the Mount Wilson Observatory. Senior Lick staff had their own houses, while single astronomers, graduate students, and computers lived in a dormitory and took meals together. Single people mixed on an informal basis more frequently than was the case in Pasadena.[110] At Williams Bay, computers at the Yerkes Observatory found rooms and boarded with families in the village; so did younger astronomers. At both Yerkes and Lick, the staff (young and old) frequently had picnics, went swimming and hiking, or played golf and tennis together.[111]

109. Martha Renner to Caroline E. Furness, 15 November 1911, p. 2. Furness Papers, VCA.

110. Fredrica Chase to Caroline E. Furness, 26 December [1905], Furness Papers, VCA.

111. Harriet McWilliams Parsons to Caroline E. Furness, 25 September 1915. Furness Papers, VCA.

Directors of the Lick and Yerkes were committed to providing recreational opportunities for the staff. Improbable as it may seem, both observatories even had a golf course, the one carved out of a mountain side and the other in a cow pasture.

The scientific activities of a computer were more or less similar from one institution to another. The factory observatories required computers to work a five-and-a-half-day week. In the beginning, Mount Wilson defined Saturday work as extending to three P.M., but soon let the computers go at noon.[112] Ruth Emily Smith (Vassar, 1906) was employed at Mount Wilson from the time she graduated until 1915. At first, she worked for Adams and Hale on various projects relating to solar physics. "It is fascinating work and there are so many different motions [when using the machine to measure the intensity of solar flocculi], and things to do that it don't [*sic*] get monotonous. Of course the work is different from anything I ever did at College, but I feel that my work there was very helpful."[113] In a few years, Smith had gained enough experience and the confidence of her male superiors so that she was placed in charge of the photographic darkroom dedicated to solar research.[114]

Harriet Parsons (Vassar, 1915) worked at Yerkes from 1915 to 1921 when she was awarded a Chicago doctorate in astronomy and returned to a position in the Vassar astronomy department. At Yerkes, computers were permitted to work toward advanced degrees. She was assigned to John A. Parkhurst, to assist in his research in photographic photometry. Parsons characterized Parkhurst as "the most patient man in America." Soon she was developing plates for Parkhurst as well as measuring extra-focal star images with a Hartmann densitometer.[115]

Fredrica Chase (Vassar, 1905) served at the Lick Observatory for two years after graduation. Her assignment was to measure positions of the asteroid Eros in order to develop more accurate knowledge of the solar parallax and, thus, the distance from the earth to the sun. Chase was delighted with conditions on Mount Hamilton. "We [her coworker was a female graduate of the Berkeley astronomy program] have our own office with a phone and the privilege of managing things in there just as we like." The office was know variously as the House of Eros, the House of Errors, or the House of Corrections.[116] If Lick computers had any energy left over

112. Ruth Emily Smith to Caroline E. Furness, 25 November 1906, pp. 1–7. Furness Papers, VCA.

113. Smith to Furness, 25 November 1906, p. 7. Furness Papers, VCA.

114. Smith to Furness, 6 February 1911, p. 3. See also Renner to Furness, 15 November 1911, p. 2. Furness Papers, VCA.

115. Parsons to Furness, 25 September 1915, p. 2. Furness Papers, VCA.

116. Chase to Furness, 26 December [1905], p. 2. Furness Papers, VCA.

at the close of the business day, they were encouraged to do piece work for the Observatory at the rate of 35¢ an hour.

Using plate-measuring machines generally meant that women had to stand rather than sit and often lean forward in order to reach the eyepiece and make adjustments. At Lick, staff astronomer Charles D. Perrine became concerned about the physical strain caused when using the Stackpole measuring engine and had a stand made to raise it to a more comfortable height.[117]

These letters contain few complaints about the duties of a computer. Phoebe Waterman, a woman of great ambition and intelligence, is an exception. She praised Mount Wilson astronomer Walter S. Adams as "simply fine to work under, very considerate and very nice [in] every way—so that we do not find the *long* hours so unbearable as they otherwise would be." To Waterman, an eight hour day was "a long time for such concentrated head work."[118] Waterman, who loved the outdoors, resented the fact that when she left the office on a winter evening, it was almost dark. She longed for weekends and the coming of spring. Waterman soon realized that she was not destined to be a computer. "The work here is interesting . . . but disappoints me a little in not opening up to individual work as I had hoped it might." Further, she did not like Adams' successor as superintendent of the computing division, Frederick H. Seares. She described Seares as a micro-manager who "likes to superintend the very details, & gives anyone working under him practically no chance for original work." Waterman hoped to "come to an understanding [with Seares] . . . when the winter's work is planned out & assigned."[119] There is no record of any such bargain and soon Waterman was on her way to Berkeley. The other computers must have felt that Waterman was different. As Ruth Emily Smith remarked in 1912, "Phoebe Waterman was down from Berkeley for the holidays. She doesn't seem very contented up there, but I guess she never is contented, no matter what she is doing."[120] Perhaps this reaction can be traced to the great difference in ambition between Waterman and the other women's college graduates working at Mount Wilson. Another Vassar graduate also had ambitions to work with telescopes rather than simply develop and measure plates taken by male astronomers. As Harriet Parsons reported from Yerkes, Parkhurst told her

117. Chase to Furness, 26 December [1905], p. 2. Furness Papers, VCA.

118. Phoebe Waterman to Caroline E. Furness, 18 January 1908, p, 1. Furness Papers, VCA.

119. Waterman to Furness, 16 September 1909, p. 1. Furness Papers, VCA. Later in the letter, Waterman suggests that she was most happy with a project Adams assigned to her where she was able to analyze and discuss photographic data on solar rotation and prepare it for publication without interference from senior male astronomers.

120. Smith to Furness, 18 January 1912, p. 3. Furness Papers, VCA.

that she could soon begin using the telescope. "Then I will feel that I'm really doing something."[121]

By far the most exciting part of a computer's life seems to have been outdoor recreation. As graduates of the women's colleges, they had considerable regard for maintaining good health through exercise and outdoor physical activity. Smith wrote that she had been up Mount Wilson twice in the space of a few weeks. One of the trips was a moonlight climb and on the second they walked up the old trail. Smith reported that she walked thirty-seven miles on the second visit. The Vassar graduate must have cut quite a figure. "I have a dandy new mountain suit, high elk skin boots that lace nearly to my knees, heavy gray blue waist [blouse], short skirt and bloomers so I can climb anywhere."[122]

When Adams relinquished supervision of the computing division, he treated the women to a party. It was a three-day trip up Mount San Antonio. All the computers and many of the observatory staff members, their wives and children joined in the adventure. Traveling by trolley and auto, they finally reached the trailhead. The first evening they toasted marshmallows and told ghost stories around the fire. Up the next day at three A.M., the party climbed toward the summit. They reached the snow line by eleven and most decided to call it quits. Smith and a few others went on across the snow fields and reached the top, where they could look across the Arizona desert and see Mount Whitney, 150 miles to the east. The second night the party fell into their blanket rolls soon after dinner. Many suffered from sunburn as a result of the light reflected from the snow. Smith reported that they looked like "lobsters for a week after."[123]

At Yerkes, swimming in Lake Geneva was one of the great attractions and provided recreation for both staff and computers. Harriet Parsons was especially charmed by the Belgian refugee astronomer George Van Biesbroeck (1880–1974), who was dubbed "the old man of the sea" because of the way his bushy white hair and beard looked as he surfaced after a dive into the lake.[124] Golf was also an important leisure time activity for Yerkes computers and astronomers. Curiously enough, the letters of male astronomers contain relatively little concerning recreation or social activities.

On at least one occasion, Ruth Emily Smith managed to work on some "original investigations" in solar physics with Professor George Abetti, "a young, brilliant and handsome Italian astronomer" who spent six months at Mount Wilson. Smith and Abetti measured solar promi-

121. Parsons to Furness, 25 September 1915, p. 3. Furness Papers, VCA.

122. Smith to Furness, 18 January 1912, p. 4. Furness Papers, VCA.

123. Smith to Furness, 28 June 1909, p. 3. Furness Papers, VCA.

124. Parsons to Furness, 25 September 1915, p. 4. Furness Papers, VCA.

nences, looking for long-term variations similar to those in the intensity of flocculi and the magnetic intensity of sunspots.[125]

Other comments about male astronomers were not, perhaps, on such a professional level. Lick Astronomer Harold K. Palmer (b. 1878) visited Mount Wilson in 1906. Smith reported that "He seems very much interested in Miss Chase, a V.[assar] C.[ollege] girl who is at Mount Hamilton."[126] Soon enough, Fredrica Chase was reporting from Mount Hamilton that she was engaged, but not to Harold Palmer! "The less fortunate one is Dr. J. H. Moore, a 'Ph.D' of Johns Hopkins in June 1903 and since then he has been in the spectroscopic work here."[127] The couple postponed their marriage until the Eros computations were completed. When she became Mrs. Moore, Fredrica resigned her post at the Lick.

Were these women lost to astronomy when they married? Traditional wisdom would probably say yes, but there is another view. William Hammond Wright, an unlikely source for positive comments on almost any subject connected with the American astronomical community, made a point of writing Armin O. Leuschner about Phoebe Waterman Haas. Leuschner had compiled a list of Berkeley astronomy doctorates and simply indicated that Mrs. Haas had left the field after her marriage. Wright did not feel this a fair representation.

Wright pointed out that Mrs. Haas observed with a 4-inch refracting telescope and for some years had contributed observations to the American Association of Variable Star Observers. Further, she was in contact with the Lick staff and kept up with the literature. Here was one woman who never exchanged her interest in astronomy for buttons and breakfasts. "She has carried her enthusiasms for, and knowledge of astronomy as a powerful cultural influence into her very charming family, and to her friends. In doing so it seems to me that she had fulfilled and is fulfilling the purposes of scientific education quite as effectively as though she were actively engaged in professional work."[128]

Women and the National Academy of Sciences

In his discussion of women in astronomy, Mount Wilson staff member Robert S. Richardson (1944: 4) noted that they stood outside the reward system of modern science. "Scientific societies being composed entirely

125. Smith to Furness, 28 June 1909, p. 1. Furness Papers, VCA. The joint publication appeared in the *Astrophysical Journal* in 1911.

126. Smith to Furness, 25 November 1906, pp. 9–10. Furness Papers, VCA.

127. Chase to Furness, [1907], p. 2. Furness Papers, VCA.

128. William Hammond Wright to Armin O. Leuschner, 23 December 1936. Department of Astronomy Papers, BLUCB. Phoebe Waterman Haas saw two sons become distinguished engineers. On the career of Phoebe Waterman Hass, see Williams (1991).

of men have been notoriously reluctant in the past to accord recognition to women." Perhaps Richardson was thinking of the National Academy of Sciences when he wrote these words. By 1944, the NAS had elected three women to membership, two of whom were still living. Only in 1981 would the Academy elect a woman astronomer to its ranks.[129]

As indicated in chapter 8, election to the Academy was seldom a straightforward process. An outstanding research record was a prerequisite, but election rested as much on the political power of patrons and supporters as on merit. Most men had a difficult time achieving election to the Academy. The process involved a campaign extending over several years. Women had an even more difficult time. The odds against them were greater, the obstacles blocking election more formidable.

A few men had special reasons for supporting the election of women to the NAS. Some of the most powerful NAS insiders, like Johns Hopkins biologist Raymond Pearl (1879–1940), were deeply upset as the Academy expanded during the interwar years. They feared that many of those elected were second-rate scientists. In discussing the nomination of physiologist Florence Sabin in 1923, Pearl admonished his friend, E. B. Wilson of the Harvard School of Public Health, "Miss Sabin is certainly as good an investigator, if not better, than anyone of the other three names proposed in the same section, and in my opinion better than at least half the other nominees offered to the Academy this year."[130] Wilson shot back. "I was the leader of a very strong minority in the American Academy [of Arts and Sciences] bent on electing women that seemed to me better scientifically than some of the men we were electing; but I got licked. The American Academy will have nothing to do with women. I am afraid that the National Academy may feel the same way."[131] Wilson went on to outline a strategy that involved the astronomers. He suggested to Pearl, "What you want to do if you want to elect Miss Sabin is to get the astronomers to run Miss Annie Cannon." Wilson believed Cannon had "made real contributions to astronomy by the intelligence and intuition with which she has classified the Harvard spectral photographs." Wilson saw no difference between "looking at a photograph and looking at the original star." Wilson ventured the opinion that Henry Norris Russell could not have achieved his great breakthrough in correlating spectral type and luminosity without Cannon's work on the Draper catalogue of stellar spectra.[132] As an outsider to astronomy, Wilson did not appreciate the

129. "Women Elected to the Academy," prepared by the Office of the Home Secretary, NAS, n.d. Copy supplied by the Office of the Archivist, NAS, 22 December 1986.

130. Raymond Pearl to E. B. Wilson, 7 March 1923. Office of the Archivist, NAS.

131. Wilson to Pearl, 9 March 1923. Office of the Archivist, NAS.

132. Ibid.

difference between the reduction and tabulation of data and their publication in catalogue form as opposed to the theoretical explanation of observational material, a difference that rested on a gender-based division of labor. Women's work and men's work were not to be rewarded in the same way.

Clearly, Wilson had a more complex agenda. If the names of three or four distinguished women scientists appeared on the preferential ballot, it "might give somebody a chance to say that they were better than the men or they would not be there and might therefore accomplish what you and [Arthur L.] Day [the home secretary of the Academy] want; namely to stop electing so many men until there are some better ones."[133]

Pearl was delighted with Wilson's reply. He had hoped to turn Wilson's attention to the political connection between current election patterns (too many second-rate men) and the possibility of using women of merit to demonstrate the need to elect better men. Pearl then reverted to the humorous mode that frequently marked the Pearl-Wilson correspondence. Both men were fond of irony, hyperbole, and caricature. "My only objection to Miss Cannon as a member of the Academy is that she is deaf." Pearl then went on, using a broad brush. "It seems to me that in the passage of time as the Academy gets larger and near its limits, we might well introduce some statement about the undesirability of electing physical defectives." From Pearl's point of view "The astronomers seem to me to be a particularly bad lot. They are either deaf, blind, or spit, or something. I suppose the reason is that they are such great men that anything like an organic balance in their makeup is not to be hoped for."[134] Perhaps, given the astronomers' tendency to nominate candidates that others felt to be scientists of lesser stature, Pearl was simply venting his anger in the form of hyperbole.

Be that as it may, in 1925 Cannon was nominated for membership in the NAS. She falls into the category of outsiders discussed in chapter 8, those who received at least one vote on the informal ballot, but no support on any formal section ballots. As far as I can tell, this was the only occasion on which Cannon was nominated. It was Harvard Observatory director Harlow Shapley who nominated Cannon. Shapley, however, was new to the Academy (elected in 1924) and had to ask Frank Schlesinger, chair

133. Ibid.

134. Pearl to Wilson, 12 March 1923. Office of the Archivist, NAS. Rossiter (1982: 286) reads this letter as if Pearl were presenting a eugenics-based argument against individuals (including Cannon) with physical defects. Given the context, especially the hyperbolic reference to the problems of other astronomers and the use of humor by Pearl and Wilson, I disagree. For example, see Wilson to Pearl, 9 March 1923, pp. 1–2, where there is a discussion of an alcoholic mathematician. This is a serious discussion as opposed to the obvious humor involved in the comments on the astronomers.

of the Astronomy Section, to explain the nomination procedures. Schlesinger provided the necessary information but cautioned Shapley. "I should like to talk over this and similar matters with you and other members of the Academy at the next meeting of the Neighborhood Club [the Neighbors]."[135] Here, perhaps for the first time, Shapley was to learn about the political power of the Neighbors.

Wilson, on the other hand, encouraged Shapley. "I can't see why the astronomers haven't made a fight for her [Cannon] before. A lot of them could never have done their work if it hadn't been for the Henry Draper Catalogue." Wilson decried the fact that Cannon had "all sorts of honors in foreign countries" but was "not even to be up before the Academy for consideration. I believe she could be elected hands down and more easily than any other astronomer not a member."[136] Although Shapley probably did not know it, he had become a pawn in the larger Wilson-Pearl game of Academy politics.

While Shapley appreciated Wilson's encouragement, he was dismayed to learn that Schlesinger and Russell "did not specially favor the candidacy of Miss Cannon. . . . Their reason seems to be that present interpretative ability rather than past accomplishment should count." Shapley was frustrated. "One European astronomer of recognized standing told me this summer than as seen from Europe Miss Cannon would rank above half a dozen of the present astronomical members of the Academy."[137] In the end, Cannon would have to settle for membership in the American Philosophical Society and an honorary degree from Oxford.

Astronomers made one other attempt to elect a woman to the NAS. From the first appearance of her name before the astronomy section in 1929, through final rejection by the Academy on the floor vote in 1943, efforts to elect Harvard astronomer Cecilia Payne-Gaposchkin extended over a period of fourteen years. Unlike Cannon, Payne-Gaposchkin was supported by all three of the generals (Shapley, Schlesinger, and Russell). At length, these powerful patrons won enough votes to place her name on the preferential ballot for consideration by the whole Academy. Three times Payne-Gaposchkin's name made it to the floor of the April meeting. Twice the balloting was closed before her name was reached. On the third occasion, the members at the April 1943 meeting rejected the Harvard astronomer for membership.

135. Harlow Shapley to Frank Schlesinger, 13 August 1925, and Schlesinger to Shapley, 2 September 1925. Astronomy Department Papers, YUA. Note that the Neighbors was sometimes called the Neighborhood Club.

136. E. B. Wilson to Harlow Shapley, 31 October 1925. Copy in the Astronomy Department Papers, YUA.

137. Shapley to Wilson, 3 November 1925. Copy in the Astronomy Department Papers, YUA. It is quite probable that Shapley is reporting his conversation with Schlesinger and Russell at the fall 1925 meeting of the Neighbors.

It took a decade for Payne-Gaposchkin's patrons to achieve her nomination by the astronomy section and even then it was close. In 1939 Payne-Gaposchkin garnered seventeen votes to sixteen for a controversial male candidate. The floor voting in 1940 closed before Payne-Gaposchkin's name was reached.[138] The same thing happened in 1941. In 1942 the astronomy section did not give any candidate a two-thirds majority on the formal ballot. As secretary of the section, Henry Norris Russell, acting on his own initiative, placed her name before the Academy for consideration in April 1943. He acted in response to a directive from NAS president Frank Jewett, suspending the bylaws governing the nomination process and allowing the Academy to elect up to thirty new members. It is not known to what degree Russell's unilateral action may have angered some members of the astronomy section. The results of the October 1942 formal section ballot saw Payne-Gaposchkin with fourteen votes and Donald Menzel, also of the Harvard College Observatory, with thirteen. It may be that gender became intertwined with various intrasectional rivalries. There is also the fact that Menzel was a fresh face in the hustings while Payne-Gaposchkin had been in contention for years. Russell clearly worked hard for her election and must have been saddened to report that the April meeting rejected Payne-Gaposchkin.[139] The repudiation of Payne-Gaposchkin was a personal defeat for the aging generals.

Although it is impossible to know the reasons for Payne-Gaposchkin's rejection by the Academy, it is interesting to note that one of her patrons, Harvard Observatory director Harlow Shapley, was sharply criticized in

138. The story can be reconstructed from the following sources. Cecilia Payne-Gaposchkin to Frank Schlesinger, 21 January 1938. Astronomy Department Papers, YUA. In this letter she thanks Schlesinger for helping with her biography and bibliography for circulation to the members of the Academy. Frederick H. Seares to "Members of the Astronomy Section of the NAS," 15 October 1940 (Astronomy Department Papers, YUA) reports that the voting closed before her name was reached at the April meeting. There is no evidence in the archival record that there was any concerted effort to make sure that the Academy did not have an opportunity to vote on the nomination in either 1940 or 1941. On Payne-Gaposchkin, see Kidwell (1987).

139. "Henry Norris Russell to Members of the Astronomy Section of the NAS," 28 October 1942. Copy in the Struve Papers, UCYOA. See also Russell to "Members of the Astronomy Section of the NAS," 15 October 1943. Adams Papers, MTWA. The rejection of an astronomer by a vote of the Academy at the April meeting was a rare event. As far as I know, it last happened in the mid-1930s, with the nomination of Seth B. Nicholson.

Blocking the election of a nominee at the April meeting was not easy. As Schlesinger once wrote in a different context, "it is extremely difficult, and even more disagreeable, to stop the election of a candidate after he has once been put among the leading fifteen on the preferential list. I have been present at many meetings of the Academy and know that unless there is a definite and important reason against the election of a candidate, he is sure to secure the necessary majority in every case." Frank Schlesinger to Robert G. Aitken, 17 November 1930. Schlesinger Papers, SALO. The context of the discussion was a plan by Aitken to reform the electoral procedures for the NAS.

connection with the preparation of biographical and bibliographical material on Payne-Gaposchkin that was circulated to Academy members. Fred Wright, home secretary of the Academy, reprimanded Shapley "on the ground that some of the statements [in the supporting materials] were laudatory." While Wright admitted that "the distinction between matters of laudatory opinion and conclusions of fact is often very closely drawn . . . an impartial observance of the spirit as well as the letter of the Bylaws requires that there be no departure from strict statements of fact." The home secretary found the following statement offensive. "Her programs in stellar photometry are probably more extensive than those of any other astrophotometrist."[140] It may be that Frederick H. Seares of Mount Wilson or Mount Wilson director Walter S. Adams complained to Wright. The long-standing tension between Harvard and Mount Wilson would only have been heightened by such remarks. Seares may well have felt that his own research in photometry, which extended over more than thirty years, was being slighted in favor of a woman who had not been in the field as long. While no archival evidence has come to light, this interpretation is highly plausible given the context and the actors.

There is a story, part of the oral tradition in Washington, concerning the centenary of the National Academy of Sciences. The occasion was a Camelot production, complete with the dashing young president and foreign dignitaries, all resplendent in academic regalia. As the Marine band played and the assembled academicians and political luminaries marched into the great marble building, someone in the crowd was heard to ask: "But where are the women?" By my count, there were five living female members of the NAS in 1963. Women were, for the most part, still outside the reward structure of American science. Clearly, this was the case for American astronomy. Politics within the astronomical community militated against the election of women and may have played a decisive role in the 1943 rejection of Payne-Gaposchkin. Only in 1981, thirty-eight years later, did the Academy elect its first woman astronomer.

THE EMPLOYMENT OF WOMEN IN THE AMERICAN ASTRONOMICAL COMMUNITY: INTERPRETATIVE CONSIDERATIONS

Differences in employment patterns between women and men in the American astronomical community are difficult to explain using neoclassical economic theory. Evidence discussed in this chapter strongly suggests

140. Fred W. Wright to Harlow Shapley, 18 November 1940. Harvard College Observatory Archives, Director's Files (Harlow Shapley), HUA. For a full statement of Wright's position, see Wright to the Chairman of Sections of the National Academy, 6 December 1933, typed postscript. Stebbins Papers, UWMA.

that there was not, at least as far as women were concerned, a single national labor market, but rather a complex dual labor market that involved several different levels. In recent years, some economists have challenged traditional theory (Doeringer and Piore 1971; Cain 1975, 1976) by developing the concept of the dual labor market. I have made use of these ideas in the discussion that follows.

The experience of women in the American astronomical community is similar to that of various racial or ethnic groups as they sought to enter the labor market. Like blacks (Baron and Hymer 1977: 96), women in many scientific communities have been relegated to positions that are "usually marginal and low paying; they require little skill or formal training." The analogy between blacks and women in the labor market is suggestive. The treatment of one group reflects racism in American culture; the treatment of the other reflects sexism. By the early nineteenth century, American culture had accepted a notion of separate spheres that characterized the abilities and social roles of women and men in diametrically opposed ways. Gender polarity became instrumental in defining woman's work in science. For American astronomy, this meant that most often women became day workers, measuring plates and reducing observations made by men. While women sometimes might be responsible for the preparation of catalogues, they were not encouraged to analyze or draw conclusions from the data. That activity was reserved for men who formulated explanatory theories and models.

By the last quarter of the nineteenth century, a dual labor market developed in American astronomy. Women entered either as computers or as faculty members at the women's colleges, or in a few private coeducational liberal arts colleges. They had little chance of moving beyond basic entry-level positions or limited institutional contexts, in the case of astronomers at the women's colleges. Their opportunities, defined either in terms of career mobility or access to power and the reward system, differed appreciably from those of many men. In the end, gender was more important that educational qualifications. Men and women of similar educational attainment generally started at different entry-level positions. Neoclassical economic theory is of little help in explaining the emergence of women's work in astronomy and the growth of gender-specific patterns of employment that developed into a dual labor market.

Dual labor market theory involves a variety of models, but the following represents the essence of the argument. Labor markets are complex and multi-dimensional. They are appear to be driven by a variety of factors, many of which are not considered by neoclassical theory. But, first, what is a labor market? "Labor markets are arenas . . . in which one or more of the following are similarly structured: employment, movement between jobs, development and differentiation of job skills, wage differences (in their own right or as functions of skills, social status, experience and other determinants). The boundaries of markets depend on a writer's interests and theoretical perspective (Miller 1982: 153)." This model is useful to social historians because it conceptualizes labor markets as a process. Institutional location and various social factors are also taken into consideration.

Basic to most dual labor market theory is the idea of *primary* and *secondary* markets. The primary labor market is defined (Miller 1982: 154) as one in which jobs possess the following traits: "relatively high wages, good working conditions, chances of advancement . . . and, above all, employment stability." Secondary markets are less well structured, have few promotion ladders that permit mobility into primary markets, and are characterized by high turnover. Secondary markets (Miller 1982: 155) often include concentrations of ethnic and racial minorities as well as young persons and women. In short, secondary markets tend to be segregated. Individuals working in a secondary labor market can seldom expect to move into a primary market. Further, dual labor market theory (Miller 1982: 154–55) suggests that primary markets can be divided into two tiers. *Higher-tier* jobs within a primary labor market are characterized by job security, autonomy, high pay and status and opportunities for promotion and mobility. *Lower-tier* jobs, while they provide security, offer only limited autonomy, lower pay, and little if any chance for mobility.

How does this model apply to the material presented in this chapter? Table 9.12 strongly suggests the existence of both primary and secondary labor markets in the American astronomical community. Table 9.12 indicates relations between first or entry-level position, level of educational attainment, status and gender. First positions are grouped into four categories. The categories are fully defined at the bottom of table 9.12. Category 1 represents the lowest observatory or academic rank. Categories 2 through 4 represent successively higher levels conforming to the traditional pattern of assistant, associate, and full professor or astronomer.

The data presented in table 9.12 illuminate relationships between status, gender, educational attainment and the ranking of the entry-level appointment. If there were a simple relationship between educational attainment, and the rank of entry-level positions, individuals with more education should receive starting positions that are higher than those for individuals with less education, without regard to status or gender. This appears *not* to be the case.

In its present form, the relationships between education and entry-level appointment are difficult to discern in table 9.12. This is due to the fact that the majority of individuals start in level-one jobs. To clarify matters, table 9.13 combines categories so that we have four educational levels, but only two grades for entry-level appointments (level one or levels

Table 9.12 First Job (by Group) by Highest Degree, by Status and Gender

	Women by Group N=359. Data missing for 67 individuals.				Rank & File by Group N=480. Data missing for 129 individuals.				Elite by Group N=119. Data missing for 51 individuals.			
Group	1	2	3	4	1	2	3	4	1	2	3	4
H.S.	264 (99%)	2	0	0	199 (94%)	7 (3%)	0	6 (3%)	8 (53%)	5 (33%)	0	2 (13%)
N=	266				212				15			
A.B.	27 (87%)	1 (3%)	0	3 (10%)	38 (66%)	7 (12%)	0	15 (25%)	14 (38%)	12 (32%)	0	11 (30%)
N=	31				60				37			
M.A.	29 (97%)	0	0	1 (3%)	40 (63%)	4 (6%)	0	20 (31%)	13 (62%)	5 (24%)	0	3 (14%)
N=	30				64				21			
Ph.D.	27 (84%)	3 (9%)	0	2 (6%)	97 (68%)	16 (11%)	2 (1%)	27 (19%)	24 (53%)	14 (31%)	0	7 (16%)
N=	32				142				45			
M.D.	0	0	0	0	1	0	0	1	0	0	0	1
N=	0				2				1			

NOTE: Groups are defined as follows. Group 1 includes the ranks/positions of instructor, computer, assistant, observer, post-doctoral fellow, research associate or volunteer assistant. Group 2 includes the ranks/positions of assistant professor, assistant astronomer, astronomer in charge of station. Group 3 includes the ranks of associate professor, associate astronomer. Group 4 includes the ranks of professor or astronomer.

Table 9.13 Highest Level of Educational Attainment by Entry-Level Position (N=958)

	Level of First Position		
	Group 1	Groups 2,3,4	**Total**
Highest level of educational attainment			
High school	471	22	**493**
Baccalaureate degree	79	49	**128**
Master's degree	82	33	**115**
Ph.D./M.D.	149	73	**222**
Total	**781**	**177**	

Table 9.14 Highest Level of Educational Attainment by Entry-Level Position for Women (N=359)

	Level of First Position		
	Group 1	Groups 2,3,4	Total
Highest level of educational attainment			
High school	264	2	**266**
Baccalaureate degree	27	4	**31**
Master's degree	29	1	**30**
Ph.D./M.D.	27	5	**32**
Total	**347**	**12**	

Table 9.15 Highest Level of Educational Attainment by Entry-Level Position for Rank and File (N=480)

	Level of First Position		
	Group 1	Groups 2,3,4	Total
Highest level of educational attainment			
High school	199	13	**212**
Baccalaureate degree	38	22	**60**
Master's degree	40	24	**64**
Ph.D./M.D.	98	46	**144**
Total	**375**	**105**	

Table 9.16 Highest Level of Educational Attainment by Entry-Level Position for the Elite (N=119)

	Level of First Position		
	Group 1	Groups 2,3,4	Total
Highest level of educational attainment			
High school	8	7	**15**
Baccalaureate degree	14	23	**37**
Master's degree	13	8	**21**
Ph.D./M.D.	24	22	**46**
Total	**59**	**60**	

two, three, and four) for the entire sample. When presented in this fashion, it becomes apparent that, overall, educational attainment predicts first job (Chi Square = 136.5, p .0001).

Table 9.14 presents the same data for women. Regardless of educational attainment, it appears that, with very few exceptions, women start in level-one positions. In fact, the first job is close to being a constant for women. Tests of significance are, of course, mute in this case, since there must be some variability in the dependent variable before a relationship can exist.

A quite different picture emerges for rank-and-file men (table 9.15). For this subsample, increasing education is related to increasingly higher levels for entry-level appointments (Chi Square = 56.14, p .0001).

The data for elite astronomers present a different picture. Inspection of table 9.16 indicates that education is not that significant for this group. About half of the elite astronomers with only a high school education start in higher positions than do about half of those with Ph.D.'s. On the other hand, elite astronomers with a bachelor's degree seem to have an advantage, while those with a master's degree do not. When taken together, the effect of education is barely significant (Chi Square = 8.31, p .05), and is driven primarily by the effect of having a bachelor's degree.

These findings are quite interesting. It seems that education provides no advantage for women in the allocation of entry-level positions, but is a significant factor for men who never attain elite status. For the elite, something other than education explains their appointment to first positions at higher entry levels. These factors (see chapter 8) include father's socioeconomic status, pre-career publications and NAS contacts (see figure 8.2 and the related discussion).

Economists Francine Blau and Carol Jusenius (1976: 192) argue that "the advancement opportunities open to workers within the enterprise are generally determined by the original entry-level job the worker has obtained." If it is the case that women are assigned different entry-level positions than men with the same level of education, then (Blau and Jusenius 1976: 195) we can infer that these patterns indicate "employer decisions to exclude women from certain entry-level positions and their associated promotion ladders and/or to promote and upgrade women more slowly than men."

Clearly, not all women were trapped in a secondary labor market. Some (often with Ph.D.'s) moved to tenure-track positions in the women's colleges, but here too, there are significant differences. While men in primary labor markets tended to be on the higher tier, women were generally on the lower. Women had job security on the lower tier of the primary labor market, but in comparison to men occupying the higher tier they had very limited autonomy, their pay was generally less than that of men, and there were virtually no mobility ladders leading out of the women's colleges to positions at public or private universities or research observatories. How many women moved from professorial rank at Smith or Vassar to a senior position at Harvard, Chicago, or California? How many directors of observatories at the women's colleges became astronomers at the Lick or Yerkes? To ask the question is to answer it.[141] Positions at the women's colleges were terminal; from those positions there was virtually no mobility, at least for female astronomers.

Limited data concerning the move to a second job indicate that women were catching up in the sense that most second positions were an improvement over entry-level jobs. This has to be understood, however, as entry into the lower tier of the primary labor market. While over 80 percent of second jobs for women were distributed across groups two through four (for definitions of groups, see table 9.12), it is clear that they were almost all located in lower tier positions in primary markets. *Future mobility would be up the academic ladder, but within the same institution.*

Further research into the dual labor market in scientific communities may illuminate some of the most interesting problems in both the history of science and women's history. As sociologists Bielby and Baron (1984: 28) suggest, "Much more has been written . . . about why employers treat men and women differently than about the extent to which they do so." Using a variety of sources, it is possible to develop quantitative data on the employment of women in scientific communities that illuminate the extent of gender-based economic and social differences.

This chapter has examined the experience of women in the American astronomical community from several perspectives. The new feminist scholarship seeks to utilize such disciplines as history, sociology and economics in order to explain some of the most complex aspects of American life. The history of science has an important contribution to make to these investigations.

This discussion has provided the materials with which to answer Matthew Arnold's question, posed when he visited Wellesley College during Annie Jump Cannon's senior year. What, indeed, were the chances for young women intending to make a career in astronomy? The answer is complex and rests on an understanding of the impact of American culture on the values and attitudes of male members of the astronomical community as well as on an understanding of the emergence of gender-based differences in scientific work and career patterns. In American astronomy,

141. Hazel Losh, who moved from an instructorship at Smith to eventual tenure at Michigan via a stop at Mount Wilson as a computer, is the only case that I know of.

the dual labor market that emerged by the end of the nineteenth century relegated women to the lower tier or secondary labor markets, thus sharply restricting their chances for mobility. At the same time men's perceptions of women as scientists denied them access to power and the reward system. However optimistic women like Cannon may have been, the answer to Arnold's question is in the negative.

10 ✦

TERMINUS AD QUEM: AMERICAN ASTRONOMY IN 1940

Chapter 2 set the stage for this book with a portrait of American astronomy in 1859. Chapter 10 rounds out the discussion with a picture of American astronomy in 1940. Quantitative and qualitative data indicate the dimensions and patterns of growth and change in the years between 1859 and the coming of World War II.

ASTRONOMY AND THE CULTURE OF AMERICAN SCIENCE

In 1859 astronomy was struggling for a place in the culture of modern science. By 1940, astronomy had established itself as the biggest of the American sciences in the age of little science. This claim does not relate to the size of the community. Numerically, chemistry and biology were far larger. The definition of Big Science used here relates to the structure and organization of American astronomy, as well as to its capitalization (Lankford and Slavings 1996). American astronomy was organized around factory observatories that produced scientific knowledge on an assembly line basis. These institutions were labor intensive, requiring a large work force ranging from Ph.D.'s to high school graduates. Factory observatories depended on expensive instrumentation. Because theirs was a data-driven science, astronomers were always planning larger telescopes and new auxiliary equipment. Factory observatories also were capital-intensive and depended on a continuous infusion of funds. Their closest competitors were the laboratories of industrial chemists organized for the discovery of new and profitable substances such as nylon or the Berkeley operation of high-energy physicist Ernest O. Lawrence, who developed the cyclotron.

Astronomy was arguably the most popular of the American sciences. Astronomical subjects almost always made good copy. During the 1930s, for example, the media covered the casting of the 200-inch mirror for the

Mount Palomar telescope and its journey from Corning, New York, to Pasadena (Woodbury 1966: chap. 28). A rich amateur tradition provided a support group for American astronomy and a pool from which new recruits were drawn. The place of astronomy in American culture did not rest on its contributions to economic growth and development. In an increasingly secular society, astronomy became a vehicle for transcendence, providing individuals access to spiritual values without encumbering dogmas.

Cast in the mold of industrial capitalism and resting on broad-based popular support, astronomy occupied a secure niche in the national scientific community and the larger American culture. And by 1940 American astronomy had reached international eminence as well. In 1859 American astronomers looked to Europe for publication opportunities and for recognition. By 1940, European astronomers were publishing in American journals and were honored by election as foreign associates of the National Academy of Sciences.[1]

AMERICAN ASTRONOMY IN 1940: DEMOGRAPHIC CONSIDERATIONS

In 1940, the American astronomical community numbered 342, a fivefold increase over 1859 (see tables 2.1 and 10.1). In 1859 virtually all American astronomers were engaged in astrometry or celestial mechanics. By 1940 just over half the community devoted its attention to astrophysics. In 1859 elite astronomers bulked large in the community (48 percent). By 1940 there was a more complex social mix including elite (17 percent), the rank and file (60 percent), and women (23 percent).

In 1859 four-fifths of the community had been born in New England and almost all the rest in the mid-Atlantic region. In 1940, over a third came from the Midwest, the rest from the mid-Atlantic region and then New England. Astronomers were also born in the South and the West (5 percent each). In 1859 a fifth of the community were non-American-born astronomers. By 1940 the figure had dropped to 15 percent. Canadians led the way, followed by astronomers from Holland, Russia, the United Kingdom, and the Scandinavian countries.

In 1940 the mean age of American astronomers was 45.5 years.[2] Predictably, members of the elite were older than the rank and file. However,

1. An examination of major journals suggests the following. In 1940, no American astronomer published in the *Astronomische Nachrichten* and two published in the *Monthly Notices of the Royal Astronomical Society* and the *Annals Astrophysique*. However, ten non-Americans published in the *Astrophysical Journal* in 1940.

2. Thronson and Lindstedt (1986) suggest that the American astronomical community in the 1980s was aging at a rapid rate. Interestingly, the community in 1940 seems to fit Thronson and Lindstedt's aging parameter. Recruitment problems through the 1920s and the Great Depression defined the age structure of the community in 1940.

Table 10.1 Fields and Statuses of Members of the American Astronomical Community in 1940 (N=342)

	Astronomy Type				
	New Astronomy	Old Astronomy	Mixed Astronomy	Unknown	**Status Total**
Status					
Elite	42	13	3	0	58
Rank & file	107	61	7	30	205
Women	31	33	7	8	79
Astronomy type total	180	107	17	38	342

the seeds of change were evident, especially if we look at some of the new entrants. In the period from 1935 to 1940, a number of young Ph.D.'s entered the community who would become major figures after mid-century. The list includes James Baker (1936), S. Chandrasekhar (1936), Carl Seyfert (1936), Jesse Greenstein (1937), Louis Henyey (1937), Martin Schwarzschild (1937), Horace Babcock (1938), Leo Goldberg (1938), Lyman Spitzer (1938), and Lawrence Aller (1943). These men would make major contributions to astronomical knowledge and often assume important administrative positions in the postwar astronomical community.

Information on the social origins of this population is fragmentary, but it suggests that for the elite, over half came from homes where the father was either a professional or a white- collar worker. The pattern is similar to that of the 1859 population. Unfortunately, missing data restrict generalization. Virtually nothing is known about the social origins of the rank and file or women.

In 1859, almost four-fifths of the American astronomical community could point to the baccalaureate as the highest level of educational attainment. By 1940 (table 10.2) more than half of the community had earned the doctorate, although there were still a number whose highest level of educational attaintment was an undergraduate degree, and 18 percent had no more than a high school diploma. More workers in the new astronomy held the Ph.D. than did those in the old astronomy.

Twelve of the 342 members of the American astronomical community in 1940 had studied in Europe, but only two (Thornton Page and Edna Carter) returned with a Ph.D. As indicated in chapter 4, American astronomers generally did not participate in the quest for European Ph.D.'s. As early as 1900, astronomy could point to a self-replicating system: American graduate schools producing American Ph.D.'s.

Table 10.2 Educational Levels of American Astronomers Active in 1940 (N=342)

Highest Earned Degree	
Ph.D.	194 (57%)
M.A.	57 (17%)
A.B.	30 (9%)
High school only	61 (18%)

In 1940 the typical American astronomer was a male, about forty-five years old, working in some branch of astrophysics, and the holder of a Ph.D. He was probably born in the Midwest and may well have done his graduate work at Chicago or Berkeley.

EMPLOYMENT PATTERNS

Tables 10.3 through 10.5 report data on the employment of American astronomers in 1940. In 1859, almost half of elite astronomers were employed by the federal government (table 2.1). This stands in marked contrast to the situation in 1940. Government science had ceased to be attractive to members of the elite. In part, this is explained by the decline of astrometry and celestial mechanics as the principal research fields of American astronomers. For elite astronomers (table 10.3) working in both the old and new astronomy, universities provided the most attractive employment sites, with private universities slightly ahead of public. Private research institutions like the Mount Wilson Observatory followed universities. But here there is an interesting difference according to astronomy type. More astrophysicists (43 percent) preferred private research institutions than did workers in the old astronomy (23 percent).

In contrast to the elite, the rank and file (table 10.4) present different employment patterns. Clearly, the rank and file provided astronomers for government research institutions such as the Naval Observatory and the Nautical Almanac Office. Rank-and-file workers in both the old and new astronomy were also found in universities, but astrophysicists preferred private universities while workers in the old astronomy were found more often in public institutions. Unlike the elite, members of the rank and file were also employed by colleges (mostly private) and some worked in private research institutions. This latter group was much smaller than the number of elite astronomers found in private research institutions.

In 1940 (table 10.5) very few women were employed by federal research institutions. Women in the new astronomy were more often (48 percent) located in university settings (most frequently private universities)

Table 10.3 Employment of Elite American Astronomers in 1940 (N=58)

	Astronomy Type			
	New Astronomy N=42	Old Astronomy N=13	Mixed Astronomy N=3	Unknown N=0
Type of Institution				
Government	1	0	0	
Public university	11	4	0	
Private university	11	5	1	
Public college	0	0	0	
Private college	1	1	1	
Private research institution	18	3	1	
Other	0	0	0	

Table 10.4 Employment of Rank-and-File American Astronomers in 1940 (N=205)

	Astronomy Type			
	New Astronomy N=107	Old Astronomy N=61	Mixed Astronomy N=7	Unknown N=30
Type of Institution				
Government	3	25	0	0
Public university	30	16	3	6
Private university	42	9	0	4
Public college	3	0	0	3
Private college	13	6	1	11
Private research institution	10	2	3	0
Other	6	3	0	6

with private research institutions coming second (38 percent). Two-thirds of the women working in the old astronomy were found in private research institutions with only 15 percent in universities. About 15 percent of the women were employed in colleges.

Employment patterns, compared with those of 1859, had changed dramatically. The emergence of the American research university (both public and private) provided a range of employment options. So did the

Table 10.5 Employment of Women in American Astronomy in 1940 (N=79)

	Astronomy Type			
	New Astronomy N=31	Old Astronomy N=33	Mixed Astronomy N=7	Unknown N=8
Type of Institution				
Government	3	1	0	0
Public university	4	3	0	1
Private university	9	2	0	2
Public college	1	0	0	0
Private college	1	3	3	4
Private research institution	12	22	4	0
Other	1	2	0	1

development of private research institutions. By 1940 the federal government no longer offered inviting employment opportunities and colleges also declined as viable employment sites. This pattern of institutionalization was an important characteristic of astronomy in comparison to the other sciences in America.

Professional activities of astronomers are closely related to employment sites. Tables 10.6 through 10.8 report data on the professional activities of American astronomers in 1940. Over a third of elite astronomers who specialized in astrophysics spent all of their time on research. If we combine this with the category of research and administration, the figure rises to over two-thirds. Elite astronomers working in astrometry and celestial mechanics fall short of these levels with a combined figure of 54 percent devoting full time to research or research and administration. If we combine categories related to teaching, approximately the same percentage of both groups of elite astronomers were involved in undergraduate and graduate education. No elite astronomer devoted full time to the classroom.

By far the most interesting data in table 10.7 concerns teaching. Fully a quarter of the rank and file devoted their energies to teaching alone while a third were full-time researchers. With categories combined, it appears that more rank-and-file astrophysicists were involved in teaching than workers in astrometry and celestial mechanics. By the same token, workers in the old astronomy devoted more of their time to research than their colleagues in astrophysics.

Table 10.8 reminds us that women working in the American astro-

Table 10.6 Professional Activities of Elite American Astronomers in 1940 (N=58)

	Astronomy Type			
	New Astronomy N=42	Old Astronomy N=13	Mixed Astronomy N=3	Unknown N=0
Activity				
Research	16	4	1	
Teaching & research	9	2	0	
Research & administration	13	3	0	
Teaching, research & administration	4	3	2	
Teaching	0	0	0	
Administration	0	1	0	
Other	0	0	0	

Table 10.7 Professional Activities of Rank-and-File American Astronomers in 1940 (N=205)

	Astronomy Type			
	New Astronomy N=107	Old Astronomy N=61	Mixed Astronomy N=7	Unknown N=30
Activity				
Research	39	27	3	0
Teaching & research	28	13	1	6
Research & administration	6	7	0	0
Teaching, research & administration	6	5	0	3
Teaching	23	8	2	18
Administration	0	0	0	1
Other	5	1	1	2

nomical community were not classroom teachers. That task was left to the rank and file (table 10.4). Women were involved in the research enterprise, working frequently in private research institutions or university settings. That almost 80 percent of the women working in astrophysics and 88 percent involved in astrometry devoted full time to research stands in

Table 10.8 Professional Activities of Women in American Astronomy in 1940 (N=79)

	Astronomy Type			
	New Astronomy N=31	Old Astronomy N=33	Mixed Astronomy N=7	Unknown N=8
Activity				
Research	24	29	3	4
Teaching & research	2	1	1	0
Research & administration	1	1	0	0
Teaching, research & administration	0	0	1	0
Teaching	3	1	2	3
Administration	0	1	0	0
Other	1	0	0	1

sharp contrast to the traditional view that, in science, women teach and men do research. A mere 11 percent of women active in the American astronomical community in 1940 were involved in teaching alone.

THE POLITICAL ECONOMY OF AMERICAN ASTRONOMY

The political economy of a scientific community brings into sharp relief the social and political processes vital for its growth and development (Kohler 1982: 1–8). Further, the political economy of a scientific community defines the operations and guides the activities of institutions, individuals, and groups that make up the community. This section examines major elements of the political economy of the American astronomical community and reviews developments since 1859.

Disputes over demarcation of boundaries and criteria for membership place severe strains on the political economy of a scientific community. The most serious demarcation problem that faced American astronomy involved the legitimation of astrophysics (chapter 3). Astrophysics was born outside the astronomical community and was legitimated only when the generic community accepted astrophysics as an appropriate scientific activity.

A second demarcation issue concerned membership in the American astronomical community. In comparison to other American scientific communities, astronomy did not emphasize academic credentialling as a primary requirement for membership (chapter 4). The Ph.D. did not become normative until the mid-twentieth century. This kept the astronomical community open to a wide variety of cognitive and technical perspectives.

The legitimation of the new astronomy and the relative ease of entry into the community suggest another element in its political economy: openness based on social, cognitive, and institutional heterogeneity. In this sense, American astronomy differed from many other sciences in America (chapter 11). The political economy of the American astronomical community reflected a diversity of scientific interests and institutions, as well as more traditional social, generational, and gender differences.

The allocation of resources is an important function of the political economy of the astronomical community. Once the legitimacy of the new astronomy had been established, another competitor joined the struggle for resources. The clash between Hale and Boss over the Carnegie Institution of Washington's (CIW) patronage suggests that, at the beginning of the twentieth century, the community had few safeguards against divisive competition for resources (chapter 7). The northeastern Neighbors group and the committee that allocated National Research Fund patronage under the supervision of Henry Norris Russell provided mechanisms to minimize conflict over scarce resources.

Market relations of the American astronomical community were complex. In 1859 the old astronomy had several markets for its products. The navy and civilians involved in ocean-born commerce relied on navigational data in the *American Ephemeris and Nautical Almanac*, based in part on observations made by astronomers at the Naval Observatory. Other astronomers constituted a secondary market.

By 1940 the old astronomy was clearly divided into an applied component that produced data useful for navigation and time-keeping and a much smaller group of astronomers working on basic research. It is important to note, however, that for both celestial mechanics and astrometry, much had changed since the mid-nineteenth century. Perhaps most significant was a feeling that there were few challenging problems left to solve. The collection and analysis of data had become routine. Second-generation mechanical computing devices increasingly took the place of human computers. Astrometric data were collected wholesale by photography. Time-keeping had been simplified by photographic instrumentation and an international shortwave radio network. The union of celestial mechanics and general relativity lay in the future, and the application of techniques of the old astronomy to the dynamics of the Milky Way galaxy was just beginning. Kinematic investigations of galaxies external to the Milky Way were also in their infancy.

In 1859 the new astronomy had no markets at all. Other scientists (especially physicists) were the first consumers of astrophysical knowledge. Langley and others tried to market solar physics as a tool for long-range weather prediction on the basis of variations in the solar constant, sun-spot cycles and the like (chapter 3; DeVorkin 1990; Eddy 1990). But these hopes were never realized. Workers in the new astronomy also turned to the literate middle class for a popular market. News of their discoveries made better copy than did mathematical studies of planetary orbits or stellar motions.

By 1940 the new astronomy had found a secure market in other astrophysicists and, to a lesser extent, scientists involved in what would now be called high-energy physics. This market was international in scope. American data were used by European astrophysical theorists, while Americans active in observational astrophysics relied on theory imported from Europe to interpret their data. By the 1920s, theoretical physicists were working on the stellar energy problem (Hufbauer 1981) and some physicists were publishing in the *Astrophysical Journal.*

The organization of research was a central element of the political economy of American astronomy. In 1859 research institutions were small-scale establishments. Eighty years later, a wide range of research institutions had been established. At the high end of the scale stood the factory observatory with its large telescope(s), staff of professional astronomers and corps of female computers as well as clerical and technical assistants, presided over by a powerful director whose responsibilities ranged from selecting research problems to securing funding. Observatories at private and public universities were generally organized on a smaller scale. Faculty were involved in both undergraduate and graduate education as well as research. Government facilities for astronomical research were much less important than in 1859 but still provided employment for astronomers, especially members of the rank and file (table 10.4). College observatories were primarily teaching institutions whose faculty sometimes were involved in routine research programs that could be carried out with limited equipment.

Further, the organization of astronomical research reflected a hierarchy arranged according to both the quality and the quantity of astronomical knowledge produced. This hierarchy also mirrored the distribution of status and power in the community as well as differential access to the reward system.

Style was also an important aspect of the political economy of American astronomy. By this I do not mean the style of scientific research (Harwood 1993; Kohler 1991c) but rather the personal styles of leading individuals as they created new institutions or acted out various roles in the community. This topic is not often discussed by historians of science (Kohler 1991b).[3]

3. I am indebted to Professor Robert E. Kohler of the University of Pennsylvania for a discussion of this problem at the Santa Fe meeting of the History of Science Society in 1993.

By the 1890s, as astronomical research in the United States expanded dramatically, the astronomer-entrepreneur appeared. The actions of these men were often cast in an imperial mold; for them science was organized on a grand scale. Pickering's insistence on imposing definitions of stellar magnitude and a system of spectroscopic classification, Hale's coup in securing Carnegie Institution funding to create the Mount Wilson Observatory or his creation of the *Astrophysical Journal*, as well as dramatic confrontations between Lowell and Campbell provide examples of the autocratic, imperial style that characterized many heads of observatories and leaders of the American astronomical community in the late nineteenth and early twentieth centuries (chapter 7).

The imperial leadership style was part and parcel of the organizational models borrowed from the larger culture. When astronomers like Pickering, Hale, and Campbell reached out and appropriated the large-scale bureaucratic model of the corporation from the industrial economy, they also took over leadership models. Complex large-scale economic institutions were managed from the top down by dynamic, often flamboyant individuals. Industrial leaders prided themselves on making the American economy one of the most productive in the world. But they operated within a fiercely competitive context. The counterparts of these captains of industry at the Mount Wilson, Harvard, or Lick observatories strove to achieve the same levels of production and often managed to reach similar levels of competition and conflict.

In an age which valued the man of action, whether a Theodore Roosevelt or an Andrew Carnegie, the leadership style of the captains of American astronomy fit comfortably with patterns in the larger American culture. And, at least by bottom-line indicators, the leaders of American astronomy succeeded. By the eve of World War I, American astronomy had reached world-class status, attracting far more Europeans for study and research in this country than Europe attracted Americans for study abroad (chapter 4).

In short, those who practiced astronomy in the imperial manner were men of unbridled ambition, vast egos, and great vision, whose projects must have seemed grandiose to contemporaries of lesser imagination. These imperialists never doubted the correctness of their views and were consummate political manipulators. When the last of these titans passed from the scene, their bureaucratic successors seemed colorless by comparison.

In both form and content, American astronomy in 1940 bore little relation to the fledgling community of 1859. Change had altered the cognitive and institutional landscape. A remarkably open and pluralistic science, American astronomy differed strikingly from many of the other sciences in America. Comparisons between astronomy and other American sciences will be developed in chapter 11.

11

ASTRONOMY COMPARED

An extensive literature on the history of American science provides rich resources for comparative analysis.[1] Of necessity, comparative history involves a level of generalization that erodes the rich particularity of the history of individual scientific disciplines (Harwood 1993: 3). There is, however, so much to be gained from comparative analysis that the risk seems worth taking. Comparative analysis highlights differences and similarities between the sciences. This, in turn, helps us recognize factors that have shaped individual scientific disciplines. By extending comparisons across national boundaries we can identify patterns (e.g., demographic, institutional, or organizational) that make astronomy in the United States different from astronomy in the United Kingdom, France, or Germany. As Charles Rosenberg points out (1979: 442), "differences between academic disciplines or among professions are, after all, at least as instructive as their similarities." Differences often stand out in high relief and can provide insights into the political economy, social structure, and development of astronomy.

IN THE AMERICAN CONTEXT: ASTRONOMY, CHEMISTRY, AND BIOLOGY

The population of scientists starred in successive volumes of J. McKeen Cattell's *American Men of Science* (1906–44) provides accordant data for demographic comparisons between astronomy, chemistry, physics, and the biological sciences (here defined as Cattell's categories of botany, zoology, and physiology). From 1906 through 1944, the last edition of *AMS*

1. The best review of the historiography of individual sciences in America is contained in Kohlstedt and Rossiter (1985: 97–209).

in which stars were awarded, 138 astronomers earned stars (Visher 1947: 11). Almost twice as many botanists were starred (257) and nearly three times as many zoologists (378). Among biologists, only the starred physiologists (115) were fewer in number than astronomers. If we combine zoologists, botanists and physiologists the number of individuals starred (750) is five and a half times that of astronomers. The number of chemists starred (468) is almost three and a half times greater than astronomers, and the number of physicists (377) over two and a half times larger. These data suggest the relative size of scientific communities in the first half of the twentieth century. Biology ranked first, followed by chemistry and then physics.

Since the early nineteenth century, scientists have made up a conspicuous if numerically small category of immigrants to the United States. Twenty-one percent of the physiologists and 19 percent of starred physicists and astronomers were non-American-born (Visher 1947: 427–75). Of chemists starred, 17 percent were non-American born. Zoologists counted only 11 percent, while botany was the most American of the biological sciences, with a mere 9 percent of its starred practitioners from abroad.[2]

The educational attainment (Visher 1947: 150–372) of starred astronomers compared with other disciplines underscores trends discussed in chapter 4. While zoologists counted more starred scientists with only a high school education (12 percent), they were followed by astronomers (9 percent). Among the other groups of starred scientists, 6 percent of physicists had no more than a high school education, botanists 5 percent, chemists 3 percent, and physiologists less than 1 percent.

Data on educational attainment are most revealing at the Ph.D. level. Forty-three percent of starred astronomers did not hold the doctorate, while the average for chemists, physicists and biological scientists was 22 percent. These data suggest that American astronomy stood in a very different relationship to the Ph.D. machine (Kohler 1990) than did many other sciences. Perhaps part of the explanation lies in the fact that the other sciences were firmly rooted in academia, while astronomy flourished in a variety of institutional niches. Further, there appear to be differences between astrometry and celestial mechanics on the one hand and astrophysics on the other. At least for those who entered the American astronomical community after 1900, practitioners of the old astronomy seemed

2. While the rise of Hitler played an important role in sending European scientists to America, the movement of astronomers to the U.S. was an established fact long before the 1930s. A quantitative and qualitative investigation of non-American-born scientists in different scientific disciplines would make an interesting study. Such an investigation should take care to identify patterns of immigration in the century before the 1930s.

Table 11.1 Comparative Institutionalization of Selected American Sciences Illustrating the Top Three Employment Sites for *AMS*-Starred Scientists in Selected Disciplines

	Astronomy	Botany	Chemistry	Physics	Physiology	Zoology
Private universities	1	2	1	1	1	1
State universities	3	1	2*	3	2	2
Private research institutions	2	3			3	3
Industry			2*	2		

*Chemists tied between state universities and industry with 23% of starred chemists in each employment site. These three categories (private and state universities and industry) account for over 80% of the employment sites for starred chemists.

SOURCE: Visher (1947: 488–97).

less inclined to take the doctorate than those who went into astrophysics.

An examination of leading employment sites for these six sciences (Visher 1947: 488–97), provides interesting perspectives (table 11.1). With one exception, *AMS* star holders preferred private universities as their first choice for employment. Botany thrived in state universities, with its second choice private universities. Botany probably reflects ties with the schools of agriculture and agricultural experiment stations associated with land grant institutions. Physiologists and zoologists found state universities most appropriate as a second employment venue, while for chemists, state universities and private industry tied for second place. The second choice for astronomers was the private research institutions such as the Mount Wilson Observatory, while physicists selected positions in industry. For both astronomers and physicists, state universities ranked third as employment sites. Botanists, physiologists, and zoologists opted for private research institutions as their third choice.

These patterns provide some indication of the ways in which the sciences were institutionalized in pre–World War II America and support Charles Rosenberg's view (1979: 445) that diversity of context is characteristic of American science in the twentieth century. Private research institutions played a more important role in astronomy than in any of the other sciences discussed here. Industrial employment was important for chemistry and physics, while all but botany preferred private to public universities as a first choice.

Even on the basis of evidence limited to elite scientists, it is clear that there were important demographic and structural differences between the sciences in America. Astronomy and physiology were the smallest of the six disciplines, but for astronomy, size and cultural significance were inversely correlated. Much larger scientific communities such as chemistry had less significance for American culture than did astronomy.[3]

Chemistry and Astronomy: Political Economy and Cultural Connections
Comparing the political economy of American chemistry and astronomy suggests interesting differences and similarities.[4] Perhaps the most obvious difference involves the centrality of industry for chemistry (Thackray et al. 1985: 83). Industrial patrons provided funds for scholarships and fellowships and supported faculty research. In addition, industry employed a large percentage of university-trained chemists, after 1920 about 70 percent (Thackray et al. 1985: 106). Chemistry demonstrated its economic value through a stream of products that improved the quality of life and stimulated economic growth and development. While astronomy played an important role in improving navigation and time keeping, it could never compete with chemistry in providing "Better Things for Better Living" (Thackray et al. 1985: 102).

Both the size and cognitive complexity of chemistry are reflected in the development of the American chemical community.[5] At the end of the nineteenth century, chemistry boasted two journals while astronomy had three. However, by the eve of World War II, while American astronomers relied on the same three journals to communicate their research, chemists could choose from eleven (Thackray et al. 1985: 439). When the Astronomical and Astrophysical Society of America (AASA) was founded (1899), chemists were attending meetings of five societies. Between 1900 and World War II they added a dozen more. By 1940 the American Chemi-

3. A complete reanalysis of Visher's data would provide important insights and help develop a deeper comparative understanding of American science in the first half of the twentieth century.

4. Beyond discussions based on the population of starred scientists, there is no attempt to compare physics and astronomy. Historians of American physics are concerned with problems so different from those considered in this book that comparisons are really not possible. As Moyer (1985:163–68) suggests, the historiography of American physics is characterized by interest in two closely related problems: the development of theoretical physics after World War I and the Manhattan Project. Few historians of physics examine nineteenth-century topics. Further, historians of physics tend to deal with the epistemic activities of elite members of the discipline. Without comparable historical data, comparative discussions degenerate into mere speculation. Empirically grounded comparative studies of astronomy and physics would be an important addition to the literature. Quantitative data on the American physics community can be found in Forman, Heilbron, and Weart (1975), Kevles (1979), Weart (1979), and Reingold and Brodansky (1985). However, these data are not easily compared with the astronomical community.

5. The American Chemical Society grew from under 2,000 in 1900 to over 25,000 in 1940 (Thackray et al. 1985: 446). This means that in 1900 the chemical community was already larger than all of astronomy during the period from 1859 to 1940.

cal Society (established almost a quarter century before the AASA) had twenty divisions. Only after World War II did the American Astronomical Society reluctantly create divisions, and the Astronomical Society of the Pacific attained national stature only in the post-war era (Bracher 1989)

If the doctorate played an ambiguous role in the formation of professional identity and the definition of membership in the astronomical community, chemists had no such problems. Thackray et al. report (1985: 147) that of Ph.D.'s in the physical sciences awarded before 1900, chemistry had the largest number (251), while astronomy Ph.D.'s amounted to only 22. In the same period, Ph.D. production in the biological sciences totalled 219 (Thackray et al. 1985: 375–77).

Chemistry was institutionalized in a number of niches (Thackray et al. 1985: 125–38). These ranged from federal, state, and local governments through industry and academe. Employment sites for chemists were almost as diverse as those available to astronomers, but the consequences of diversity differed between the two communities. For chemists, industry became the major employer. This guaranteed that chemistry would have a significant applied component. Except for astronomers associated with the federal government, astronomy could make no claims as an applied science.

Many more high school students were exposed to chemistry than to astronomy. Between 1890 and the mid-1930s, approximately 9 percent of American high school students took courses in chemistry. While there are no figures for astronomy, there is little evidence to suggest that it had a comparable position in the secondary science curriculum.[6] Generally, high school students chose from courses in biology, physics, and chemistry, with biology being the most popular (Thackray et al. 1985: 304). Only after the Sputnik crisis of the 1950s did funding become available to equip some high schools with telescopes and planetariums.

Given the number of secondary school students who studied chemistry, one might assume that literate Americans would be sensitive to the activities of chemists and have a general idea of what the science of chemistry entailed. In fact, during these years, the educated middle class was indifferent to chemistry (Thackray et al. 1985: 82). Physics, biology, and the health sciences received far more attention than chemistry (Thackray et al. 1985: 76) .

It is precisely here that we find important differences between chemistry and astronomy. Of the sciences in America, astronomy has the deepest cultural roots. This state of affairs is not of recent origin (chapter 2)

6. For an analysis of astronomy in the secondary school curriculum at the end of the nineteenth century, see Jordan D. Marché, "The Planetarium in America. 1930–1970: A Social History." Ph.D. dissertation, Indiana University. Forthcoming.

but can be traced at least to the early nineteenth century. Public support for astronomy in the pre–Civil War era rested on its spiritual and cultural value, not on any material contributions astronomers might make to American life. Apparently Americans valued astronomy as a way of achieving spiritual transcendence. Even in the pre–Civil War era of keen sectarian competition, Protestants of whatever denomination could agree that astronomy was a powerful teacher of spiritual values. As Gilded Age materialism threatened to erode evangelical religion, the glories of the heavens served to remind Americans that there were higher values than the "Bitch goddess of success."

In short, Emerson was wrong. In an age before megalopolis obliterated the night, the average man and woman did indeed know the names of many stars and constellations and took comfort from the spiritual messages they read in the starry heavens. The willingness of individuals, from those who contributed pennies to build the Cincinnati Observatory to wealthy philanthropists, to support astronomy (Miller 1970: chap. 5) provides ample proof of the relationship between astronomy and American cultural values.

An examination of nineteenth-century newspapers suggests that the building of observatories received wide coverage and that competition between observatories was a favorite subject of reporters by the 1890s (chap. 7).[7] More important, with the establishment of the *Sidereal Messenger* (1882), America had its first modern popular astronomical journal. In time the *Sidereal Messenger* became *Popular Astronomy* (1893), and in the 1930s periodicals such as *The Sky* and *The Telescope* appeared. Later (1941) these last two publications merged to form *Sky & Telescope*.

The growth of amateur astronomy points to the development of a new element in the political economy of American astronomy. Confrontations between amateurs and professionals (Lankford 1981a; 1981b), were eventually superseded by cooperation on terms dictated by professionals. The founding in 1911 of the American Association of Variable Star Observers and the American Meteor Society indicates the emerging division of labor (Rothenberg 1981). Amateurs provided reliable observational data that professionals then interpreted. Amateurs were also innovators in the development of new instrumentation. For example, after World War I, the Reverend Joel Metcalf fabricated state-of-the-art photographic refractors for the Harvard College Observatory, and California amateurs built the first Schmidt cameras in the United States (Williams 1993).

7. Newspaper clippings in the Lick and Yerkes Observatory archives attest to these contentions. At the Lick there is also documentation of media treatment of observatory politics as California newspapers took sides in dealing with the directorship of Edward S. Holden (Osterbrock 1984b).

In the early twentieth century, the amateur telescope-makers movement emerged. Nourished by the editor of *Scientific American* (Williams 1992), telescope-making clubs appeared, and grinding six-inch (or larger) parabolic mirrors occupied the spare time of countless Americans (Willard 1976). A few amateurs used their instruments for research in variable star astronomy, study of the moon and planets, or a search for comets.

Astronomy clubs proliferated in the early years of the twentieth century and institutionalized the culture of amateur astronomy. Defined in part by a longing for transcendence and in part by the desire for technical mastery, the culture of amateur astronomy reflected a mixture of the spiritual and material. Perhaps the best source for understanding the rich culture of amateur astronomy is the myriad poems that appeared in amateur journals and books through the 1940s, a body of material that awaits careful analysis. Most of the verse is doggerel, but its content suggests key elements in the culture of amateur astronomy: knowledge of God and the beauty of the sidereal universe, as well as the ability to fabricate instruments that will provide pleasing views of the heavens. A third element in the culture of amateur astronomy in the United States was the desire of some amateurs to contribute data useful to professionals or to produce technical innovations that enhanced the professional's ability to make observations. Amateur astronomy in America was virtually an exclusively male culture. Only after mid-century did women appear in amateur organizations.

Professional astronomers encouraged the amateurs. Many were willing to speak to amateur groups and some, for example, Shapley and Menzel at Harvard, Struve at Yerkes, and Russell at Princeton, wrote popular accounts of the latest research in astronomy for amateur consumption. Between the two world wars, planetariums opened in New York, Chicago, Pittsburgh, Philadelphia, and Los Angeles. These institutions provided centers for amateur activities and disseminated knowledge of astronomy to the general public.

In a larger sense, amateurs provided a support group for professional astronomers. This aspect of the political economy of astronomy would take on added significance after World War II as an important element in a political climate that made possible the creation of a system of national observatories and space-based instruments paid for by federal tax dollars. And, of course, professionals were always on the lookout for bright young people who might be recruited from the ranks of amateurs and encouraged to prepare for careers in astronomy.

No equivalent of the amateur tradition in astronomy existed in the field of chemistry. While high school students might study chemistry and even find Gilbert chemistry sets under the Christmas tree, there were few amateur chemistry clubs, or popular journals devoted to chemistry,[8] and no national amateur organizations whose members made contributions to experimental chemistry or provided new instrumentation for use in the laboratory. Laypersons did not seek to develop technical skills or find spiritual solace in chemistry. While the chemical laboratory might be a symbol in the popular press (LaFollette 1990:112), it lacked deep personal meaning for any but the professional. Compared to a solar eclipse or meteor shower, the carbon atom held little popular appeal. The complex nature of the amateur tradition in American astronomy and the manner in which it linked the astronomical community to the larger culture suggest one of the most important ways in which astronomy differed from the other physical sciences in the United States.

Biology and Astronomy: Political Economy and the Growth of New Sciences

Parallels between the rise of astrophysics and classical genetics are fascinating.[9] Although the transformation of American astronomy started forty years before changes in biology, it can be argued that morphology and physiology were the equivalent of the old astronomy (celestial mechanics and astrometry) and that classical genetics provides an analogue to astrophysics. In each case the emergence of the new science was marked by a defining moment. For astronomy it was the publication of the experiments of Bunsen and Kirchhoff in 1859 and for biology the recovery of the work of Gregor Mendel in 1900 (Bowler 1983).

A comparison of the origins of astrophysics and classical genetics provides illuminating perspectives. Astrophysics in the United States was a self-conscious creation involving political, social, and epistemic motives detailed in chapter 3. Development of the new astronomy was contingent on the vision and ability of its founders and a first generation of followers.[10]

8. In the 1940s Helen Miles Davis edited *Chemistry Magazine* for Science Service and developed a number of experimental chemistry kits. But the circulation of the magazine was small and its impact not equivalent to that of *Popular Astronomy* or *Sky & Telescope*.

9. A rich literature explores the rise of classical genetics in the United States. Much of this scholarship centers on a significant historiographical discussion. Was the emergence of classical genetics a revolutionary (Allen 1978) or an evolutionary (Maienschein 1991) process? That American biology was transformed after 1900 is not in dispute. The tempo and mode of change and the motives of its leaders are, however, the subject of considerable debate. Maienschein (1985) provides a critical review of the literature and the original essays in Rainger et al. (1988, 1991) discuss major historiographical issues. The spring 1981 issue of the *Journal of the History of Biology* contains a number of important papers debating the Allen thesis together with a reply from Allen.

10. This interpretation stands in contrast to the whiggish practitioner-perspective that portrays the emergence of astrophysics as little more than a rational division of labor within astronomy driven by epistemic considerations (Struve and Zebergs 1962; Meadows, 1984a, 1984b; Osterbrock 1984a).

The rise of classical genetics in the United States presents a different situation. Current historiography suggests that, rather than a revolt against earlier practices and theories, there was a gradual extension of biological research into new territory. This process involved the reconceptualization of old questions and was directed by respected scientists, virtually all of whom where products of American Ph.D. programs (Rosenberg 1976: 207; Rainger et al. 1988: 5). The contrast with American astrophysics could not be greater. The fathers of astrophysics frequently lacked academic training and often worked in locations on the periphery of the American astronomical community.

Morphology and physiology were related both to the tradition of field naturalists and the laboratory approach to biology (Allen 1978). Classical genetics, however, was clearly located in the laboratory. Like genetics, astrophysics inclined toward the experimental sciences. In the strictest sense, no branch of astronomy can be called an experimental science (Harwit 1981: 4–5), but it should be emphasized that the new astronomy took great pains to develop instruments and techniques for gathering physical data. From new ways of measuring the sun's heat to narrow passband solar spectroscopy or high dispersion stellar spectroscopy, astrophysicists did indeed carry out experiments. Rather than the direct manipulation of objects under examination, an obvious impossibility, the experimental tradition in astrophysics emphasized innovative ways of collecting data of interest.

Through World War II, astrophysics in the United States can best be understood as a branch of experimental spectroscopy. Bunsen and Kirchhoff provided powerful empirical laws that permitted spectroscopists to reconstruct physical conditions in the sun and stars, but only in the 1930s did theoretical physicists succeed in explaining the source of stellar energy (Hufbauer 1991: 96–112). The American astrophysical community begin to produce theorists in significant numbers only after World War II.[11] In genetics, however, the situation was different. Both Mendelian and Darwinian theory quickly came to play an important role in the new science.

The early histories of astrophysics and genetics are sharply contrasting and the differences can be traced to the political economy of each discipline. A comparatively large number of consumers, located in several different markets, used knowledge produced by geneticists. These markets included scientists in various academic departments, agricultural colleges and agricultural experiment stations, as well as private breeders. There was, in short, more demand for the products of geneticists than for those of astrophysicists.

11. This point is controversial. Slavings and I plan to address the problem using bibliometric analysis. We are at work on a study of problem selection in American astronomy, 1859–1940.

The founders of classical genetics can be characterized as *insiders.* They were securely located in the context of academic science. The originators of astrophysics were *outsiders*, free to select topics that interested them rather than problems dictated by traditions in which they had been socialized as undergraduate or graduate students. In short, the founders of astrophysics were not inhibited by professional training. Their research was driven by imagination and opportunity (chapter 3).

The genetics community came together rapidly after 1900. Within a few years there was a critical mass of credentialled scientists working in a variety of academic settings (Maienschein 1991: 73–176). Fifteen years after the recovery of Mendel's work, Thomas Hunt Morgan and his team were able to publish *The Mechanism of Mendelian Heredity.* In astrophysics the situation was quite different. Fifteen years after the publication of the work of Bunsen and Kirchhoff the number of astrophysicists was still very small (table 3.1) and no classic monograph appeared, blending theory and observational data. Indeed, it would be difficult to say when, before 1940, the equivalent of such a monograph was published by an astronomer in the United States.[12]

Research in genetics payed off more rapidly than in astrophysics. Work in Morgan's fly room at Columbia, and many other laboratories and research stations, linked Darwinian evolution and Mendelian genetics. By 1915 geneticists had accomplished what Darwin failed to achieve: a quantitative theory that explained variation.

Genetics became institutionalized in American science more rapidly than astrophysics. The new area of inquiry first took root in agricultural colleges and agricultural experiment stations at land grant universities as well as among private breeders (Sapp 1983: 337–41), but quickly spread from the agricultural campus to academic departments. In the beginning, astrophysics flourished in amateur observatories (Lankford 1981b). Then a few academic institutions such as Dartmouth and the University of Western Pennsylvania saw their astronomers take up research in solar physics. Unlike genetics, astrophysics long remained a research rather than a teaching field. The new astronomy departments at Berkeley and the University of Chicago concentrated on celestial mechanics and astrometry (chapter 4), leaving astrophysics to the off-campus observatory. At Harvard, for example, astrophysics was not taught until the 1930s.

Biology and astronomy provide interesting comparisons. Biology, a much larger and more complex community, tended to concentrate in university settings. On the campus, however, there was remarkable diversity. Biologists were found in colleges of arts and sciences (most often departments of botany and zoology) as well as in colleges of agriculture and

12. Cecilia Payne's (Gaposchkin) 1925 Harvard dissertation, "Stellar Atmospheres," would be a prime candidate for such a monograph, but its immediate impact was negligible.

experiment stations. Medical schools employed biologists in departments ranging from anatomy and microbiology to biochemistry and physiology. Indeed, one early twentieth-century commentator described the sweep of the biological sciences as encompassing fields from paleontology to pharmacology (Appel 1988: 110). Biologists of whatever persuasion early became involved with the Ph.D. machine, and the doctorate (except for field naturalists) became normative in the biological sciences much earlier than in astronomy. Like astronomy, some fields of the biological sciences included a rich amateur tradition.[13]

Compared to astronomy, the biological community was a sprawling enterprise with many specialist research communities. By the eve of World War I, there were at least eight professional societies plus an umbrella organization, while biologists published in at least nine journals (Appel 1988). Astronomers seemed content with one professional organization and three journals. The multiplication of journals and professional societies in biology continued unabated between the wars.

Public policy provides another point of comparison. The biological sciences in America have important policy connections. Public health and medicine, the food supply and agriculture, social policy and eugenics—these linked biology to government at the local, state, and national levels. Astronomy had no such policy connections. Over the years, astronomers played a significant role in the development of modern navigation and time-keeping, were early collectors of meteorological and geophysical data, and assisted in trigonometric and topographic surveys, but these activities provided no permanent links to public policy. The Naval Observatory and the Nautical Almanac Office stand as lone examples of the connection between astronomy and the national government. State and local governments may have paid observatories to provide accurate time, but this was in no way equivalent to biologists' connections with state departments of agriculture or public health.

For many sciences there are linkages between public policy and patronage. The modern state has become a primary consumer of scientific knowledge. Through a network of departments, agencies, and bureaus, federal, state, and local governments provide patronage for biologists (Dupree 1957b). Government patronage for the biological sciences has increased over the years, but only after World War II did astrophysics benefit from federal largesse.

13. The amateur naturalist in America has yet to be studied in detail. On ornithology see Ainley (1987). Evans and Evans (1970) provide a fascinating account of William Morton Wheeler as he moved from the status of an amateur naturalist to a professional zoologist who made major contributions to the study of animal behavior. A comparative discussion of the amateur tradition in biology and astronomy would make an exciting contribution to the literature.

These comparisons between biology and astrophysics are suggestive. Astrophysics developed before the advent of the Ph.D. machine, while genetics developed in the context of the research university. The political economy of the biology community insured ready markets for genetics research. Astrophysics suffered from underdeveloped markets. Genetics was able to assimilate theory much more rapidly than astrophysics. American astrophysics remained an observational science long after genetics developed a balance between theory and laboratory work.

CROSS-CULTURAL PERSPECTIVES: ASTRONOMY IN THE UNITED KINGDOM, FRANCE, GERMANY, AND THE UNITED STATES

This is a preliminary reconnaissance dealing with large-scale structures. Demography, the organization and institutionalization of astronomy, patronage and the rough contours of careers are the focus of attention. The goal is to understand major differences and similarities in several national contexts. The section concludes with a discussion of national styles in astronomical research and the impact of the *Carte du Ciel.*

Given the paucity of monographic studies, the discussion relies on quantitative data drawn from three surveys by the staff of the Belgian Royal Observatory at Uccle. The size and composition of the astronomical community and the number and type of observatories in each country are of special interest. But first we must consider an important technical issue: the distribution of large telescopes in the period 1859–1940.

Large Telescopes: Europe and America

The rise of American astronomy to world-class status can not be explained simply as a consequence of superior instrumentation. As table 11.2 suggests, the perception that America had more large telescopes than Europe is an error.

Of the sixty-seven large telescopes constructed during this eighty year period, 26 (39 percent) were in the United States while 41 (61 percent) were in Europe. In the category of large refractors, Europe had four instruments with objectives between thirty and thirty-nine inches while the U.S. had but two. Of course, America boasted the largest working refractor in the world (the Yerkes 40-inch), but only at the highest end of the scale did America outdistance Europe. Europe had five reflectors with mirrors from 40 to 59 inches in diameter, while at the next level (mirrors from 60 to 99 inches) the distribution was reversed. The Hooker telescope at the Mount Wilson Observatory was the largest reflector in the world. In both large refractors and reflectors, the U.S. was out-gunned by approximately 3:2. America led only at the highest end of the scale.

American leadership in observational astronomy and astrophysics af-

Table 11.2 Large Telescopes, 1859–1940: Europe and America

	Refractors N=37*			Relectors N=30*	
	USA (N=14)	Europe (N=23)		USA (N=12)	Europe (N=18)
Aperture (Inches)			Aperture (Inches)		
20–29	11	19	30–39	5	12
30–39	2	4	40–59	1	5
40	1		60–99	5	1
			100	1	

*Includes telescopes at U.S. overseas stations and the Royal Observatory at the Cape of Good Hope.

SOURCE: Gingerich (1984: Aiii–Avi).

ter 1900 cannot be explained on the basis of telescopes alone. Generous patronage and the organization of astronomical research in factory observatories are at least as important as telescope size or the location of observatories.[14]

Astronomers and Astronomical Institutions

In 1886 the Belgian Royal Observatory at Uccle published a report on world observatories (Lancaster 1886).[15] The most striking inference to be drawn from the 1886 data (table 11.3) concerns the commanding position of American astronomy. Both in number of observatories and personnel, the United States surpassed the United Kingdom, France, and Germany. American astronomy was firmly rooted in the context of higher education with fifteen observatories located in universities and fourteen in colleges.[16]

14. By the same token, it is unacceptable to argue that astronomical seeing (i.e., steadiness and transparency of the atmosphere) at American observatories was superior to conditions in Europe. For every Mount Wilson or Mount Hamilton there were the observatories at *Pic du Midi* or Meudon in France, Max Wolf's Astronomical Institute at Heidelberg, or Schiaparelli's observatory in Milan. Neither technological determinism nor determinism resting on climatic conditions is an acceptable historical explanation for the development of American astronomy after the middle of the nineteenth century.

15. Lancaster (1886) was based on library research rather than self-reporting questionnaires, as would later be the case (Stroobant et al. 1907, 1931). The 1886 study probably underestimates personnel. For example, there are no computers listed in the U.K., but we know they were on staff at Greenwich by this date.

16. There is something of a paradox here. When considered from a domestic point of view, American astronomy is striking because of its institutional diversity. Astronomers were found in a variety of institutional settings. In comparison with Europe, however, the connection between astronomy and the expansion of higher education in America appears paramount.

Table 11.3 Astronomy in the United States, the United Kingdom, France, and Germany in 1886

	USA	UK	France	Germany
Total observatories	**40**	**32**	**16**	**26**
Total excluding amateurs	39	20	15	19
National	2	4	13	7
College/University	29	8	0	9
Private Research	8	8	2	3
Amateur	1	12	1	7
Total personnel	**128**	**70**	**63**	**81**
Astronomers	89	59	54	52
Professors	29	11	6	21
Computers	10	0	3	8

SOURCE: Lancaster (1886).

NOTE: In this and other tables based on the Uccle surveys (Stroobant et al. 1907, 1931), amateurs are not counted as personnel. For the U.S., directors of college observatories are counted as professors. For Germany, *dozenten* are included under the category of professor. Private research observatories range from Mount Wilson and the Dudley in the U.S. through German institutions such as Boothkamp were H. C. Vogel and O. Lohse started their careers to the observatory of William Huggins in London or that of Lord Rosse at Birr Castle. Amateur institutions range from well equipped and productive observatories like that of W. H. Steavenson in England to the country parson with a three-inch refractor who listed the moon and planets as his areas of special interest. Organizations such as the *Recheninstitut*, *Bureau des Longitudes*, and the Nautical Almanac Offices in the UK and the US are counted as national observatories.

Second came private research institutions like the observatories of Henry Draper and Lewis M. Rutherfurd or the Dudley Observatory at Albany. National and amateur operations are at the bottom of the list. Amateur observatories would increase in number over the years, but in 1940 there were still only two national astronomical institutes in the United States. Astronomy in the U.K. was institutionalized in amateur observatories, universities, and private research institutions as well as national observatories. This pattern reflects the amateur tradition in British science. Beyond institutions such as the Royal Greenwich Observatory, there was very limited government support for astronomy. The expansion of higher education in the U.K. came later and was on a much smaller scale than the American experience; astronomy in the U.K. did not benefit in similar ways. The astronomical community in the U.K. was smaller than in either the U.S. or Germany.

On the Continent, French astronomy ranked behind Germany and the U.K. France had more national observatories, but the peculiarities of

Table 11.4 Astronomy in the United States, the United Kingdom, France, and Germany in 1907

	USA	UK	France	Germany
Total observatories	**102**	**119**	**43**	**45**
Total excluding amateurs	86	31	20	32
National	2	5	14	7
College/University	71	12	6	20
Private Research	13	14	0	5
Amateur	16	88	23	13
Total personnel	**294**	**161**	**150**	**152**
Astronomers	153	85	90	112
Professors	53	15	16	35
Computers	88	61	44	5

SOURCE: Stroobant et al. (1907).

French higher education meant that there were no observatories formally associated with universities. Indeed, from the Napoleonic era through the 1890s, France had no university system comparable to those of Germany, the U.K. or the U.S. Germany ranked third in observatories and personnel. As in the U.S., German astronomy was rooted in the university context, but there were more national observatories than in either the U.S. or the U.K. The number of professors of astronomy stood just behind the US.

Twenty-two years later the staff at Uccle produced a second report (Stroobant et al. 1907) on world astronomy (table 11.4).[17] Interesting changes occurred between the first and second Uccle reports. In 1907, the American lead in both observatories and personnel remained undiminished. There were more than twice as many observatories in the U.S. than in Germany, the closet competitor. The same held true for personnel. American astronomy continued to cluster in universities (38 observatories) and colleges (33 observatories). Both amateur and private research institutions increased in number, but in the U.S., private research institutions were by far the more important.

The U.K. and Germany were virtually tied for second place in terms of observatories and personnel, but the institutionalization of astronomy differed. In Germany, higher education was the primary location, followed by amateur institutions, national observatories and private research insti-

17. These data, if used with care (Stroobant et al. 1907:3–4), are valuable for comparative purposes. In the case of amateurs in the U.K., there may be a tendency to over-representation. But this should not negate our appreciation of the continuation of the amateur tradition in British science into the Edwardian era.

Table 11.5 Astronomy in the United States, the United Kingdom, France, and Germany in 1931

	USA	UK	France	Germany
Total observatories	**90**	**52**	**43**	**46**
Total excluding amateurs	77	26	14	31
National	2	3	6	7
College/University	64	16	5	17
Private Research	11	7	3	7
Amateur	13	26	29	15
Total personnel	**378**	**108**	**123**	**177**
Astronomers	170	76	72	131
Professors	104	17	10	26
Computers	104	15	41	20

SOURCE: Stroobant et al. (1931).

tutions. In the U.K. amateur observatories led the way, followed by private research institutions, observatories connected with institutions of higher education and national observatories. While the U.K. slightly exceeded Germany in total personnel, Germany numbered more professors and astronomers. The U.K. took a commanding lead in computers.

France, with a system of modern higher education scarcely a decade old, lagged behind in a number of ways. There were fewer observatories connected with institutions of higher learning and the size of the community placed French astronomy last, just behind Germany. France, however, still led in national observatories and ranked second in amateur institutions.

Almost a quarter century later, the Uccle staff carried out a third survey (Stroobant et al. 1931). During the intervening years, war and economic dislocation brought many changes, but it is remarkable how similar the 1907 and 1931 data are (table 11.5). The U.S. had solidified its position, possessed of the largest and most diverse astronomical community. In Europe, Germany had pulled ahead of the U.K. in number of observatories, while France lost ground. Germany led France and the U.K. in total personnel, showing a deficit only in computers. France still had the smallest number of university-related observatories, while national observatories had declined by more than 50 percent. In the U.S., forty-three universities and twenty-one colleges reported associated observatories. America led the way in all categories of personnel.

Data reported in the Uccle surveys are instructive. The growth of American astronomy was well under way in the 1880s. Even at this early

Table 11.6 Participation in the International Astronomical Union, 1922 and 1938

	Rome (1922)		Stockholm (1938)	
	Commission Presidents N=32	Delegates N=83	Commission Presidents N=36	Delegates N=284
U.S.	11 (34%)	11 (13%)	12 (33%)	62 (22%)
U.K.	8 (25%)	18 (22%)	4 (11%)	43 (15%)
France	8 (25%)	12 (14%)	5 (14%)	22 (8%)
Germany*			0	11 (4%)

*Germany was excluded from the IAU and all other international scientific organizations after World War I.

SOURCE: Fowler (1922) and Oort (1939).

date, Astronomy in the U.S. outdistanced its counterparts in the U.K., France, and Germany in many respects. The most obvious advantage of European astronomy lay in the number of national observatories. America's most obvious advantage involved linkages between astronomy and the expansion of higher education. Before we examine each country in detail, there is one further point: participation in the international astronomical community.

While American astronomers attended most of the international meetings connected with the *Carte du Ciel*, no American observatory took part in the project. The International Union for Cooperation in Solar Research, on the other hand, was an American creation and served as a vehicle for imposing American standards in spectroscopy and photometry on the international community. After World War I, the *Carte* and the Solar Union were subsumed as part of a larger entity: the International Astronomical Union (IAU). A comparison of participation in the IAU by astronomers from the U.S., the U.K., France, and Germany is instructive. Let us look at the first (1922) meeting of the IAU and the final prewar meeting of the Union in 1938 (table 11.6).

At both meetings, Americans dominated the Union with a third of commission presidencies. It was in commissions that the work of the IAU was accomplished. Commissions brought together specialists who decided on technical standards, nomenclature, and the like. Their decisions were virtually binding on all astronomers. In absolute terms, the number of American delegates increased between 1922 and 1938. Neither France nor the U.K. could match the U.S. in commission presidents or delegates. It appears that from its inception, American astronomers assumed a commanding position in the IAU. German scientists were excluded from the international scientific community for most of this period, so we can not compare their participation.

Conditions That Affected the Development of Astronomy

The following section examines conditions that affected the development of astronomy in four national contexts. The discussion is exploratory. In the history of modern astronomy, there is virtually no historiographical tradition dealing with topics like this.

Science and Higher Education The expansion of higher education in the U.S. and Europe proceeded in very different ways. There was no connection in America between higher education and the civil service, nor did universities prepare students for examinations that determined admission to the professions (Ringer 1979a: 248). In the U.S. there was freedom for institutions of higher education to develop in response to a wide range of imperatives that reflected the needs of the community, the vision of powerful educational leaders, or the agendas of state and federal governments (Herbst 1983). Further, higher education in America was the beneficiary of generous private, local, state, and federal patronage. In short, in the American system of higher education, there was room for both the church college and the research university.

In the century after 1830, British universities grew from four to twenty-two (Perkin 1983: 209), but this expansion did not produce anything comparable to the American research university. In the U.K. the expansion of higher education was driven by the demands of empire and economic growth (Rothblatt 1983: 132). Within British higher education, most Dons continued to teach and the D.Phil. (the equivalent of the American Ph.D.) never became a prerequisite for a faculty position (Ringer 1979a: 217). British philanthropists were much less generous than their American cousins in supporting higher education (Rothblatt 1983: 140). While the observatories at Cambridge and Oxford were important research institutions and the Cavendish Laboratory at Cambridge became the center of British experimental physics, for the most part the growth of nineteenth-century British science took place outside the universities (Ringer 1979a: 212).

In France, the situation was so different that it is impossible to make direct comparisons with the rest of Europe or the United States. Between the abolition of the universities of the *ancien régime* and the creation of the modern French system of higher education in 1896, higher education in France was virtually unique. Created by Napoleon I, the *Université Impériale* divided the nation into twenty-six districts (reduced to seventeen by 1870). In each district there was a faculty of medicine, law, arts and science, and theology. In truth, these faculties were little more than loose confederations without institutional or collegial identity. Their primary function was to examine candidates for degrees. Neither research nor instruction occupied a great deal of their time (Moody 1978: 24; Ringer 1979a: 115). In metropolitan centers like Paris, seeking multiple academic posts (*cumul*) and preparing popular lectures for delivery to crowds of up to a thousand paying spectators consumed the energy of many scientists (Fox 1976; Fox and Weisz 1980).

The Napoleonic system of *Grandes écoles* (e.g., the *École Normale Supérieure*, 1808) was designed to prepare civil servants, engineers, military officers and other experts for state service. These institutions were geared to instruction, not research. The Observatory, the *Collège de France*, and the *Muséum d'historie naturelle* were the primary research institutions in Paris. The French Academy of Sciences stimulated research through a system of prizes and rewarded individuals with various honors as well as membership for the chosen few.

Fox (1976) argues that when the golden age of Napoleonic science came to an end with the passing of the astronomers Laplace and Delambre, the chemist Gay-Lussac, and biologists Cuvier and Lamarck, their successors turned away from the laboratory to the drawing room in search of political patronage and to the crowded lecture hall in search of income. This happened at the very moment Justus von Liebig was opening his chemical institute at the University of Giessen—a milestone event in the development of scientific research in the German university system (Fox 1976: 29). Later in the century, Pasteur and the Curies (Fox 1976: 34–35) should be viewed as exceptions that prove the rule: they stood outside the system of French higher education.

Astronomy in France was institutionalized either in national observatories or in private research institutions and amateur establishments. Without access to graduate students and located in institutional contexts in which a competitive research ethic was not paramount, French astronomers devoted themselves to routine astrometric research during much of the nineteenth century. Just as thermodynamics and electromagnetism developed, for the most part, outside the French physics community, so astrophysics emerged with only one significant French participant: the solar physicist Pierre Jules César Janssen.

Higher education in Germany and the United States proved hospitable to astronomy, but there were differences in both degree and kind. Certain elements in the culture of German higher education militated against specialization. Specialization conflicted with the idea of *Bildung*, a cluster of deeply cherished values that characterized the culture of German academic life (Ringer 1979a; Harwood 1993: 23). Among other things, *Bildung* meant a well-rounded education that produced individuals with a wide knowledge in the humanities (especially the classics in literature and philosophy), as well as mathematics and the sciences (Shaffer 1990). Generalists came closer to attaining the deep meaning of *Bildung* than did specialists. Specialization, with its emphasis on research and the application of knowledge to real-world problems, threatened the moral foundations of the German university. A product of early nineteenth-century romantic idealism, the German university was committed to a balance between teaching and research, while remaining aloof from the practical uses of knowledge (Lundgreen 1980).

Specialization was of concern for political reasons as well. German universities were organized around faculties and research institutes, not departments (Jungnickel and McCormmach 1986 1: chaps. 1, 4, 9; Ringer 1979b). Research institutes were generally the domain of a single professor, with his assistants and students. Specialization entailed the multiplication of institutes and this, in turn, meant a drain on the resources of the ministry of education that paid the bills, as well as changes in the political balance of the faculty. The creation of new institutes had to be approved by a university faculty and the ministry.[18]

In America, specialization was a key ingredient in the expansion of higher education. Public and private universities included a wide range of departments and research facilities. Specialization became the hallmark of American academic science; it was also sometimes encouraged by state patronage for reasons of public policy. This was the case for biology, but not astrophysics. In comparison to the situation in Germany, American universities had more flexibility in accommodating new specialties. American universities were organized around departments and it was the university president rather than the faculty and ministry of education that made decisions about educational policy. University presidents could create new departments or add to established units. Presidents controlled resources and defined institutional goals (Harwood 1993: 156ff; Veysey 1965).

The expansion of academic astronomy from the 1880s reflects the ability of American university presidents to respond to the emergence of astrophysics. To be sure, the process was contingent, depending on local circumstances and personalities. Not all university presidents were as expansionist as Chicago's William Rainey Harper; not all astronomers were as entrepreneurial as George Ellery Hale. But where the chemistry was right, a team like Harper and Hale could create a Yerkes Observatory. A few years later a sympathetic administration encouraged the revitalization of astronomy at Michigan. These examples can be multiplied (chapter 4).

Between 1860 and 1930, enrollments in American higher education increased twenty-two fold. This compares to a growth factor of eleven for

18. Lundgreen (1980: 315) suggests that by the eve of World War I there were almost 5,000 technical institutes associated with German universities.

the U.K. and of eight for Germany. In the U.S., faculty size kept pace with student enrollments (Jarausch 1983: 13). These patterns reflect the expansion of graduate education as well (Harwood 1993: 142–48). While there were periods of oversupply in the production of graduates in the German university system (Titze 1983: 58), it would appear that American colleges and universities did not face similar fluctuations.

By the early twentieth century, German higher education seems to have reached a plateau. Faculty size and financial resources grew much more slowly than student enrollment. Harwood (1993: 145) reports that in 1910–11 the total annual expenses for the top twelve U.S. universities were twice the total budget for all twenty-one German universities. During the interwar years, American higher education fared much better than its German equivalent.

For American astronomy, the lesson is clear. The development of the research university provided a growing market for astronomers (chapters 5–6). Lundgreen (1983:156) reports that in 1864 there were twelve professors of astronomy in the German university system that totalled twenty institutions. In 1931 the number of professors stood at fourteen. The 1864 figure, however, should be augmented by eight *ausserordentliche professoren* and two *privatdozenten*. By 1931 the figures for these two groups of secondary faculty stood at seven each. Thus it is not clear that astronomy was represented in all of the German universities. It may be that astronomy was more widely represented in the American system of higher education than in the German.[19]

Access to careers in astronomy must have differed greatly between the U.K., France, Germany and the U.S. Patterns of employment were very dissimilar, and market forces shaped career opportunities differently. But before we consider the nature of careers, it may be useful to look at patronage.

Patronage Before World War II, American astronomy was the beneficiary of several forms of patronage. As astronomy became institutionalized in the context of higher education, institutional budgets provided operating funds for observatories and salaries for astronomers and professors. To be sure, there was an overlap between private philanthropy and institutional budgets. In many cases (e.g., the Lick Observatory of the Uni-

19. In comparing the institutionalization of astronomy in the American and German systems of higher education, a perplexing question concerns the location of astrophysics. The Potsdam Astrophysical Observatory is well known as the center of observational astrophysics in Germany. But where was the subject being taught? Where could students study astrophysics at the graduate level? Lundgreen (1983: 159) reports a single *ausserordentliche professor* teaching astrophysics as a member of the physics faculty at Berlin, but the position apparently did not survive World War I. Data are not readily available for other universities. This problem deserves investigation.

versity of California or the Yerkes Observatory of the University of Chicago) private philanthropy played a central role in the founding of the institution and its continued operation. Indeed, the Harvard College Observatory relied completely on gifts and bequests and was financially independent of the university.

Private research institutions, ranging from the observatories of wealthy individuals in the nineteenth century to the Mount Wilson Observatory were, as the term suggests, privately funded. Philanthropists financed not only telescopes and the buildings to house them, but paid staff salaries, provided for institutional maintenance and the like, as well as gifts for specific projects. At the beginning of the twentieth century there was a discernible shift in patronage as the new philanthropic foundations entered the scene and funded astronomy on a scale that few individuals could match. Individual philanthropy, however, was never extinguished by the great foundations.

As solar physicist Samuel P. Langley suggested in the 1880s (chapter 3), there were significant differences between funding the old and the new astronomy. Federal dollars supported research in astrometry and celestial mechanics at the U.S. Naval Observatory and the Nautical Almanac Office, while astrophysics, which did not benefit from government support until after 1945, relied on private donors and foundations. France, Germany, and the U.K., each supported one major astrophysical research center: Meudon in France, Potsdam in Germany, and South Kensington in the U.K.

Scientific societies (the National Academy of Sciences, the American Philosophical Society, or the American Academy of Arts and Sciences) provided grants that enabled individual astronomers to purchase auxiliary equipment, hire assistants, and begin new projects. The sums were small, but their impact on astronomical research was significant.

In the U.K., institutions of higher education played a very limited role in funding astronomy. After 1889, government patronage for university science increased (Brock 1976: 184). Government support for individual projects was channeled through the Royal Society. Between the 1907 and 1931 Uccle reports on world astronomy, private research institutions in the U.K. declined significantly. This reflects changing patterns of patronage. Arguably, government-funded observatories such as the Royal Greenwich Observatory remained the strongest and best financed in the U.K. before World War II.

Before the end of the nineteenth century, astronomy in France was not a university-based science. It was rooted in government institutions like the Paris Observatory and its several stations. Insofar as private research institutions reflect a level of individual philanthropy, France must have been far behind the U.S. or other European nations (see table 11.3). It may well be, however, that the prize system of the French Academy of Sciences provided more support for certain kinds of astronomical research than did other European academies.

Germany astronomy was rooted in the university system, but national observatories also played an important role. Since the state financed both the universities and national astronomical research facilities such as the *Recheninstitut*, it is obvious that government patronage was decisive for the development of astronomy in Germany.

What can be concluded from this overview? Clearly the state played a more important role in German and French astronomy than it did in either the U.K. or the U.S. In the U.K., universities appear to have been less important in providing patronage for astronomy than in the U.S. The pluralistic system of patronage for American astronomy helps explain its remarkable development after the middle of the nineteenth century.

Careers In the U.S. there were a number of well-defined paths leading to careers in astronomy (chapters 5–6). As American higher education expanded, market conditions favored the intending astronomer. In the U.K. markets were less elastic. The expansion of higher education provided some posts, but it is not clear whether these compensated for the loss of jobs as national and private observatories declined after 1907. Until the end of the nineteenth century, the astronomical career in France was located in either national or private research observatories; the new university system seems to have provided relatively few positions before World War II. German higher education offered fewer careers in astronomy than did higher education in the U.S. National observatories remained important employers before World War II, while private research institutions registered modest gains.

With overall demand much greater in the U.S. than in Europe, it is not surprising that the American astronomical community should have been larger than European communities. Nor is it surprising that European astronomers immigrated to the U.S., drawn by the promise of employment. The expansion of American higher education can be identified as the primary driver, but at a deeper level the cultural significance of astronomy must be taken into consideration. This deep cultural interest must have had something to do with the willingness of American colleges and universities to support the subject.

Amateurs The status of amateurs in Europe differed from their status in the U.S. In the U.K. and on the Continent, wealthy amateurs built observatories and hired astronomers to carry out research programs. Other affluent amateurs constructed or purchased their own equipment and with it made valuable contributions to the science. While there were amateur observatories in the U.S. in which important research was accomplished, there was no analogue to the tradition of well-to-do individuals establishing observatories and paying the salaries of astronomers.[20]

Given the dramatic growth of astronomy in the U.S., amateurs soon found they could not command the resources available to professionals. By the end of the nineteenth century, amateurs occupied a secondary position in the political economy of American astronomy. However, amateurs in the U.S. were instrumental in fostering a cultural climate that encouraged investment in astronomy.

When comparing amateurs in these four national contexts, we can imagine a continuum, with the activities of a wealthy European elite at one extreme and the American amateur building a 6-inch telescope in the basement at the other. American amateurs played a more active role in popularizing astronomy than did counterparts in Europe. In so doing, amateurs provided important links between astronomy and the larger society.

National Styles in Science: Astronomy and Astrophysics in Europe and the United State

National contexts play a significant role in defining the problems scientists select and the ways in which they organize and carry out research (Kohler 1991b, 1991c; Harwood 1993). Far from transcending national boundaries (Simpson 1962: 1–14), science seems to follow the political map. This section compares national styles in astronomy and astrophysics in Europe and America. The goal is to identify and explain major differences in problem selection and research practice.

From the 1850s through the 1930s the growth of European astronomy was dominated by astrometry. While astrophysics became part of the European research agenda, it was secondary to more traditional topics. In the U.S., observational astrophysics played a central role in the expansion of astronomy. Both astrometry and celestial mechanics, however, remained viable research fields.

These differences are reflected in the organization and practice of astronomical research. In Europe, even where astrophysical concerns were primary, research was organized in traditional ways. The transit-circle mentality (chapter 3) predominated and astrophysics was as much a cus-

20. Roslyn House Observatory, built by Gustavus Wynne Cook in Wynnewood, Pennsylvania, is perhaps the best example of a wealthy American constructing a private observatory and employing professional astronomers (e.g., Orren Mohler) to make observations. The observatory eventually was given to the University of Pennsylvania. The McMath-Hulbert Solar Observatory, which became part of the University of Michigan in the early 1930s, is a more complex case. The founders remained actively connected with the observatory even after it was taken over by the university.

tom product as astrometry or celestial mechanics. European astronomers lavished a great deal of individual time and energy on large-scale projects in observational astrophysics. In America these activities were carried out wholesale by photography, and the data were reduced and analyzed by assembly line methods in factory observatories (Lankford 1994; Lankford and Slavings 1996).

The growth of European astronomy can be linked to a research tradition in astrometry that entailed large, expensive, labor-intensive projects. The first of these great projects was the *Bonner Durchmusterung* (1859–62), a catalogue and chart giving the positions and magnitudes of over 300,000 stars between the pole and two degrees south of the celestial equator. In 1886 a supplementary catalogue carried the study to -23°, adding 133,000 stars. The *Astronomische Gesellschaft* soon organized a program that divided the sky into zones for the observation of stars to the ninth magnitude with transit circles. Fifteen European and two American observatories were involved in the AG project. The transits of Venus (1874, 1882) prompted European astronomers to mount expeditions to the far ends of the earth to observe these events in order to develop a more precise measure of the sun's distance (Lankford 1987c). The European tradition of large-scale astrometric research culminated in the *Carte du Ciel*, a project that photographed the sky from pole to pole and produced both a catalogue and chart. Eventually the *Carte* involved twenty-four observatories stretching from England to India, from Helsinki to Sydney. Thirteen of these institutions were in Europe, with the Paris Observatory serving as headquarters for the project (Lankford 1984a).

Astrometry played a significant part in American astronomy but it did not drive the growth of the science. A number of institutions carried out important astrometric research including the Naval Observatory, the Dudley, Allegheny after Langley, Yale, and Virginia. Any observatory with a large long-focus refractor devoted some time to double-star work (e.g., Yerkes, Lick and the USNO) while institutions such as Swarthmore and Allegheny specialized in the photographic study of stellar parallaxes. Successful astrometric operations resembled work in observational astrophysics, organized along factory lines.

Astrometric research drove the institutional growth of European astronomy, providing the justification for increased resources, including budgets, staff, and instruments. This pattern left little room for other forms of astronomical investigation. As Dieter Herrmann (1984:32) concludes, in late nineteenth-century Europe "the growing tradition of positional astronomy had reached maturity; it had indeed become an overgrown part of the astronomer's life and only grudgingly allowed room for new ideas." The cognitive pluralism and institutional diversity that marked American astronomy provided greater flexibility. New research universities were open to both astrophysics and astrometry; so were private research institutions.

Astronomical practice differed between the U.S. and Europe. At Harvard, Pickering organized his staff for efficient production. In its first year, for example, the number of stars classified for the Henry Draper Memorial spectroscopic survey totalled 8,000. During the second year, the number tripled to 24,000 stars (Jones and Boyd 1971: 235–36). This stands in sharp contrast to the work of European astronomers like William Huggins, who lavished years of study on the spectrum of a few stars (Jones and Boyd 1971: 234; Becker 1993).

British astronomers were sharply critical of American methods (Jones and Boyd 1971: 272–76). They responded negatively when Pickering announced that Harvard's new 24-inch Bruce photographic telescope could photograph the whole sky in about 2,000 plates, thus rendering the *Carte*, with its estimated 88,000 plates, obsolete. German observers were critical of American photometric research because of its scale and assembly-line methods (Jones and Boyd 1971: 342). Frequently, (Jones and Boyd 1971: 209) transatlantic disputes boiled down to the fact that European astronomers disapproved of the assembly-line approach to research.

Europe had many more national observatories (tables 11.3, 11.4 and 11.5) than the U.S. Most of these institutions concentrated on astrometry. France led the way in national observatories. Germany ranked second and England stood in third place. In the category of computers, however, the U.S. never lost its lead after 1886. The factory organization of American astronomy is clearly mirrored in these data.

Jonathan Harwood has drawn fruitful distinctions between research traditions. Harwood suggests (1993: 189ff.) that in the German genetics community there was a comprehensive research tradition as well as a pragmatic tradition and that these traditions attracted very different kinds of scientists in terms of political values and social backgrounds. Available data do not permit analysis similar to Harwood's, but we can suggest that the European tradition of astrometry was comprehensive in that it strove to produce inclusive catalogues to a given limiting magnitude. American astrophysics, on the other hand, was more pragmatic. Convenient targets of opportunity were eagerly grasped by early workers (chapter 3). Even Pickering's vast surveys contained elements of pragmatism. They were designed to take advantage of new forms of organization and production (chapter 7), as well as American technical skills in fabricating photographic instrumentation.

Differences in national research traditions point to disparities in the way astrophysics developed in Europe and America. In Europe, astrophysics tended to be theoretical rather than observational. Theory, by definition, seeks to achieve the highest level of comprehensiveness and extensibility (Ziman 1984: 13–33). On the other hand, a pragmatic, essentially observational research tradition is, at best, involved with low-level explanation.

In America, astrophysics was observational and descriptive, rooted in a pragmatic research tradition. In Europe astrophysics was theoretical and comprehensive.[21] The *Handbuch der Astrophysik* (Eberhard et al. 1928–36) offers a test of these generalizations. In thirty-six chapters, these seven volumes cover the subject as understood in the 1930s. European astronomers wrote thirty-two chapters and American astronomers contributed four. All of the American chapters deal with observational astrophysics.[22] They are descriptive rather than theoretical. Of the chapters by European astronomers, eight deal with theory, twelve discuss observational astrophysics, and twelve are devoted to methods and instrumentation. The strong emphasis on theory, methods, and instrumentation suggest European concerns for a comprehensive approach to astrophysical research, rooted, perhaps, in the transit circle mentality.

The Carte du Ciel *and the Political Economy of European Astronomy*

Initially, the *Carte* involved eighteen observatories (nine in Europe) and was the most ambitious international program in astronomy undertaken before World War II. Several institutions withdrew or did not complete their assignments and six additional observatories stepped in to help. In this way four additional European institutions became involved bringing the number of European institutions to thirteen. No American observatory took part in the *Carte*.

The project was developed by transit circle astronomers who had little idea of the potential of astronomical photography. Their plan of work involved standardized instrumentation (all participants would use identical long-focus photographic refractors). The plates were to be taken in two series: one long exposure (20 minutes) for the chart and one of 6 minutes for astrometric measurement.[23] Zones were so arranged that there would be overlapping plates, and each zone required on average about 1,200 plates for one series. The grand total for eighteen observatories taking both series came to 88,216 plates.

The great irony of the project is that the international committee directing the *Carte* froze instrumentation and research design at the very

21. Of course, fine-grained analysis will show differences between European nations. Lacking adequate instrumentation, the U.K. developed astrophysical theory early. After World War I, quantum mechanics may have had a more significant impact on European astrophysical theory than on British workers. The question deserves investigation.

22. American contributions include S. A. Mitchell on solar eclipses (4: 231–358); R. H. Curtiss on classification and description of stellar spectra (5, pt. 1: 1–108); H. Shapley on star clusters (5, pt. 2: 698–773) and H. D. Curtis on nebulae (5, pt. 2: 774–936).

23. Lankford (1984a: 29–39) provides an introduction to the history of the *Carte*. The topic deserves a book.

time astronomical photography was developing exponentially (Lankford 1984a). Instruments created by Pickering at Harvard and, later, Schlesinger at Allegheny would have made it possible to complete the work much more rapidly. But once the *Carte* leadership defined the project, uniformity of instrumentation and method was sacrosanct. Just as with transit circle work, observations had to be internally accordant. No changes were permitted, even though technical developments soon rendered the *Carte* as originally planned obsolete. One of the most telling actions of the *Carte* participants was a lengthy and sometimes heated debate over the stability of photographic emulsions for astrometric measurement. The debate was conducted from a transit circle perspective and gave little consideration to the photographic plate as a physical or chemical object.[24]

The *Carte* was an enormously expensive and time-consuming project. The leadership estimated that participating observatories could finish in twenty years. In fact, the final zones were completed well after World War II. British astronomer Herbert Hall Turner (1912b) has left a detailed record of Oxford's participation in the *Carte*. His data suggest just how critical the project was in defining the political economy of European astronomy.

Turner (1912b: 62ff.) reported that each participating observatory had at least a half-million star images to measure for position and that each image had to be measured twice. At Oxford a staff of four or five assistants worked for ten years measuring plates, each one of which averaged four to five hundred images. Printing the catalogue, Hall estimated, would take another five years. He did not indicate the time required to collect observational data, but given variable weather conditions, imperfections in some of the photographic plates, and problems with the telescope, five years might not have been an excessive estimate. Turner reported the weight of the photographic plates for the Oxford zone to be about three tons and the paper required for printing the Oxford catalogue at approximately two tons.

Once committed to the observation, measurement, reduction, and publication of the catalogue and chart for a *Carte* zone, an observatory dedicated most of its resources (financial and personnel) for at least twenty years. This meant there was little flexibility to engage in other research activities and certainly no room for major new initiatives.

Numerical data are available only for Oxford.[25] The telescope was a

24. The debate can be followed in the pages of the *Observatory* and the *Monthly Notices of the Royal Astronomical Society* from the late 1880s.

25. Turner's figures are in pounds sterling, but he indicates a figure of $5.14 to the pound in 1912 (Turner 1912b: 71). Converted into 1912 dollars, the sums are then multiplied by 149, the Consumer Price Index (1982=100), in order to express the value in 1990 dollars. See U.S

gift to the university, its value estimated at £600 ($460,000 in 1990 dollars). Maintenance of the observatory and salaries for assistants over twenty years was placed at £13,000—almost 10 million 1990 dollars. Turner's salary for the same time period was £18,000 (almost 14 million 1990 dollars). The project was aided by a government grant of £1,200 channeled through the Royal Society. This amounts to just under 1 million 1990 dollars. Turner suggested that printing costs for the Oxford portion of the *Carte* would be covered by this grant. Plates cost 1.5 million 1990 dollars. In the end, Oxford stopped with the catalogue plates and never attempted to secure the second series for the chart. For this reason (and because the observatory carried out other investigations as well), Turner concluded (1912b: 73) that the Oxford portion of the *Carte* probably cost £20,000. This expenditure amounts to over 15 million 1990 dollars.

If we assume that the same figure applies to the other seventeen institutions assigned *Carte* zones, plus the six that subsequently took over the work of observatories that dropped out, the final figure expressed in 1990s dollars is in excess of 300 million.[26] Data are not available that would allow us to examine the *Carte* as a percentage of the budget of a single observatory or of a national astronomy budget.

The *Carte* is not to be understood simply in pounds sterling, but in terms of constraints it imposed on the political economy of European astronomy. Resources were channeled into the *Carte* that might have gone to other projects. For more than three decades, directors and astronomers at thirteen leading European observatories devoted a substantial portion of their resources to the project. In terms of individual careers and the direction of astronomical research, astrometry as represented by the *Carte* virtually defined European astronomy after 1890.

At the same time, Americans were moving rapidly into observational astrophysics. Since U.S. observatories did not cooperate with the *Carte*, the political economy of American astronomy was not constrained by the project. Resources could be directed to the construction of instruments and the creation of observatories dedicated to astrophysical investigation. Careers were not centered on astrometry; the political economy of American astronomy provided resources and opportunities for research programs in both astrometry and observational astrophysics. Recruits entering the field had a choice.

Department of Commerce, Bureau of the Census (1975: 210–11; 1992: 469). I am indebted to Professor Virgil Norton, Department of Agricultural Economics, West Virginia University, and to Dean Mary McPhail Gray, Kansas State University, for aid in developing an accordant series from which to derive a value for the CPI in 1990 that could be used to convert 1912 dollars.

26. Almost all of the observatories that dropped out of the *Carte* project (e.g., Potsdam) made considerable investments of time and money before giving up.

In Europe, astronomers interested in astrophysics had few opportunities to engage in observational work. The political economy of European astronomy was largely defined by astrometric concerns exemplified by the *Carte*. Ministries of education and observatory directors were not in a position to secure funds for the construction of instrumentation for observational astrophysics. The budgets of university and national observatories reflected commitments to astrometric research and there was little possibility of directing funds into observational astrophysics.

Given these circumstances, there was really only one alternative: theory. Pens, pencils, and paper were a good deal cheaper than large telescopes and associated auxiliary equipment. And Europe had an advantage over the United States: a rich tradition in theoretical physics.

More than fifty years later, Otto Struve (1943: 474–75) recalled that commitments to astrometry (ranging from the catalogues of the *Astronomische Gesellschaft* to the *Carte*) did "much harm by virtually killing the ambitions and scientific aspirations of hundreds of younger astronomers in Europe." Large astrometric projects that involved a great deal of routine did not "attract a brilliant young man intent upon making a name for himself." Struve concluded that "some of the more independent persons became discouraged and gave up the study of astronomy" while others, "less independent . . . became mere office holders." The Yerkes director ended his homily with a personal recollection. Returning from the 1913 meeting of the *Astronomische Gesellschaft*, Struve (1943: 475) eavesdropped on a conversation in a German railway carriage. The speaker was a young astronomer who had been employed in a routine transit circle program. "His appointment had expired on December 31, and he was telling with considerable delight how at the exact second of midnight he had interrupted the transit observations of a star and written *finis* in the official record-book." While Struve achieved maximum rhetorical effect with this story, it would have meant little had not similar tales been a part of the oral tradition of the astronomical community.

In sum, Americans had access to the field of observational astrophysics, while European astronomers faced serious obstacles. The political economy of European astronomy offered few opportunities to engage in work similar to that carried out by many Americans. For the ambitious young European, theoretical astrophysics was virtually the only way to approach the field.

WHAT IS AMERICAN ABOUT AMERICAN ASTRONOMY?

This final section addresses a question of considerable historiographical interest: what is American about American astronomy? After detailed comparative analysis, we are in a position to deal with this problem. Five distinctive features combine to provide an answer. American astronomy can be characterized by its cognitive pluralism and institutional diversity, assembly-line techniques for the mass production of scientific knowledge in factory observatories, the role of women in the community, American nonparticipation in the *Carte du Ciel*, and a commitment to observational astrophysics. Of course, these elements existed in other national astronomical communities. It is a matter of degree and of the complex symbiotic relationship between these factors that gives astronomy in America its unique character.

Cognitive pluralism and institutional diversity are hallmarks of American astronomy. Astronomers were found in both public and private colleges and universities. In comparison to Europe, American astronomy benefited from the expansion of higher education that started in the late nineteenth century. University-based astronomers taught freshmen and supervised Ph.D. candidates as well as conducting research. Astronomers who devoted themselves exclusively to research clustered in private research institutions or were employed by the federal government. The community was clearly aware of institutional diversity and recognized the implied hierarchy. Research output was essential to gaining peer recognition and access to the reward system, but some institutions offered better opportunities than others. The most prestigious positions were in observatories and astronomy departments with the greatest potential for enabling astronomers to carry out significant research.

Cognitive pluralism is evident in the division between the old and the new astronomy. Fine-grained analysis suggests that within these broad divisions there were many choices. Within astrometry, for example, one could specialize in transit circle work, the study of stellar parallaxes, or dynamical studies of binary stars. Astrophysicists could choose between topics in solar or stellar spectroscopy, work in stellar photometry, or investigate the distribution, distance, and structure of nebulae and star clusters. In turn, each of these categories could be further subdivided.

A complex pattern of patronage mirrored the cognitive pluralism of American astronomy. Before World War II, federal patronage supported the old astronomy while astrophysics depended on private patrons and foundations. Individual benefactors contributed to the building of large telescopes and observatories to house them. They also supplied funds for salaries and the acquisition of axillary equipment. Trust funds of learned societies provided small grants and prizes that supported research in astronomy. In the twentieth century, the new philanthropic foundations became major patrons of large-scale research institutions such as the Mount Wilson Observatory. American astronomy may or may not have been funded at higher levels than its European counterparts. It is, however, clear that forms of patronage were different.

Heterogeneity was manifested in other ways as well. Membership in the American astronomical community remained relatively open in comparison to other scientific communities in the U.S. or to European science. The Ph.D. did not become normative until mid-century. This meant that recruitment proceeded differently in comparison to the American chemical or biological communities. It may be that astronomy's openness to non-Ph.D.'s guaranteed that the science remained receptive to diverse intellectual and technical stimuli that otherwise would have been unavailable. There are, after all, constraints as well as benefits imposed by graduate training.

These patterns of institutional diversity and cognitive pluralism guaranteed that conflict often predominated over consensus. Further, institutional and cognitive heterogeneity ensured that however powerful astrophysics became within American astronomy, it could never completely overshadow astrometry or celestial mechanics.

In the U.S., both the old and the new astronomy developed factory methods for the acquisition, reduction, analysis, and publication of data. Factory observatories were hierarchically organized. A powerful director controlled operations. The division of labor in a factory observatory was organized around status and gender. Male astronomers developed research programs in consultation with the director and made observations with the help of male assistants. Female assistants reduced data and sometimes prepared it for publication. Analysis and interpretation were generally the province of male astronomers.

Like their counterparts in the national economy, directors of factory observatories were always on the lookout for new technologies that would make the production of knowledge more cost-effective. From telescopes to measuring engines, these institutions were technology-intensive. New research programs almost always demanded new instrumentation. These developments, in turn, often required additions to the female work force. As with the national economy, there was a tendency toward a community-wide (i.e., national) division of labor. Some factory observatories specialized in solar physics or stellar spectroscopy and photometry, while others were dedicated to the study of stellar parallaxes or solar system astronomy.

There were significant differences in scale and organization between American and European observatories. These differences translated into more complex administrative challenges as well as the greater administrative freedom and flexibility available to American directors. Superficial similarities must not blind us to these institutional realities. The role of observatory director was substantially different in the U.S. American directors presided over complex institutions including both teaching and research activities and often had administrative responsibility for an overseas research station. In Europe, directors were often state officials in a sense unknown in the U.S. As state officials they were free from the burdens of fund-raising. Few American directors ever enjoyed this luxury. American directors were also responsible for the image and reputation of their institutions and were often involved in acrimonious political controversy.

The number of women in the American astronomical community marks one of the most significant points of contrast between Europe and the U.S. Between 1859 and 1940, fully a third of the community was female. Most were assistants—women who made possible the production of astronomical knowledge on an assembly-line basis at factory observatories. Both astrometry and observational astrophysics owe a great deal to female assistants who labored over columns of astrometric data, spectrograms, and photometric plates. Professors at the women's colleges played an important role, providing graduates to work in factory observatories and stimulating interest in astronomy among undergraduates. It is, perhaps, this concentration of women that is the most unique aspect of American astronomy.

Another important element in American astronomy is something that never happened: no observatory in the U.S. took part in the *Carte du Ciel*. Because of this nonhappening, the budgets of major American observatories were not encumbered by an expensive project that severely restricted new research initiatives for many decades. Non-participation in the *Carte* had a profound effect on American astronomy as contrasted with Europe.

Finally, there is the American commitment to observational astrophysics. This was central to the Hale program (chapter 3) which envisioned solar and stellar spectroscopy as an extension of experimental spectroscopy. After 1920, European developments in theoretical physics transformed the field. Innovation in observational astrophysics in America, however, involved little more than the quest for spectroscopes of higher resolution and telescopes with greater light-gathering power. Astrophysical theory was, for the most part, an exotic European import.

A curious symbiotic relationship developed after 1920. No American student went to Europe to earn a degree in theoretical astrophysics. Rather European theoreticians came to America to give seminars and lectures for the instruction of their American colleagues and to examine rich collections of observational data preserved in the plate vaults of American observatories.

Taken together, these five characteristics suggest an answer to the question: What is American about American astronomy? These elements developed and combined in distinctive ways to give American astronomy its particular character.

This story ends in 1940, as the American astronomical community was about to go to war (DeVorkin 1982b; Kidwell 1992). After 1945, epistemic, political, and social developments transformed American astronomy. Distinctions between the old and the new astronomy blurred as a result of these changes. Following the war, new patrons appeared in the form of the military. In the 1950s, the National Science Foundation became a major source of funds, and plans were developed for national optical and radio observatories. Astronomy moved from being a beneficiary of military largesse to the status of a player in Cold War diplomacy (Doel, forthcoming). Cold War tensions led to the founding of the National Aeronautics and Space Administration. Soon astronomers were involved in space-based astronomy (Tatarewicz 1990; Doel 1996), as whole new portions of the electromagnetic spectrum became available for investigation. In the 1960s, astronomers developed priorities for federal funding based on community-wide consensus (Greenstein 1972). This reflected important changes in political economy. An ever-widening variety of careers became available to astronomers, while political patronage and power within the community assumed new forms. The demographic composition of astronomy was also changing, as physics Ph.D.'s migrated into the field. Even the staid American Astronomical Society lost its club-like flavor and became a large-scale bureaucratic organization. In the 1960s, television provided coverage of rocket launchings and moon landings and in the 1970s astronomy became popular fare on the new educational TV channels. In the 1980s and 90s the electronic media covered planetary missions.

When Harlow Shapley died in the early 1970s, American astronomy bore little resemblance to the scientific community he entered sixty years earlier. Never again could a group like the generals (chapter 7) exert so much power and control. One of the most striking results of the postwar expansion of astronomy was a political economy so complex that the science became a federation of specialist research communities and powerful interest groups within communities. Each specialist community or interest group had its own leaders and ambassadors to the generic astronomical community, as well as to the national and international scientific communities. Both the generic community and specialist research communities worked closely with the federal government, the science bureaucracy, as well as the legislative and executive branches. In both scale and structure, this was, indeed, a far cry from the American astronomical community of Shapley's youth.

But these are topics for further research and other books.

✦

REFERENCES

Abbot, C. G. 1962. "Samuel E. Mitchell." *Biographical Memoirs of the National Academy of Sciences* 36: 254–76. New York, Columbia University Press.

Abbott, A. 1988. *The Structure of Professions: An Essay on the Division of Expert Labor*. Chicago, University of Chicago Press.

Adams, W. S. N.d. . "Autobiographical Notes." Typescript. National Academy of Sciences, Deceased Members Files.

———. 1941. "George Ellery Hale." *Biographical Memoirs of the National Academy of Sciences* 21: 181–241. Washington, D. C., National Academy of Sciences.

Ainley, M. G. 1987. "Field Work and Family: North American Women Ornithologists, 1900–1950." In *Uneasy Careers and Intimate Lives: Women in Science, 1789–1979*, ed. P. G. Abir-Am and D. Outram, 60–76. New Brunswick: Rutgers University Press.

Alexander, A. F. O'D. 1962. *The Planet Saturn: A History of Observation, Theory, and Discovery*. London, Faber and Faber.

Allen, G. E. 1978. *Life Sciences in the Twentieth Century*. Cambridge, Cambridge University Press.

———. 1979. "The Rise and Spread of the Classical School of Heredity, 1910–1930: Development and Influence of the Mendelian Chromosome Theory." In *The Sciences in the American Context: New Perspectives*, ed. N. Reingold, 209–28. Washington, D.C., Smithsonian Institution Press.

Allison, P. D. 1980. *Processes of Stratification in Science*. New York, Arno Press.

Allison, P. D., J. S. Long, and T. K. Krauze. 1982. "Cumulative Advantage and Inequality in Science." *American Sociological Review* 47: 615–25.

Althauser, R. P., and A. L. Kalleberg. 1981. "Firms, Occupations, and the Structure of Labor Markets: A Conceptual Analysis and Research Agenda." In *Sociological Perspectives on Labor Markets*, ed. I. E. Berg, 120–49. New York, Harcourt, Brace and Jovanovich.

Angier, N. 1988. *Natural Obsessions: The Search for the Oncogene*. Boston, Houghton Mifflin.

Anonymous. 1886. "Report of the AAAS Meeting." *The Observatory* 9: 366–67.

———. 1893. "Editorial." *The Observatory* 16: 122–23.

———. 1896. "RAS Discussion." *The Observatory* 19: 221–23.

———. 1897. "RAS Discussion." *The Observatory* 20: 85–87.

———. 1901. "On the Assignment of the Nomenclature and the Formation of a New Catalogue of Variable Stars." *Astronomical Journal* 22: 77–81.

———. 1903. "Review of Clerke, *Problems in Astrophysics*." *Astrophysical Journal* 18: 156–158.

———. 1905. "Comparison of the 36-inch and 40-inch Refractors." *The Observatory* 28: 74.

———. 1913. *The Observatory* 46: 264.

Appel, Toby A. 1988. "Organizing Biology: The American Society of Naturalists and Its 'Affiliated Societies,' 1883–1923." In *The American Development of Biology*, ed. R. Rainger, K. Benson, and J. Maienschein, 87–120. Philadelphia, University of Pennsylvania Press.

Astronomischer Jahresbericht. 1899–1940. W. Wislicenus et al., eds. Berlin, Astronomischen Gesellschaft.

Astrophysical Journal Board. 1896. "Minutes of a Meeting of the Editorial Board of the Astrophysical Journal at the Fifth Avenue Hotel, New York on November 2, 1894." Typescript copy. UCYOA, Directors Papers.

Babbitt, M. K. 1912. *Maria Mitchell as Her Students Knew Her*. Poughkeepsie, NY., n.p.

Babcock, H. W. 1975–77. *Oral History Interview*. New York, American Institute of Physics, Center for the History of Physics.

Baldassare, M. 1992."Suburban Communities." In *Annual Review of Sociology*, ed. J. Blake and J. Hagen, 18: 475–95.

Barnard, J. 1968. "Community Disorganization." In *International Encyclopedia of the Social Sciences*, ed. D. L. Sills, 3: 156–74. New York, Macmillan.

Barnes, S. B., and R. G. A. Dolby. 1970. "The Scientific Ethos: A Deviant Viewpoint." *Archives Européennes de Sociologie* 11: 3–25.

Baron, H. M., and B. Hymer. 1977. "The Dynamics of the Dual Labor Market." In *Problems in Political Economy: An Urban Perspective*, ed. D. M. Gordon, 94–101. Lexington, MA, D. C. Heath.

Basalla, G., ed. 1968. *The Rise of Modern Science: External or Internal Factors?* Lexington, D. C. Heath.

Bausch, J. L., ed. 1988. *Sky & Telescope Cumulative Index, Volumes 1–70, 1941–1985*. Cambridge, Sky Publishing Co.

Becker, Barbara J. 1992. "The Reception of Kirchhoff's Spectroscopic Methods in England: The Case of William Huggins." Paper presented at the Washington meeting of the History of Science Society, December 1992.

———. 1993." Eclecticism, Opportunism, and the Evolution of a New Research Agenda: William and Margaret Huggins and the Origins of Astrophysics." Doctoral dissertation, The Johns Hopkins University.

Ben-David, J. 1971. *The Scientist's Role in Society: A Comparative Study*. Englewood Cliffs, Prentice-Hall.

———. c. 1982. "Norms of Science and the Sociological Interpretation of Scientific Behavior." In *Scientific Growth: Essays on the Social Organization and Ethos of Science,* ed. G. Freudenthal, 469–84. Berkeley, University of California.

Ben-David, J., and R. Collins. 1966. "Social Factors in the Origin of a New Science: The Case of Psychology." *American Sociological Review* 31: 451–65.

Berendzen, R. 1974. "Origins of the American Astronomical Society." *Physics Today* 27: 32–39.

Berendzen, R., and M. T. Moslen. 1972. "Manpower and Employment in American Astronomy." In *Education in and History of Modern Astronomy*, ed. R. Berendzen, 46–65. New York, New York Academy of Sciences.

Berkhofer, R. 1969. *A Behavioral Approach to Historical Analysis*. New York, The Free Press.

Bielby, W. T. and J. N. Baron. 1984. "A Woman's Place Is with Other Women: Sex Segregation within Organizations." In *Sex Segregation in the Workplace*, ed. B. F. Reskin, 27–55. Washington, D.C., National Academy of Sciences.

Blau, F. D. and C. L. Jusenius. 1976. "Economists' Approaches to Sex Segregation in the Labor Market: An Appraisal." *Signs* 1: 181–99.

Bloor, D. 1992. "Left and Right Wittgensteinians." In *Science as Practice and Culture,* ed. A. Pickering, 266–82. Chicago, University of Chicago Press.

Böeme, G. 1977. "Models of the Development of Science." In *Science, Technology and Society: A Cross-Disciplinary Perspective*, ed. I. Spiegel-Rösing and D. J. DeS. Price, 319–54. London, Sage.

Boss, B. 1920. "Biographical Memoir of Lewis Boss." *Biographical Memoirs of the National Academy of Sciences* 9: 239–60. Washington, D.C., National Academy of Sciences.

———. 1968. *History of the Dudley Observatory*. Albany, Dudley Observatory.

Boss, L. 1895. "The New Dudley Observatory." *Astronomical Journal* 14: 169–75.

Bowler, P. J. 1983. *The Eclipse of Darwinism: Anti-Darwinian Evolution Theories in the Decades Around 1900*. Baltimore, The Johns Hopkins University Press.

———. 1989. *Evolution: The HIstory of an Idea*. Berkeley, University of California Press.

Bracher. K. 1989. "The Stars for All: A Centennial History of the Astronomical Society of the Pacific." *Mercury* 18: 1–43.

Braudel, F. 1980. *On History*. Chicago, University of Chicago Press.

Broad, W. J. 1980. "History of Science Losing Its Science." *Science* 207: 389.

Brock, W. H. 1976. "The Spectrum of Science Patronage." In G. L'E. Turner, ed., *The Patronage of Science in the Nineteenth Century*, 173–206. Leyden, Noordhoff International.

Brouwer, D. 1945. "Frank Schlesinger." *Biographical Memoirs of the National Academy of Sciences* 24: 105–44. Washington, D. C., The National Academy of Sciences.

Brown, E. W. 1916. "Biographical Memoir of George W. Hill." *Biographical Memoirs of the National Academy of Sciences* 8: 275–309. Washington, D.C., National Academy of Sciences.

Bruce, R.V. 1987. *The Launching of Modern American Science, 1846–1876*. New York, Knopf.

Brush, S. G. 1979. "Looking Up: The Rise of Astronomy in America." *American Studies* 20: 41–67.

Byrd, M. E. 1886. "Popular Fallacies about Observatories." *Sidereal Messenger* 5: 263–266.

Cahnman, W. J., and A. Boskoff, eds. 1964. *Sociology and History: Theory and Research*. New York, The Free Press.

Cain, G. G. 1975. "The Challenge of Dual and Radical Theories of the Labor Market to Orthodox Theory." *The American Economic Review* 65: 16–22.

———. 1976. "The Challenge of Segmented Labor Market Theories to Orthodox Theory." *Journal of Economic Literature* 14: 1215–57.

Campbell, W. W. 1908. "Comparative Power of the 36-inch Refractor of the Lick Observatory." *Popular Astronomy* 16: 560–62.

———. 1924. "Simon Newcomb: A Biographical Memoir." *Biographical Memoirs of the National Academy of Sciences* 17: 1–18. Washington, D. C., National Academy of Sciences.

Carnegie Institution of Washington 1904. *Yearbook*. Washington, D.C.

Caswell, A. 1860. "Address of the President of the Association." *Proceedings of the American Association for the Advancement of Science for 1859*, 1–26. Cambridge, Mass, Allen and Farnham.

Cattell, J. McK., ed. 1906–44. *American Men of Science*. New York, Bowker.

Chandler, A. D., Jr. 1962. *Strategy and Structure: Chapters in the History of Industrial Enterprise*. Cambridge: MIT Press.

Chandler, A. D., Jr., and Louis Galambos 1970. "The Development of Large-Scale Economic Organizations in Modern America." *Journal of Economic History* 30: 201–217.

Chandler, S. C. 1888a. "Catalogue of Variable Stars." *Astronomical Journal* 8: 81–92.

———. 1888b. "On the Colors of Variable Stars." *Astronomical Journal* 8: 137–140.

———. 1893. "Second Catalogue of Variable Stars." *Astronomical Journal* 13: 89–91.

———. 1894. "Supplement to the Second Catalogue of Variable Stars." *Astronomical Journal* 14: 81–92.

———. 1895. "Revised Supplement to Second Catalogue of Variable Stars." *Astronomical Journal* 15: 81–85.

Chandrasekhar, S. 1977. *Oral History Interview*. New York, American Institute of Physics, Center for the History of Physics.

Chester, C. M. 1903. *Report of the Superintendent of the United States Naval Observatory for the Fiscal Year Ending 30 June 1903*. Washington, D. C., Government Printing Office.

Chubin, D. E. 1976. "The Conceptualization of Scientific Specialties." *Sociological Quarterly* 17: 448–76.

Clegg, S. R. 1989. *Frameworks of Power*. London, Sage.

Clerke, A. M. 1887. *A Popular History of Astronomy During the Nineteenth Century*. 1st edition. London, Adam and Charles Black.

———. 1903. *Problems in Astrophysics*. London, Adam and Charles Black.

———. 1908. *A Popular History of Astronomy During the Nineteenth Century*. 4th edition. London, Adam and Charles Black.

Cochrane, R. C. 1978. *The National Academy of Sciences: The First Hundred Years, 1863–1963*. Washington, D. C., National Academy of Sciences.

Cohen, I. B. 1985. *Revolution in Science*. Cambridge, Harvard University Press.

Cohen, J., and P. Cohen. 1975. *Applied Multiple Regression/Correlation Analysis for the Behavioral Sciences*. Hillsdale, N. J., Lawrence Erlbaum Associates.

Cole, A. H. 1959. *Business Enterprise in its Social Setting*. Cambridge, Harvard University Press.

Cole, J. R. 1979. *Fair Science: Women in the Scientific Community*. New York, The Free Press.

Cole, J. R., and S. Cole. 1973. *Social Stratification in Science*. Chicago, University of Chicago.

Cole, S., and J. R. Cole. 1967. "Scientific Output and Recognition: A Study in the Operation of the Reward System in Science." *American Sociological Review* 32: 377–90.

Collins, N. W. 1983. *Professional Women and Their Mentors: A Practical Guide to Mentoring for the Woman Who Wants to Get Ahead*. Englewood Cliffs, Prentice Hall.

Collins, R. 1975. *Conflict Sociology: Toward an Explanatory Science*. New York, Academic Press.

Committee on the Status of Women, American Astronomical Society 1988. "Issues and Questions: An Open Meeting of the CSWA." Mimeographed.

Comstock, G. C. 1922. "Biographical Memoir of Benjamin A. Gould." *Biographical Memoirs of the National Academy of Sciences* 17: 155–80. Washington, D. C., National Academy of Sciences.

Cott, N. 1977. *The Bonds of Womanhood: "Women's Sphere" in New England, 1780–1835*. New Haven, Yale University Press.

Cozzens, S. F., and T. F. Gieryn, eds. 1990. *Theories of Science in Society*. Bloomington, Indiana University Press.

Crane, D. 1965. "Scientists at Major and Minor Universities: A Study in Productivity and Recognition." *American Sociological Review* 30: 699–714.

———. 1970. "The Academic Marketplace Revisited: A Study of Faculty Mobility Using the Cartter Ratings." *American Journal of Sociology* 75: 953–64.

Crosland, M. 1979. "From Prizes to Grants in the Support of Scientific Research in France in the Nineteenth Century: The Montyou Legacy." *Minerva* 17: 355–80.

Crowe, M. J. 1986. *The Extraterrestrial Life Debate, 1750–1900: The Plurality of Worlds from Kant to Lowell*. Cambridge, Cambridge University Press.

Curti, M. E. 1959. *The Making of an American Community: A Case Study of Democracy in a Frontier County*. Stanford, Stanford University Press.

Danto, A. C. 1968. *Analytical Philosophy of History*. Cambridge, Cambridge University Press.

Davis, A. F. 1973. *An American Heroine: The Life and Legend of Jane Addams*. New York, Oxford University Press.

Davis, H. S. 1898. "Women Astronomers, 1750–1890." *Popular Astronomy* 6: 211–26.

Davis, L. E., and D. J. Kevles. 1974. "The National Research Fund: A Case Study in the Industrial Support of Science." *Minerva* 12: 207–20.

DeVorkin, D. H. 1977. "W. W. Campbell's Spectroscopic Study of the Martian Atmosphere." *Quarterly Journal of the Royal Astronomical Society* 18: 37–53.

———. 1980. *Preliminary Finding Aid to the Lick Observatory Archives*. New York, American Institute of Physics.

———. 1981. "Community and Spectral Classification in Astrophysics: The Acceptance of E. C. Pickering's System in 1910." *Isis* 72: 29–49.

———. ed. 1982a. *The History of Modern Astronomy and Astrophysics: A Selected, Annotated Bibliography*. New York, Garland Publishing.

———. 1982b. "An Astronomer Responds to War: Otto Struve and the Yerkes Observatory During World War II." *Minerva* 18: 595–623.

———. 1984a. "Stellar Evolution and the Origin of the Hertzsprung-Russell Diagram." In *Astrophysics and Twentieth-Century Astronomy to 1950*, ed. O. Gingerich, 4A: 90–108. Cambridge, Cambridge University Press.

———. 1984b. "The Harvard Summer School in Astronomy." *Physics Today* 37: 48–55.

———. 1990. "Defending a Dream: Charles Greeley Abbot's Years at the Smithsonian." In *Two Astronomical Anniversaries: Harvard College Observatory and the Smithsonian Astrophysical Observatory*, ed. O. Gingerich and M. A. Hoskin, 121–36. Cambridge, Harvard-Smithsonian Center for Astrophysics.

———. 1996. "Astrophysics." In *The History of Astronomy: An Encyclopedia*, ed. J. Lankford. New York, Garland.

Dexter, F. B. 1912. "Denison Olmsted." *Biographical Sketches of the Graduates of Yale College with Annals of the College History* 6: 592–600. New Haven, Yale University Press.

Diamond, S. 1955. *The Reputation of the American Businessman*. Cambridge, Harvard University Press.

Dick, S. J. 1983. "How the United States Naval Observatory Began, 1830–1865." In *Sesquicentennial Symposia of the United States Naval Observatory*, ed. S. J. Dick and L. E. Doggett, 167–181. Washington, D. C., USNO.

DiMaggio, P., and W. Powell. 1983. "The Iron Cage Revisited: Institutional Isomorphism and Collective Rationality in Organizational Fields." *American Sociological Review* 48: 147–60.

Doel, R. E. 1996. *Solar System Astronomy in America: Patronage, Communities, and Interdisciplinary Research, 1920–1960*. New York, Cambridge University Press.

———. Forthcoming. "Diplomatic Constraints on American Science: The Cold War Relations between the U.S., Soviet, and Chinese Astronomers, 1950–1961."

Doeringer, P. B., and M. J. Piore, eds. 1971. *Internal Labor Markets and Manpower Analysis*. Lexington, D. C. Heath.

Doggett, L. E. 1996. "Nautical Almanac Offices." In *The History of Astronomy: An Encyclopedia*, ed. J. Lankford, Garland.

Dreyer, J. L. E. 1923. "The Decade 1830–1840."In *History of the Royal Astronomical Society, 1820–1920*, ed. Dreyer and H. H. Turner, 50–81. London, Royal Astronomical Society.

Dupree, A. H. 1957a. "The Founding of the National Academy of Sciences: A Reinterpretation." *Proceedings of the American Philosophical Society* 101: 434–40.

———. 1957b. *Science in the Federal Government: A History of Policies and Activities to 1940*. Cambridge, Harvard University Press.

———.1979. "The National Academy of Sciences and the American Definition of Science." In *The Organization of Knowledge in Modern America, 1860–1920*, ed. A. Oleson and J. Voss, 342–63. Baltimore, The Johns Hopkins University Press.

Eastman, J. R. 1892. "The Neglected Field of Fundamental Astronomy." *Proceedings of the American Association for the Advancement of Science, Forty-Third Meeting*, ed. Secretary of the AAAS, 17–32. Salem, AAAS.

Eberhard, G., A. Kohlschütter, and H. Ludendorff, eds. 1928–1936. *Handbuch der Astrophysik*. 7 vols. Berlin, Springer.

Eddy, J. A. 1990. "Founding the Astrophysical Observatory: The Langley Years." In *Two Astronomical Anniversaries: Harvard College Observatory and the Smithsonian Astrophysical Observatory*, ed. O. Gingerich and M. A. Hoskin, 111–20. Cambridge, Mass., Harvard-Smithsonian Center for Astrophysics.

Edge, D. O. 1979. "Quantitative Measures of Communication in Science: A Critical Review." *History of Science* 17: 102–34.

Edge, D. O., and M. J. Mulkay. 1976. *Astronomy Transformed: The Emergence of Radio Astronomy in Britain*. New York, John Wiley and Sons.

Eggers, S. 1995. "The Making of a Scientific Ph.D. in 1900: Caroline Ellen Furness." Undergraduate Honors Thesis, University of Missouri-Columbia. Department of History.

Eiseley, L. 1958. *Darwin's Century: Evolution and the Men Who Discovered It*. Garden City, Doubleday.

Elliott, C. A. 1982. "Models of the American Scientist: A Biographical and Summary View." *Isis* 73: 77–93.

Evans, D.S. 1944. "The Future of Astronomy." *The Observatory* 65: 237–41.

Evans, M. A., and H. E. Evans 1970. *William Morton Wheeler, Biologist*. Cambridge, Mass., Harvard University Press.

Featherman, D. L. 1981. "Social Stratification and Mobility: Two Decades of Cumulative Social Research." *American Behavioral Scientist* 24: 364–85.

Fisch, R. 1977. "The Psychology of Science." In *Science, Technology and Society: A Cross-Disciplinary Perspective*, ed. I. Spiegel-Rösing and D. J. DeS. Price, 227–318. London, Sage.

Flam, F. 1991. "Still a 'Chilly Climate' for Women?" *Science* 252 5013): 1604–6.

Fleming, W. P. 1893. "A Field for Women's Work in Astronomy." *Astronomy and Astrophysics* 12: 683–89.

Fletcher, R. S. 1971. *A History of Oberlin College from Its Foundation Through the Civil War*. New York, Arno Press.

Fogel, R. W. 1974. *Time on the Cross: The Economics of American Negro Slavery*. Boston, Little, Brown.

Forbes, E. G., A. J. Meadows, and D. Howse, eds. 1975. *Greenwich Observatory: The Royal Observatory at Greenwich and Herstmonceaux, 1675–1975*. 3 vols. London, Taylor and Francis.

Forbes, G. 1916. *David Gill: Man and Astronomer*. London, John Murray.

Forman, P. 1991. "Independence, Not Transcendence, for the Historian of Science." *Isis* 82: 71–86.

Forman, P., J. L. Heilbron, and S. R. Weart. 1975. "Physics *circa* 1900: Personnel, Funding, and Productivity of the Academic Establishments." *Historical Studies in the Physical Sciences* 5: 3–185.

Fowler, A., ed. 1922. *Transactions of the International Astronomical Union*. Vol. I. London: Imperial College.

Fox, M. F. 1983. "Publication Productivity Among Scientists: A Critical Review." *Social Studies of Science* 13: 285–305.

Fox, R. 1976. "Scientific Enterprise and the Patronage of Research in France, 1800–1870." In *The Patronage of Science in the Nineteenth Century*, ed. G.L'E. Turner, 9–52. Leyden, Noordhoff International.

Fox, R., and G. Weisz, eds. 1980. *The Organization of Science and Technology in France, 1808–1914*. Cambridge, Cambridge University Press.

Fricke, W. 1970. "Friedrich Wilhelm Bessel." *Dictionary of Scientific Biography*, ed. C. C. Gillispie, 2: 97–102. New York, Charles Scribner's Sons.

From, W. H. 1968. "Occupations and Careers." In *International Encyclopedia of the Social Sciences,* ed. D. L. Sills, 11: 245–53. New York, Macmillan.

Frost, E. B. 1910. "Biographical Memoir of Charles A. Young." *Biographical Memoirs of the National Academy of Sciences* 7: 91–114. Washington, D. C., National Academy of Sciences.

———. 1926. "Edward Emerson Barnard." *Memoirs of the National Academy of Sciences* 21: 1–23. Washington, D. C., Government Printing Office.

———. 1933. *An Astronomer's Life*. Boston, Houghton Mifflin.

Galambos, L. 1983. "Technology, Political Economy, and Professionalization: Central Themes of the Organizational Synthesis." *Business History Review* 57: 471–93.

Galison, P. 1983. "Re-reading the Past from the End of Physics: Maxwell's Equations in Retrospect." In *Functions and Uses of Disciplinary Histories,* ed. L. Graham et al., 35–51. Dordrecht, Reidel.

Gaston, J. 1978a. *The Reward System in British and American Science.* New York, Wiley.

———. 1978b. "The Norm of Universalism." *Sociological Inquiry* 48: 3–4.

Geiger, R. L. 1986. *To Advance Knowledge: The Growth of American Research Universities, 1900–1940.* New York, Oxford University Press.

Gerstner, P. A. 1976. "The Academy of Natural Sciences of Philadelphia, 1812–1850." In *The Pursuit of Knowledge in the Early American Republic: American Scientific and Learned Societies from Colonial Times to the Civil War*, ed. A. Oleson and S. C. Brown, 174–93. Baltimore, The Johns Hopkins University Press.

Gieryn, T. F. 1980. "Patterns in the Selection of Problems for Scientific Research: American Astronomers, 1950–1975." Ph.D. dissertation, Columbia University.

Gill, D. 1891. "An Astronomer's Work in a Modern Observatory." *The Observatory* 14: 335–341; 370–411.

Gingerich, O., ed. 1984. *Astrophysics and Twentieth-Century Astronomy to 1950. The General History of Astronomy.* Vol. 4A of *The General History of Astronomy*, M. Hoskin series ed. Cambridge, Cambridge University Press.

Gingerich, O., and M. A. Hoskin, eds. 1990. *Two Astronomical Anniversaries: The Harvard College Observatory and the Smithsonian Astrophysical Observatory.* Cambridge, Harvard-Smithsonian Center for Astrophysics.

Goldberg, L. 1977. "Donald H. Menzel." *Sky & Telescope* 53: 244–51.

———. 1978. *Oral History Interview.* New York, American Institute of Physics, Center for the History of Physics.

Goldfarb, S. 1969. "Science and Democracy: A History of the Cincinnati Observatory, 1842–1872." *Ohio History* 78: 172–78; 222–28.

Gordon, L. D. 1990. *Gender and Higher Education in the Progressive Era*. New Haven, Yale University Press.

Gouldner, A. W. 1957. "Cosmopolitans and Locals: Towards an Analysis of Latent Social Roles." *Administrative Science Quarterly* 2: 281–306, 444–80.

Greene, J. C. 1976. "Science, Learning and Utility: Patterns of Organization in the Early American Republic." In *The Pursuit of Knowledge in the Early American Republic: American Scientific and Learned Societies from Colonial Times to the Civil War*, ed. A. Oleson and S. C. Brown, 1–20. Baltimore, The Johns Hopkins University Press.

Greene, M. 1988. *A Science Not Earthbound: A Brief History of Astronomy at Carleton College*. Northfield, Minn., Carleton College.

Grove-Hills, E. H. 1923. "The Decade 1850–1860." In *History of the Royal Astronomical Society, 1820–1920*, ed. J. L. E. Dreyer and H. H. Turner, 110–28. London, Royal Astronomical Society.

Greenstein, J. L. 1972. *Astronomy and Astrophysics for the 1970s*. Astronomy Survey Committee of the Committee on Science and Public Policy, National Academy of Sciences. Washington, D.C., NAS.

———. 1977–1978. *Oral History Interview*. New York, American Institute of Physics, Center for the History of Physics.

Guralnick, S. M. 1979. "The American Scientist in Higher Education, 1820–1910." In *The Sciences in the American Context: New Perspectives*, ed. N. Reingold, 99–142. Washington, D. C., Smithsonian Institution Press.

Hagstrom, W. O. 1965. *The Scientific Community*. New York, Basic Books.

———. 1971. "Inputs, Outputs and the Prestige of University Science Departments." *Sociology of Education* 44: 375–97.

Hale, G. E. 1895. "The *Astrophysical Journal*." *The Astrophysical Journal* 1: 80–84.

———. 1897. "The Aim of the Yerkes Observatory." *Astrophysical Journal* 6: 310–321.

———. 1905a. "The Development of a New Observatory." *Publications of the Astronomical Society of the Pacific* 17: 41–52.

———. 1905b. "The Solar Observatory of the Carnegie Institution of Washington." *Astrophysical Journal* 21: 151–72.

———. 1908. *The Study of Stellar Evolution: An Account of Some Recent Methods of Astrophysical Research*. Chicago, University of Chicago Press.

Hall, A. 1908. *An Astronomer's Wife: The Biography of Angeline Hall*. Baltimore, Nunn and Co.

Hargens, L. L., N. C. Mullins, and P. K. Hecht. 1980. "Research Areas and Stratification Processes in Science." *Social Studies of Science* 10: 55–74.

Hargens, L. L., and D. H. Felmlee. 1984. "Structural Determinants of Stratification in Science." *American Sociological Review* 49: 685–97.

Hargens, L. L., and W. O. Hagstrom. 1967. "Sponsored and Contest Mobility of American Academic Scientists." *Sociology of Education* 40: 24–38.

Harwit, M. 1981. *Cosmic Discovery: The Search, Scope, and Heritage of Astronomy*. New York, Basic Books

Harwood, J. 1993. *Styles of Scientific Thought: The German Genetics Community, 1900–1933*. Chicago, The University of Chicago Press.

Hawley, A. H. 1950. *Human Ecology: A Theory of Community Structure*. New York, Ronald.

Hearnshaw, J. B. 1986. *The Analysis of Starlight: One-Hundred and Fifty Years of Astronomical Spectroscopy*. Cambridge, Cambridge University Press.

———. 1996. "Photometry, Astronomical." In *The History of Astronomy: An Encycolpedia,* ed. J. Lankford. New York, Garland.

Herbst, J. 1983. "Diversification in American HIgher Education." In *The Transformation of the Higher Learning, 1860–1930,* K. H. Jarausch ed., 196–206. Chicago, University of Chicago Press.

Hergert, P. 1977. *Oral HIstory Interview*. New York, American Institute of Physics, Center for the History of Physics.

———. 1978. "Armin O. Leuschner." *Biographical Memoirs of the National Academy of Sciences* 49: 129–47. Washington, D. C., National Academy of Sciences.

Herrmann, D. B. 1971. "B. A. Gould and His Astronomical Journal." *Journal for the History of Astronomy* 2: 98–108.

———. 1984. *The History of Astronomy from Herschel to Hertzsprung*. Trans. K. Krisciunas. Cambridge, Cambridge University Press.

Hetherington, N. S. 1983. "Mid-Nineteenth Century American Astronomy: Science in a Developing Nation." *Annals of Science* 40: 61–80.

———. 1988. *Science and Objectivity: Episodes in the History of Astronomy*. Ames, Iowa State University Press.

Hodge, J. E. 1977. "Charles Dillon Perrine and the Transformation of the Argentine National Observatory." *Journal for the History of Astronomy* 8: 12–25.

Hoffleit, D. 1979. *Oral HIstory Interview*. New York, American Institute of Physics, Center for the History of Physics.

———. 1991. "The Evolution of the Henry Draper Memorial." *Vistas in Astronomy* 34: 107–62.

———. 1992. *Astronomy at Yale, 1701–1968. Memoirs of the Connecticut Academy of Arts and Sciences*, 23(1992).

Holden, C. 1989. "Chauvinism in Nobel Nominations." *Science* 243: 471.

Holden, E. S. 1881a. "Astronomy." *Annual Report of the Smithsonian Institution*, 191–230. Washington, D.C.

———. 1888b. "The Ring Nebulae in Lyra." *Monthly Notices of the Royal Astronomical Society* 48: 383–88.

———. 1897. "Mr. Lowell's Observations of Mercury and Venus." *Publications of the Astronomical Society of the Pacific* 9: 92–93.

Holden, E. S., and W. Winlock. 1880. "Astronomy." *Annual Report of the Smithsonian Institution*, 183–219. Washington, D.C.

———. 1882. "Astronomy." *Annual Report of the Smithsonian Institution*, 277–324. Washington, D. C.

———. 1883. "Astronomy." *Annual Report of the Smithsonian Institution*, 365–441. Washington, D. C.

———. 1884. "Astronomy." *Annual Report of the Smithsonian Institution*, 159–213. Washington, D. C.

Holland, L. 1990. "Which Scientists Might Be Honored With The Nobel Prize?" *The Scientist* 1 October): 17–18.

Horowitz, H. L. 1987. *Campus Life: Undergraduate Culture from the End of the Eighteenth Century to the Present.* New York, Knopf.

Hoyt, W. G. 1976. *Lowell and Mars.* Tucson, University of Arizona Press.

Hufbauer, K. 1981. "Astronomers Take Up the Stellar Energy Problem, 1917–1920." *Historical Studies in the Physical Sciences* 11: 277–303.

———. 1991. *Exploring the Sun: Solar Science Since Galileo.* Baltimore, The Johns Hopkins University Press.

Huggins, W. 1900. "The New Astronomy: A Personal Retrospect." In *Essays in Astronomy,* ed. E. S. Holden, 441–52. New York, Appleton and Co.

Hutchins, R. M. 1947. "The Yerkes Observatory." *Science* 106 (2749): 195–96.

Jackson, D. N., and J. P. Rushton, eds. 1987. *Scientific Excellence: Origins and Assessments.* Beverly Hills, Sage.

James, E. T., J. W. James, and P. S. Boyer, eds. 1971. *Notable American Women, 1607–1950: A Biographical Dictionary.* 3 vols. Cambridge, Harvard University Press.

James, M. A. 1987. *Elites in Conflict: The Antebellum Clash Over the Dudley Observatory.* New Brunswick, Rutgers University Press.

Jarausch, K. H. 1983. *The Transformation of Higher Learning, 1869–1930.* Chicago, University of Chicago Press.

Jardine, N. 1994. "A Trial of Galileo." *Isis* 85: 279–83.

Jencks, C., et al. 1972. *Inequality; A Reassessment of the Effect of Family and Schooling in America.* New York, Basic Books.

Jenkins, R. V. 1977. *Images and Enterprise: Technology and the American Photographic Industry, 1839–1925.* Baltimore, The Johns Hopkins University Press.

Jones, B. Z., and L. G. Boyd. 1971. *The Harvard College Observatory: The First Four Directorships, 1839–1919.* Cambridge, Harvard University Press.

Joy, A. H. 1958. "Walter S. Adams." *Biographical Memoirs of the National Academy of Sciences* 31: 1–31. New York, Columbia University Press.

Jungnickel, C., and. R. McCormmach. 1986. *Intellectual Mastery of Nature: Theoretical Physics from Ohm to Einstein.* 2 vols. Chicago, University of Chicago Press.

Keeler, J. E. 1897. "The Importance of Astrophysical Research and the Relation of Astrophysics to Other Physical Sciences." *Astrophysical Journal* 6: 271–88.

Kendall, P. M. 1896. *Maria Mitchell: Life, Letters, and Journals*. N.p., Books for Libraries Series. Reprint 1977.

Kerckhoff, A. C. 1984. "The Current State of Social Mobility Research." *The Sociological Quarterly* 25: 139–53.

Kevles, D. J. 1968. "George Ellery Hale, the First World War, and the Advancement of Science in America." *Isis* 59: 427–37.

———. 1979. "The Physics, Mathematics, and Chemistry Communities: A Comparative Analysis." In *The Organization of Knowledge in Modern America, 1860–1920*, ed. A. Oleson and J. Voss, 139–72. Baltimore, The Johns Hopkins University Press.

———. 1978. *The Physicists: The History of a Scientific Community in Modern America*. New York, Knopf.

Kidwell, P. A. 1986. "E. C. Pickering, Lydia Hinchman, Harlow Shapley and the Beginning of Graduate Work at the Harvard College Observatory." *Astronomy Quarterly* 5: 157–71.

———. 1987. "Cecilia Payne-Gaposchkin: Astronomy in the Family." In *Uneasy Careers and Intimate Lives: Women in Science, 1789–1979*, ed. P. G. Abir-Am and D. Outram, 216–38. New Brunswick, Rutgers University Press.

———. 1990. "Three Women of Astronomy." *American Scientist* 78: 244–51.

———. 1992. "Harvard Astronomers in World War II: Disruption and Opportunity." In *Science at Harvard University*, ed. C. Elliott and M. Rossiter, 285–302. Bethlehem, Pennsylvania, Lehigh University Press.

Knapp, R. H., and H. B. Goodrich. 1952. *Origins of American Scientists: A Study Made under the Direction of a Committee of the Faculty of Wesleyan University*. New York, Russell and Russell.

Knorr-Cetina, K., and M. J. Mulkay, eds. 1973. *Science Observed: Perspectives on the Social Study of Science*. London, Sage.

Knorr-Cetina, K. D., R. Krohn, and R. Whitley, eds. 1981. *The Social Process of Scientific Investigation*. Dordrecht, Reidel.

Kohler, R. E. 1980. "Letter." *Science* 207: 934–35.

———. 1982. *From Medical Chemistry to Biochemistry: The Making of a Biomedical Discipline*. Cambridge: Cambridge University Press.

———. 1990. "The Ph.D. Machine: Building on the Collegiate Base." *Isis* 81: 638–62.

———. 1991a. *Partners in Science: Foundations and Natural Scientists, 1900–1945*. Chicago, University of Chicago Press.

———. 1991b. "Drosophila and Evolutionary Genetics: The Moral Economy of Scientific Practice." *History of Science* 29: 335–75.

———. 1991c. "Systems of Production: Drosophila, Neurospora, and Biochemical Genetics." *Historical Studies of the Biological and Physical Sciences* 22: 87–130.

———. 1994. *Lords of the Fly: Drosophila Genetics and the Experimental Life.* Chicago, University of Chicago Press

Kohlstedt, S. G. 1976. *The Formation of the American Scientific Community: The American Association for the Advancement of Science, 1848–1860.* Urbana, University of Illinois Press.

———. 1978a. "Maria Mitchell: The Advancement of Women in Science." *New England Quarterly* 51: 39–63.

———. 1978b. "In from the Periphery: American Women in Science, 1830–1880." *Signs* 4: 81–96.

Kohlstedt, S. G., and M. W. Rossiter, eds. 1985. *Historical Writing on American Science. Osiris*, 2nd series, vol. 1.

Kragh, H. 1987. *An Introduction to the Historiography of Science.* Cambridge, Cambridge University Press.

Krisciunas, K. 1992. "Otto Struve, 1897–1963." *Biographical Memoirs of the National Academy of Sciences* 61: 351–87. Washington, D.C., The National Academy.

Kuhn, T. S. 1970. *The Structure of Scientific Revolutions.* Chicago, University of Chicago Press.

———. 1977. *The Essential Tension: Selected Studies in Scientific Tradition and Change.* Chicago, University of Chicago Press.

LaFollette, M. C. 1990. *Making Science Our Own: Public Images of Science, 1910–1955.* Chicago, University of Chicago Press.

Lancaster, A. 1886. "Liste Générale Observatoires et des Astronomes, des Sociétés et des Revues Astronomiques." *Annuaire de L'Observatoire Royal De Bruxelles* 54: 444–577.

Langley, S. P. 1879. "The Recent Progress of Solar Physics." *Popular Science Monthly* 16: 1–11.

———. 1884. "The New Astronomy." *The Century Illustrated Monthly Magazine* 28: 712–26; 922–36.

———. 1892. *The New Astronomy.* Boston, Houghton, Mifflin.

Lankford, J. 1979. "Amateur Versus Professional: The Transatlantic Debate Over the Measurement of Jovian Longitude." *Journal of the British Astronomical Association* (October 1979), 574–82.

———. 1980. "A Note on T. J. J. See's Observations of the Craters on Mercury." *Journal for the History of Astronomy* 11: 129–32.

———. 1981a. "Amateurs Versus Professionals: The Controversy Over Telescope Size in Late Victorian Science." *Isis* 72: 11–28.

———. 1981b. "Amateurs and Astrophysics: A Neglected Aspect in the Development of a Scientific Specialty." *Social Studies of Science* 11: 275–303.

———. 1984a. "The Impact of Astronomical Photography to 1920." In *Astrophysics and Twentieth-Century Astronomy to 1950,* ed. O. Gingerich, 4A:16–39. Cambridge, Cambridge University Press.

———. 1984b. "In Search of Henry Fitz, Early American Telescope Maker." *Sky & Telescope* 66: 214–18.

———. 1984c. "Margaret W. Rossiter's *Women Scientists in America* Considered from a Sociology of Science Perspective." *Isis* 75: 192–95.

———. 1985. "How Not to Run an International Halley Watch." *Sky & Telescope* 68: 530–33.

———. 1987a. "Charting the Southern Sky." *Sky & Telescope* 74: 243–46.

———. 1987b. "Private Patronage and the Growth of Knowledge: The J. Lawrence Smith Fund of the National Academy of Sciences, 1884–1940." *Minerva* 25: 269–81.

———. 1987c. "Photography and the Nineteenth Century Transits of Venus." *Technology & Culture* 28: 648–57.

———. 1988. "Amateurs: Astronomy's Enduring Resource," *Sky & Telescope* 75: 482–83.

———. 1994. "Women and Women's Work at Mt. Wilson Observatory Before World War II." In *The Earth, the Heavens and the Carnegie Institution of Washington*, G. Good. ed. *History of Geophysics* 5: 125–27. Washington, D.C., American Geophysical Union.

Lankford, J., ed. 1996. *The History of Astronomy: An Encyclopedia.* New York, Garland.

Lankford, J. and R. L. Slavings. 1990. "Gender and Science: The Experience of Women in American Astronomy, 1859–1940." *Physics Today* 43: 58–65.

———. 1996. "The Industrialization of American Astronomy, 1880–1940." *Physics Today* 49: 34–41.

Latour, B., and S. Woolgar. 1979. *Laboratory Life: The Social Construction of Scientific Facts.* Beverly Hills, Sage.

LeGrand, H. 1988. *Drifting Continents and Shifting Theories: The Modern Revolution in Geology and Scientific Change.* Cambridge: Cambridge University Press.

Leonard, W. L., ed. 1921. *Percival Lowell: An Afterglow.* Boston, Richard G. Badger.

Levinson, D. J., et al. 1978. *The Seasons of a Man's Life.* New York, Knopf.

Lewis-Beck, M. S. 1990. *Applied Regression: An Introduction.* Beverly Hills, Sage.

Lindeman, E. C. 1931. "Community." In *Encyclopedia of the Social Sciences*, ed. E. R. Seligman, 4: 102–5. New York, Macmillan.

Long, J. S. 1978. "Productivity and Academic Position In the Scientific Career." *American Sociological Review* 43: 889–908.

Long, J. S., P. D. Allison, and R. McGinnis. 1979. "Entrance Into the Academic Career." *American Sociological Review* 44: 816–30.

Long, J. S., and R. McGinnis. 1981. "Organizational Context and Scientific Productivity." *American Sociological Review* 46: 422–42.

Loomis, E. 1851. *The Recent Progress of Astronomy: Especially in the United States.* New York, Harper and Brothers.

———. 1856. "Astronomical Observatories in the United States." *Harper's New Monthly Magazine* 13: 25–52.

Lowell, P. 1905. "Comparison Charts of the Region Following Delta Ophiuchi." *Monthly Notices of the Royal Astronomical Society* 66: 57–58.

Lubbock, C. A., ed. 1933. *The Herschel Chronicle.* Cambridge, Cambridge University Press.

Lukes, S. 1974. *Power: A Radical View.* London, Macmillan.

Lundgreen, P. 1980. "The Organization of Science and Technology in France: A German Perspective." In *The Organization of Science and Technology in France, 1808–1914.* In R. Fox and G. Weisz, eds., 311–32. Cambridge, Cambridge University Press.

———. 1983. "Differentiation in German Higher Education." In *The Transformation of Higher Learning, 1860–1930,* K. H. Jarausch ed., 149–79. Chicago, University of Chicago Press.

Lynd, R. S., and H. Lynd 1929. *Middletown: A Study in Contemporary American Culture.* New York, Harcourt, Brace.

———. 1937. *Middletown in Transition: A Study in Cultural Conflicts.* New York, Harcourt, Brace.

MacLeod, R. 1977. "Changing Perspectives on the Social History of Science." In *Science, Technology and Society: A Cross-Disciplinary Perspective,* ed. I. Spiegel-Rösing and D. J. DeS. Price, 149–96. London, Sage.

Maienschein, J. 1981. "Shifting Assumptions in American Biology: Embryology, 1890–1910." *Journal of the History of Biology,* 14: 89–113.

———. 1985. "History of Biology." In *Historical Writing on American Science,* ed. S. G. Kohlstedt and M. W. Rossiter, 147–62. *Osiris,* 2d ser., 1.

———. 1991. *Transforming Traditions in American Biology, 1880–1915.* Baltimore, The Johns Hopkins University Press.

Marks, C. 1991. "The Urban Underclass." In *Annual Review of Sociology* ed. W. R. Scott and J. Blake, 17: 445–66.

Marsden, B. G. 1971. "Seth Carlo Chandler." *Dictionary of Scientific Biography,* ed. C. C. Gillispie 3: 194. New York, Charles Scribner's Sons.

Marx, K. 1976. *Capital.* Harmondsworth, Penguin.

Maunder, E. W. 1900. *The Royal Observatory Greenwich.* London, Religious Tract Society.

McCormmach, R. 1966. "Ormsby MacKnight Mitchel's *Sidereal Messenger,* 1846–1848." *Proceedings of the American Philosophical Society* 110: 35–47.

McGucken, W. 1969. *Nineteenth Century Spectroscopy: Development of the Understanding of Spectra, 1802–1897.* Baltimore, The Johns Hopkins University Press.

Meadows, A. J. 1970. *Early Solar Physics.* Oxford, Pergamon Press.

———. 1984a. "The Origins of Astrophysics." In *Astrophysics and Twentieth-Century Astronomy to 1950*, ed. O. Gingerich, 4A: 3–15. Cambridge, University of Cambridge Press.

———. 1984b. "The New Astronomy." In *Astrophysics and Twentieth-Century Astronomy to 1950,* ed. O. Gingerich, 4A: 59–72. Cambridge, University of Cambridge Press.

Menard, H. W. 1971. *Science: Growth and Change.* Cambridge, Harvard University Press.

Menzel, D. H., ed. 1962. *Selected Papers on Physical Processes in Ionized Plasmas.* New York, Dover.

Mermin, N. D. 1989. "What's Wrong with These Prizes?" *Physics Today* 42: 9–11.

Merrill, P. W. 1923. "A Research Career in Astronomy." *Science* 57 (1480): 546–48.

Merton, R. K. 1936. "Civilization and Culture." *Sociology and Social Research* 21: 1–30.

———. 1938. "Science and the Social Order." *Philosophy of Science* 5: 321–37.

———. 1939. "Science and the Economy of Seventeenth Century England." *Science and Society* 3: 3–27.

———. 1942. "The Normative Structure of Science." Reprinted in *The Sociology of Science: Theoretical and Empirical Investigations,* ed. N. W. Storer, 267–78. University of Chicago Press, 1973.

———. 1957. "Priorities in Scientific Discovery." Reprinted in *The Sociology of Science: Theoretical and Empirical Investigations*, ed. N. W. Storer, 286–323. Chicago, University of Chicago Press, 1973.

———. 1960. "'Recognition' and 'Excellence': Instructive Ambiguities." Reprinted in *The Sociology of Science: Theoretical and Empirical Investigations*, ed. N. W. Storer, 419–38. Chicago, University of Chicago Press, 1973.

———. 1968. "The Matthew Effect in Science." Reprinted in *The Sociology of Science: Theoretical and Empirical Investigations,* ed. N. W. Storer, 439–59. Chicago, University of Chicago Press.

———. 1970. *Science, Technology and Society in Seventeenth-Century England.* New York, Harper & Row.

———. 1988. "The Matthew Effect in Science, II: Cumulative Advantage and the Symbolism of Intellectual Property." *Isis* 79: 606–23.

Merton, R. K., and P. A. Sorokin. 1935. "The Course of Arabian Intellectual Development, 700–1300 A.D." *Isis* 22: 516–24.

Merton, R. K., and A. Thackray. 1972. "On Discipline-Building: The Paradoxes of George Sarton." *Isis* 63: 473–95.

Merton, R. K., and H. Zuckerman. 1971. "Institutionalized Patterns of Evaluation in Science." Reprinted in *The Sociology of Science: Theoretical and Empirical Investigations,* ed. N. W. Storer, 460–96. Chicago, University of Chicago Press, 1973.

———. 1972. “Age, Aging and Age Structure in Science.” Reprinted in *The Sociology of Science: Theoretical and Empirical Investigations*, ed. N. W. Storer, 497–559. Chicago, University of Chicago Press, 1973.

Meyer, J. W., and B. Rowan. 1977. “Institutionalized Organizations: Formal Structure as Myth and Ceremony.” *American Journal of Sociology* 83: 340–63.

Miller, H. S. 1970. *Dollars for Research: Science and its Patrons in Nineteenth Century America*. Seattle, University of Washington Press.

Miller, R. K. 1982. “Labor Market Structure and Career Occupational Status Mobility: A Theoretical Model.” *Sociological Inquiry* 52: 152–62.

Miller, W., ed. 1952. *Men in Business: Essays in the History of Entrepreneurs*. Cambridge, Harvard University Press.

Mitroff, I. 1974a. “Norms and Counter-Norms in a Select Group of the Apollo Moon Scientists: A Case Study of the Ambivalence of Scientists.” *American Sociological Review* 39: 579–95.

———. 1974b. *The Subjective Side of Science: A Philosophical Inquiry into the Psychology of the Apollo Moon Scientists*. New York, American Elsevier.

Moody, J. N. 1978. *French Education Since Napoleon*. Syracuse, Syracuse University Press.

Morrell, J., and A. Thackray. 1981. *Gentlemen of Science: The Early Years of the British Association for the Advancement of Science*. New York, Oxford University Press.

Moyer, A. E. 1985. “History of Physics.” In *Historical Writing on American Science*, ed. S. G. Kohlstedt and M. W. Rossiter, 163–82. *Osiris*, 2nd series, vol. 1.

———. 1992. *A Scientist’s Voice in American Culture: Simon Newcomb and the Rhetoric of Scientific Method*. Berkeley: University of California Press.

Mukerji, C. 1989. *A Fragile Power: Scientists and the State*. Princeton, Princeton University Press.

Mulkay, M. J. 1977. “Social Studies of Science: Disciplinary Perspectives.” In *Science, Technology and Society: A Cross-Disciplinary Perspective*, ed. I. Spiegel-Rösing and D. J. DeS. Price, 93–148. London, Sage.

———. 1980. “Sociology of Science in the West.” *Current Sociology* 28: 1–116.

———. 1991. *Sociology of Science: A Sociological Pilgrimage*. n.p., Open University Press.

Musto, D. F. 1968. “Yale Astronomy in the Nineteenth Century.” *Ventures* 8: 7–18.

National Science Foundation, Division of Policy and Planning Analysis, Office of Planning and Resources Management. 1982. *Studies of Scientific Disciplines: An Annotated Bibliography*. Washington, D. C., National Science Foundation.

Newcomb, S. 1882. “Introduction.” *Astronomical Papers of the American Ephemeris and Nautical Almanac* 1:vii-xiv. Washington, D. C., Government Printing Office.

———. 1903. *Reminiscences of an Astronomer*. Boston, Houghton Mifflin.

Newell, H. F. 1923. "The Decade 1860–70." In *The History of the Royal Astronomical Society, 1820–1920*, ed. J. L. E. Dreyer and H. H. Turner, 129–66. London, The Royal Astronomical Society.

Norberg, A. L. 1974. "Simon Newcomb and Nineteenth Century Positional Astronomy." Ph.D. dissertation, University of Wisconsin at Madison.

———. 1978. "Simon Newcomb's Early Astronomical Career." *Isis* 69: 209–25.

Novick, P. 1988. *That Noble Dream: The "Objectivity Question" and the American Historical Profession*. Cambridge: Cambridge University Press.

Oort, J., ed. 1939. *Transactions of the International Astronomical Union*. Vol. 6. Cambridge: Cambridge University Press.

Osterbrock, D. E. 1984a. *James E. Keeler: Pioneer American Astrophysicist and the Early Development of American Astrophysics*. Cambridge, Cambridge University Press.

———. 1984b. "The Rise and Fall of Edward S. Holden." *Journal for the History of Astronomy* 15: 81–127; 151–76.

———. 1985. "The Quest for More Photons." *Astronomy Quarterly* 5: 87–95.

———. 1986. "Failure and Success: Two Early Experiments with Concave Gratings in Stellar Spectroscopy." *Journal for the History of Astronomy* 17: 119–29.

———. 1990. "Armin O. Leuschner and the Berkeley Astronomy Department." *Astronomy Quarterly* 7: 95–115.

———. 1993. *Pauper and Prince: Ritchey, Hale and Big American Telescopes*. Tucson, University of Arizona Press.

Osterbrock, D. E., J. R. Gustafson, and W. J. S. Unruh. 1988. *Eye on the Sky: Lick Observatory's First Century*. Berkeley, University of California Press.

Osterbrock, D. E., and P. K. Seidelmann. 1987. "Paul Hergert." *Biographical Memoirs of the National Academy of Sciences* 57: 59–86. Washington, D.C., National Academy of Sciences.

Pais, A. 1986. *Inward Bound: Of Matter and Forces in the Physical World*. Oxford, The Clarendon Press.

Park, R. E. 1925. *The City*. Chicago, University of Chicago Press.

——— 1952. *Human Communities: The City and Human Ecology*. Glencoe, The Free Press.

Parsons, T. 1949. *The Structure of Social Action: A Study in Social Theory with Special Reference to a Group of Recent European Writers*, Glencoe, The Free Press.

Parsons, T., and E. A. Shills., eds. 1951. *Toward A General Theory of Action*. Cambridge, Harvard University Press.

Patterson, E. 1983. *Mary Somerville and the Cultivation of Science, 1815–1840*. The Hague, Martinus Nijhoff.

Paul, E. R. 1993. *The Milky Way Galaxy and Statistical Cosmology, 1890–1924*. Cambridge, Cambridge University Press.

———. 1996. "Netherlands, Astronomy in." In *The History of Astronomy: An Encyclopedia*, ed. J. Lankford. New York, Garland.

Pavalko, R. M. 1987. *Sociology of Occupations and Professions*. Itasca, Ill., F. E. Peacock.

Payne-Gaposchkin, C. 1984. *An Autobiography and Other Recollections*. Cambridge, Cambridge University Press.

Pelletier, M. 1990. *La Carte de Cassini: l'extraordinaire aventure de la Carte de France*. Paris, Presse de l'Ecole national des ponts et chaussées.

Perkin, H. 1983. "The Pattern of Social Transformation in England." In *The Transformation of Higher Learning, 1860–1930*, ed. K. H. Jarausch, 207–18. Chicago: University Chicago Press.

Perry, R. B. 1949. *Characteristically American*. New York, Knopf.

Peterson, C. J. 1989. "The United States Navy Corps of Professors of Mathematics." *Griffith Observer* 54: 2–11, 14.

———. 1990. "A Very Brief Biography and Popular Account of the Unparalleled T. J. J. See." *Griffith Observer* 54: 2–5.

Pickering, A. 1984. *Constructing Quarks: A Sociological History of Particle Physics*. Chicago, University of Chicago Press.

———, ed. 1992. *Science as Practice and Culture*. Chicago, University of Chicago Press.

Pickering, E. C. 1913. "Discussion at RAS Meeting." *The Observatory* 36: 285–86.

Pierson, G. W. 1952. *Yale College: An Educational History*. 2 vols. New Haven, Yale University Press.

Piore, M. J. 1975. "Notes for a Theory of Labor Market Stratification." In *Labor Market Segmentation*, ed. R. Edwards et al., 125–50. Lexington, D. C. Heath.

———. 1977. "The Dual Labor Market: Theory and Implications." In *Problems in Political Economy: An Urban Perspective*, ed. D. M. Gordon, 93–97. Lexington, D. C. Heath.

Plotkin, H. 1978. "Astronomers versus the Navy: The Revolt of American Astronomers over the Management of the United States Naval Observatory, 1877–1902." *Proceedings of the American Philosophical Society* 122: 385–99.

———. 1993. "Harvard College Observatory's Boyden Station in Peru: Origin and Formative Years, 1879–1898." In *Mundialización de la ciencia y cultura nacional. Actas del Congreso Internacional "Ciencia, descubrimiento y mundo colonial,"* ed. A. Lafuente, A. Elena, and M. L. Ortega, 689–705. Madrid, Doce Calles.

Price, D. J. DeS. 1963. *Little Science, Big Science*. New York, Columbia University Press.

Pyenson, L. 1977. "Who the Guys Were: Prosopography in the History of Science." *History of Science* 15: 155–88.

Rainger, R., K. R. Benson, and J. Maienschein, eds. 1988. *The American Development of Biology.* New Brunswick, Rutgers University Press.

———1991. *The Expansion of American Biology*, New Brunswick, Rutgers University Press.

Redman, R.W. 1945. "Post-War Astronomy." *The Observatory* 66: 153–59.

Reingold, N. 1964. *Science in Nineteenth Century America: A Documentary History.* Chicago, University of Chicago Press.

———. 1976. "Definitions and Speculations: The Professionalization of Science in America in the Nineteenth Century." In *The Pursuit of Knowledge in the Early American Republic: American Scientific and Learned Societies from Colonial Times to the Civil War*, ed. A. Oleson and S. C. Brown, 33–69. Baltimore, The Johns Hopkins University Press.

———. 1977. "The Case of the Disappearing Laboratory." *American Quarterly* 29: 79–101.

———. 1979. "National Science Policy in a Private Foundation: The Carnegie Institution of Washington." In *The Organization of Knowledge in Modern America, 1860–1920*, ed. A. Oleson and J. Voss, 313–41. Baltimore, The Johns Hopkins University Press.

———. 1991. "Introduction." *Science, American Style*, 1–10. New Brunswick, Rutgers University Press.

———. 1991a. "Science, Scientists and Historians of Science." In *Science American Style*, ed. N. Reingold, 379–88. Rutgers, Ruthers University Press.

Reingold, N., and J. N. Brodansky 1985. "The Sciences, 1850–1900: A North Atlantic Perspective." *Biological Bulletin* 168: 44–61.

Reskin, B. F. 1977. "Scientific Productivity and the Reward Structure of Science." *American Sociological Review* 42: 491–504.

———. 1979. "Academic Sponsorship and Scientists' Careers." *Sociology of Education* 52: 129–46.

Richardson, R. S. 1944. *Astronomy: The Distaff Side.* Leaflets of the Astronomical Society of the Pacific, vol. 181. San Francisco, Astronomical Society of the Pacific.

Rigaux, F., ed. 1959. *Les Observatoires Astronomiques et les Astronomes.* Brussels, Observatoire Royal de Belgique/International Astronomical Union.

Riley, S., and D. Wrench. 1985. "Mentoring Among Women Lawyers." *Journal of Applied Social Psychology* 15: 376–85.

Ringer, F. K. 1979a. *Education and Society in Modern Europe.* Bloomington, Indiana University Press.

———. 1979b. "The German Academic Community." In *The Organization of Knowledge in Modern America, 1860–1920*, ed. A. Oleson and J. Voss, 409–29.

Roe, A. 1951. "A Psychological Study of Physical Scientists." *Psychological Monographs* 43: 121–239.

———. 1953a. "A Psychological Study of Eminent Psychologists and Anthropologists and a Comparison with Biologists and Physical Scientists." *Psychological Monographs* 67. Whole number 2.

———. 1953b. *The Making of a Scientist*. New York, Dodd, Mead.

Rosenberg, C. 1976. *No Other Gods: On Science and American Social Thought*. Baltimore, The Johns Hopkins University Press.

———. 1988. "Woods or Trees? Ideas and Actors in the History of Science." *Isis* 79: 565–70.

———. 1979. "Toward an Ecology of Knowledge: On Discipline, Context, and History." In *The Organization of Knowledge in Modern America, 1860–1920*, A. Oleson and J. Voss, eds. 440- 55. Baltimore, The Johns Hopkins University Press.

Rosenberg, R. 1982. *Beyond Separate Spheres: Intellectual Roots of Modern Feminism*. New Haven, Yale University Press.

Rossiter, M. W. 1982. *Women Scientists in America: Struggles and Strategies to 1940*. Baltimore, The Johns Hopkins University Press.

Rothblatt, S. 1983. "The Diversification of HIgher Education in England." In *The Transformation of Higher Learning, 1860–1930*, ed. K. H. Jarausch, 131–48. Chicago, University of Chcago Press.

Rothenberg, M. 1974. "The Education and Intellectual Background of American Astronomers." Ph.D. dissertation, Bryn Mawr College.

———. 1981) "Organization and Control: Professionals and Amateurs in American Astronomy, 1889–1918." *Social Studies of Science* 11:305–25.

———. 1990. "Patronage of the Harvard College Observatory, 1839–1851." In *Two Astronomical Observatories: The Harvard College Observatory and the Smithsonian Astrophysical Observatory*, ed. O. Gingerich, 37–46. Cambridge, Harvard-Smithsonian Center for Astrophysics.

Royal Society of London. 1867–1925. *Catalogue of Scientific Papers*. London, Eyre and Spottiswoode.

Rudolph, F. 1962. *The American College and University: A History*. New York, Knopf.

Rufus, W. C. 1951. "The Astronomy Department and Astronomical Observatories at Ann Arbor." In *The University of Michigan: An Encyclopedic Survey*, ed. W. B. Shaw, 2: 442–76. Ann Arbor, University of Michigan.

Sampson, J. A., and E. S. C. Weiner. eds. 1989. *The Oxford English Dictionary*. Oxford, The Clarendon Press.

Sampson, R. A. 1923. "The Decade 1840–1850." In *History of the Royal Astronomical Society, 1820–1920*, ed. J. L. E. Dreyer and H. T. Turner, 82–109. London: Royal Astronomical Society.

Sapp, J. 1983. "The Struggle for Authority in the Field of Heredity, 1900–1932: New Perspectives on the Rise of Genetics." *Journal of the History of Biology* 16: 311–42.

Sassen, S. 1990) "Economic Restructuring and the American City." In *Annual Review of Sociology*, ed. W. R. Scott and J. Blake, 16: 465–90.

Sawyer, J. E. 1958. "Entrepreneurial Studies: Prospects and Directions, 1948–1958." *Business History Review* 32: 434–43.

Schaffer, S. 1988. "Astronomers Mark Time: Discipline and the Personal Equation." *Science in Context* 2: 115–45.

Schlesinger, F. 1926. "Letter." *Science* 64 (1668): 596.

———. 1940. "William L. Elkin." *Biographical Memoirs of the National Academy of Sciences* 18: 175–88. Washington, D.C., National Academy of Sciences.

Servos, John W. 1990. *Physical Chemistry from Oswald to Pauling: The Making of a Science in America*. Princeton, Princeton University Press.

Shaffer, E. S. 1990. "Romantic Philosophy and the Organization of the Disciplines: The Founding of the Humboldt University of Berlin." In *Romanticism and the Sciences*, A. Cunningham and N. Jardine, eds. 38–54. Cambridge, Cambridge University Press.

Shapin, S. 1982. "History of Science and Its Sociological Reconstructions." *History of Science* 20: 157–211.

———. 1991. "Discipline and Bounding: The History and Sociology of Science as Seen through the Externalism-Internalism Debate." *Conference on Critical Problems and Research Frontiers in History of Science and History of Technology, Madison, Wis., History of Science Society*, 203–27. Madison, History of Science Society.

———. 1993. Book Review. *Isis* 84:623–24.

———. 1994. *A Social History of Truth: Civility and Science in Seventeenth-Century England*. Chicago, University of Chicago Press.

Shapin, S., and A. Thackray. 1974. "Prosopography as a Research Tool in the History of Science." *History of Science* 12: 1–28.

Shapiro, H. D. 1976. "The Western Academy of Natural Science of Cincinnati and the Structure of Science in the Ohio Valley, 1810–1850." In *The Pursuit of Knowledge in the Early American Republic: American Scientific and Learned Societies from Colonial Times to the Civil War*, ed. A. Oleson and S. C. Brown, 219–47. Baltimore, The Johns Hopkins University Press.

Shapley, H. 1966. *Oral History Interview*. New York, American Institute of Physics, Center for the History of Physics.

———. 1969. *Through Rugged Ways to the Stars: Reminiscences of an Astronomer*. New York: Charles Scribner and Sons.

Sheehan, W. 1988. *Planets and Perception: Telescopic Views and Interpretations, 1609–1909*. Tucson, University of Arizona Press.

Shorter, E. 1971. *The Historian and the Computer: A Practical Guide*. Englewood Cliffs, Prentice-Hall.

Simpson, G. E., and J. M. Yinger 1953. *Racial and Cultural Minorities: An Analysis of Prejudice and Discrimination*. New York, Harper.

Smith, R. W. 1982. *The Expanding Universe: Astronomy's "Great Debate," 1900–1931*. Cambridge, Cambridge University Press.

———. 1989. *The Space Telescope: A Study of NASA, Science, Technology and Politics*. Cambridge: Cambridge University Press.

———. 1991. "A National Observatory Transformed: Greenwich in the Nineteenth Century." *Journal for the History of Astronomy* 22: 5–20.

Sokal, M. M. 1984. "The Gestalt Psychologists in Behavioralist America." *American Historical Review* 89: 1240–63.

Spitzer, A. B. 1973. "The Historical Problem of Generations." *American Historical Review* 78: 1353–85.

Stearns, P. J. 1985. "Social History: A Progress Report." *Journal of Social History* 19: 319–23.

Stein, M. R. 1960. *The Eclipse of Community: An Interpretation of American Studies*. Princeton, Princeton University Press.

Stetson, H. T. 1928. "The Students' Astronomical Observatory of Harvard University, 1903–1928." *Popular Astronomy* 36: 589–96.

Stimpson, D., ed. 1962. *Sarton on the History of Science*. Cambridge, Harvard University Press.

Stone, L. 1972. "Prosopography." In *Historical Studies Today*, ed. F. Gilbert and S. R. Graubard, 107–40. New York, W. W.Norton.

Stone, O. 1887. "Photographers versus Old Fashioned Astronomers." *Sidereal Messenger* 6: 1–4.

Storer, N. W. 1966. *The Social System of Science*. New York, Holt, Rinehart and Winston.

———, ed. 1973) *The Sociology of Science: Theoretical and Empirical Investigations*. Chicago, University of Chicago Press, 1973.

Stroobant, P., J. Delvosal, H. Philippot, E. Delporte, and E. Meelin. 1907, rev. ed. 1931, supplement, 1936. *Les Observatoires Astronomiques et les Astronomes*. Brussels, Observatoire de Belgique.

Struve, O. 1942. "Astronomy Faces the War." *Popular Astronomy* 50: 465–72.

———. 1943. "Fifty Years of Progress in Astronomy." *Popular Astronomy* 51: 469–81.

Struve, O., and V. Zebergs 1962. *Astronomy of the Twentieth Century*. New York, Macmillan.

Tatarewicz, J. N. 1990. *Space Technology and Planetary Astronomy*. Bloomington, Indiana University Press.

Taylor, P. 1992. "Community." In *Keywords in Evolutionary Biology*, ed. E. F. Keller and E. A. Lloyd, 52–60. Cambridge, Harvard University Press.

Thackray, A., J. L. Sturchio, P. T. Carroll, and R. Budd 1985. *Chemistry in America, 1876–1976: Historical Indicators.* Dordrecht, Reidel.

Thomas, W. I. 1966. *W. I. Thomas on Social Organization and Social Personality: Selected Papers.* Chicago, University of Chicago Press.

Thronson, H. A., and S. L. Lindstedt. 1986. "The Graying of American Astronomy." *Publications of the Astronomical Society of the Pacific* 98: 941–47.

Thurow, L. C. 1975. *Generating Inequality.* New York, Basic Books.

Tickamyer, R. R., And C. M. Duncan. 1990. "Poverty and Opportunity Structure in Rural America." In *Annual Review of Sociology*, ed. W. R. Scott and J. Blake, 16: 67–86.

Titze, H. 1983. "Enrollment Expansion and Academic Overcrowding in Germany." In *The Transformation of Higher Learning, 1860–1930*, K. H. Jarausch, ed., 57–88. Chicago, University of Chicago Press.

Tobey, R. C. 1971. *The American Ideology of National Science, 1919–1930.* Pittsburgh, University of Pittsburgh Press.

Traweek, S. 1988. *Beamtimes and Lifetimes: The World of High Energy Physics.* Cambridge, Harvard University Press.

Turner, G. L'E., ed. 1976. *The Patronage of Science in the Nineteenth Century.* Leyden, Noordhoff.

Turner, H. H. 1891. "Oxford Notebook." *The Observatory* 14: 189–92.

———. 1902. *Modern Astronomy: Being Some Account of the Revolution of the Last Quarter Century.* Westminster, U.K., A. Constable and Co.

———. 1912a. "Obituary of W. P. Fleming." *Monthly Notices of the Royal Astronomical Society* 72: 261–64.

———. 1912b. *The Great Star Map.* London, John Murray.

U.S. Department of Commerce, Bureau of the Census. 1973. *1970 Census of Population, Vol. 1, Characteristics of the Population, Pt. 1, Survey of Population, Sec. 1.* Washington, D.C., Government Printing Office.

———. 1975. *Historical Statistics of the United States, Colonial Times to the Present.* Washington, D.C., Government Printing Office.

———.1992. *Statistical Abstract of the United States 1992.* Washington, D.C., Government Printing Office.

Van de Kamp, P. 1977–1979. *Oral History Interview.* New York, American Institute of Physics, Center for the HIstory of Physics.

Van Den Bos, W. H. 1958. "Robert Grant Aitken." *Biographical Memoirs of the National Academy of Sciences* 32: 1–30. New York, Columbia University Press.

Van Helden, A. 1984a. "Telescope Building, 1850–1900" In *Astrophysics and Twentieth-Century Astronomy to 1950*, ed. O. Gingerich, 4A: 40–58. Cambridge, Cambridge University Press.

———. 1984b. "Building Large Telescopes, 1900–1950." In *Astrophysics and*

Twentieth-Century Astronomy to 1950, ed. O. Gingerich, 4A: 134–52. Cambridge, Cambridge University Press.

Veysey, L. R. 1965. *The Emergence of the American University*. Chicago, University of Chicago Press.

Visher, S. S. 1947. *Scientists Starred, 1903–1943, in "American Men of Science": A Study of Collegiate and Doctoral Training, Birthplace, Distribution, Backgrounds and Developmental Influences*. Baltimore, The Johns Hopkins University Press.

Vlachy, J. 1985. "Scientometric Analysis in Physics: A Bibliography of Publication, Citation, and Mobility Studies." *Czechoslovak Journal of Physics* B35: 1389–1436.

Von Konkoly, N. 1887. *Practische Ankitung zur Himmelsphotographie*. Halle, Wilhelm Knapp.

Wali, K. C. 1991. *Chandra: A Biography of S. Chandrasekhar*. Chicago, University of Chicago Press.

Wallis, R., ed. 1979. *On the Margins of Science: The Social Construction of Rejected Knowledge*. Keele, University of Keele.

Walton, J. 1993. "Urban Sociology: The Contribution and Limits of Political Economy." In *Annual Review of Sociology*, ed. J. Blake and J. Hagen, 19: 301–20.

Warner, D. J. 1979. "Astronomy in Antebellum America." In *The Sciences in the American Context: New Perspectives*, ed. N. Reingold, 55–76. Washington, D.C., Smithsonian Institution Press.

Weart, S. R. 1979. "The Physics Business in America, 1919–1940: A Statistical Reconnaissance." In *The Sciences in the American Context: New Perspectives*, ed. N. Reingold, 295–358. Washington, D. C., Smithsonian Institution.

Weart, S. R., and D. H. DeVorkin. 1981. "Interviews as Sources for the History of Modern Astrophysics." *Isis* 72: 471–77.

Weber, M. 1947. *The Theory of Social and Economic Organization*. London, Routledge and Kegan Paul.

Wecter, D. 1941. *The Hero in America: A Century of Hero Worship*. New York, Charles Scribner's Sons.

Welther, B. 1966. "The Cult of True Womanhood, 1820–1860." *American Quarterly* 18: 151–74.

Westfall, R. S. 1985. "Science and Patronage: Galileo and the Telescope." *Isis* 76: 11–30.

White, L. D. 1958. *The Republican Era, 1869–1901: A Study in Administrative History*. New York, Macmillan.

Whitford, A. E. 1962. "Joel Stebbins." *Biographical Memoirs of the National Academy of Sciences* 49: 293–316. Washington, D.C., National Academy of Sciences.

Willard, B. C. 1976. *Russell W. Porter: Arctic Explorer, Artist, Telescope Maker.* Freeport, Maine, Bond Wheelwright.

Williams, T. R. 1989. "Contributions of Amateurs to Astronomy, 1889–1898." *Publications of the Astronomical Society of the Pacific* 101:888.

———. 1991. "Phoebe Haas—An AAVSO Volunteer." *Journal of the American Association of Variable Star Observers* 20: 18.

———. 1992. "Albert Ingalls and the ATM Movement." *Sky & Telescope* 83: 140–43.

———. 1993. "Springfield in the West—Amateur Telescope Making in Southern California, 1930–1955." *Bulletin of the American Astronmical Society* 25: 931.

Winlock, W. 1885. "Astronomy." *Annual Report of the Smithsonian Institution,* 343–456. Washington, D.C.

———. 1887. "Astronomy." *Annual Report of the Smithsonian Institution,* 99–187. Washington, D.C.

———. 1888. "Astronomy." *Annual Report of the Smithsonian Institution,* 125–216. Washington, D.C.

———. 1890. "Astronomy." *Annual Report of the Smithsonian Institution,* 121–82. Washington, D.C.

———. 1892. "Astronomy." *Annual Report of the Smithsonian Institution,* 681–774. Washington, D.C.

Wolf, C. 1902. *Histoire de l'Observatoire de Paris de sa création à 1793.* Paris, Gauthier-Villars.

Woodbury, David O. 1966. *The Glass Giant of Palomar.* New York, Dodd, Mead.

Woolgar, Steve. 1982. "Laboratory Studies: A Comment on the State of the Art." *Social Studies of Science* 12: 481–98.

Wright, H. 1949. *Sweeper in the Sky: The Life of Maria Mitchell, First Woman Astronomer in America.* New York, Macmillan.

———. 1966. *Explorer of the Universe: A Biography of George Ellery Hale.* New York, E. P. Dutton.

Wright, H., J. N. Warnow, and C. Weiner, eds. 1972. *The Legacy of George Ellery Hale: The Evolution of Astronomy and Scientific Institutions in Pictures and Documents.* Cambridge, Massachusetts Institute of Technology Press.

Wrong, D. H. 1988. *Power: Its Forms, Bases, and Uses.* Chicago, University of Chicago Press.

Yeo, R. 1989. "Science and Intellectual Authority in Mid-Nineteenth Century Britain: Robert Chambers and *Vestiges of the Natural History of Creation.*" In *Energy and Entropy: Science and Culture in Victorian Britain,* ed. P. Brantlinger, 1–27. Bloomington, Indiana University Press.

Young, C. A. 1881. *The Sun.* New York, Appleton.

———. 1891. "Address at the Dedication of the Kenwood Observatory." *Sidereal Messenger* 10: 312–21.

———. 1902. *Manual of Astronomy: A Text Book*. New York, Ginn and Co.

Ziman, J. 1984. *An Introduction to Science Studies: The Philosophical and Social Aspects of Science and Technology*. Cambridge, Cambridge University Press.

Zuckerman, H. 1970. "Stratification in American Science." *Sociological Inquiry* 40: 235–57.

———. 1977." Deviant Behavior and Social Control in Science." In *Deviance and Social Change*, ed. E. Sagarian, 87–138. Beverly Hills, Sage.

✦

INDEX